Infosys Science Foundation Series

Infosys Science Foundation Series in Mathematical Sciences

The **Infosys Science Foundation Series in Mathematical Sciences,** a Scopus-indexed book series, focuses on high-quality content in the domain of pure and applied mathematics, biomathematics, financial mathematics, operations research and theoretical computer science. With this series, Springer hopes to provide readers with monographs, textbooks, edited volume, handbooks and professional books of the highest academic quality on current topics in relevant disciplines. Literature in this series will appeal to a wide audience of researchers, students, educators and professionals across the world.

Kalyan Chakraborty · Shigeru Kanemitsu ·
Takako Kuzumaki

Modular Relations and Parity in Number Theory

Springer

Kalyan Chakraborty
Department of Mathematics
SRM University, Andhra Pradesh
Amaravati, Andhra Pradesh, India

Shigeru Kanemitsu
Mathematical Research Center
Shandong University
Jinan, Shandong, China

Takako Kuzumaki
Faculty of Engineering
Gifu University
Gifu, Japan

ISSN 2363-6149 ISSN 2363-6157 (electronic)
Infosys Science Foundation Series
ISSN 2364-4036 ISSN 2364-4044 (electronic)
Infosys Science Foundation Series in Mathematical Sciences
ISBN 978-981-96-6470-2 ISBN 978-981-96-6471-9 (eBook)
https://doi.org/10.1007/978-981-96-6471-9

Mathematics Subject Classification: 30-XX, 32-XX, 37J38, 42A16, 94A11, 11Rxx, 14G10, 11-XX, 33-XX, 65D20, 26-XX, 37K20, 44A20, 11E45, 11Mxx, 37C30

This Springer imprint is published by the registered company Springer Nature Singapore Pte Ltd.
The registered company address is: 152 Beach Road, #21-01/04 Gateway East, Singapore 189721, Singapore

Dedicated to Professor Dr. Jianya Liu on his 60th birthday, with great respect and friendship

Preface

This book intends to provide the reader with the ways of finding and thinking on research problems as well as enough frontier knowledge of zeta-functions and modular and parity relations—odd and even. It can be observed that most of far-reaching results incorporate the modular relation (in the wide sense, to be explained below) in some or other form and in some cases, parity classification shows its effect, giving a decisive result. We shall exhibit manifestation of these two principles by ample examples.

Our description is rather unique in providing the process from finding possible problems to treating them from a scratch. We try to give intermediate process of working with concrete examples and worked-out exercises, incorporating them into more general unified theorems. This will show how to proceed from some naive observations to more high brow achievements. Such a process is hidden in published papers or even books which state the final results and proofs, never showing the intermediate process of thinking. It is hoped that this exposure of the secret will be beneficial to younger researchers who are eager to establish their research career. For advanced readers, mere browsing through general results will be of use.

One of the most influential and creative source of ideas lies in the **modular relations**, the correspondence between the transformation formula (of modular form-like objects) and the functional equation (of the corresponding zeta-functions).

The origin of this correspondence goes back to Riemann (1859) and has been developed by Hecke (1936), Hecke (1983) and Bochner (1951), Chandrasekharan and Narasimhan, M. Knopp, B. Berndt et al. These are culminated in Kanemitsu and Tsukada (2014) as the modular relations (as a memorial to Bochner) in the wide sense, **equivalent expressions to the functional equation**.

The naming **Riemann-Hecke-Bochner-correspondence** of modular relations by Knopp (2000) expresses the historical events well. In this respect we also refer to the modular relation as RHB correspondence or just RHB.

In Kanemitsu and Tsukada (2014) the underlying principle of modular relations has been elucidating those important results which were obtained hitherto by the methods that do not look like modular relations because of the incorporated extra ingredients. In early chapters of this book, we shall introduce the notion of cosmic

Table 1 Parity and cosmic dual force

Objects	Even	Odd
Character	Even	Odd
Quadr. field	Real	Imaginary
Function	Clausen	Bernoulli polyn.
Class no.	Max. real subfld	Relative class no.

Table 2 Transform and its inversion

Objects	Easier	Wanted
Process	Transform	Inversion
RHB corr.	Mellin transf.	Inverse Mellin
Fourier-Bessel	Boundary Lerch	Perturbed Dirichlet ser.
Diff., int.	Diff.	Int.

dual force by which we mean loose classification of the whole into even and odd part: W = E + O to elucidate on aspects of such cases which do not look like RHB. This cosmic dual force classification has been perceived in an implicit way in many literature but not in an explicit way but partially formulated in de la Vallée Poussin (1896), de la Vallée Poussin (1897), Schnee (1930), Ishibashi and Kanemitsu (1999), Milnor (1983).

There are many objects which belong to the odd part. Imaginary quadratic field, relative class number, Maillet-Demyanenko determinants, linear independence of cotangent values (Chowla's problem), Riemann's posthumous fragment, etc. (Table 1).

In a wide sense, a.a. processes are to make the object easier and then return to the original (Table 2).

Weil (1971), p. 1 states "A rather full but somewhat longwinded function-theoretic treatment of that case (the classical modular functions and modular forms) was given in 1890–1892 by R. Fricke and F. Klein; the arithmetical aspects, which are intimately tied up with the theory of complex multiplication, were considered by H. Weber in his Algebra, vol, III (for a modern treatment, cf. a forthcoming book by G. Shimura). **The relation between modular forms and Dirichlet series with functional equations was discovered by Hecke**, whose epoch-making work during the years 1930–1940, based on that discovery and that of the "Hecke operators", brought out completely new aspects of a theory which many mathematicians would have regarded as **a closed chapter long before**."

Here Shimura's book is Shimura (1971) and the relation is the RHB correspondence for the Hecke type functional equation.

"He not only believes, but positively knows, that such and such apparently arbitrary arrangements of the matter constitute and alone constitute the true beauty. His reasons, however, have not yet been matured into expression. It remains for a more profound analysis than the world has yet seen, fully to investigate and express them"—The domain of Arnheim by E. A. Poe.

We would like to express our hearty thanks to Prof. Jay Mehta for his devoted help in typesetting and comments and to Messrs. Shamim Ahmad, Bharath Kumar Dhamodharan and Suresh Dharmalingam for their thoughtful support during the whole process and patience.

Amaravati, India — Kalyan Chakraborty
Jinan, China — Shigeru Kanemitsu
Gifu, Japan — Takako Kuzumaki

Contents

Chapter 1
Introduction and Preliminaries

1.1 Notation, Terminology, and Short-Hand

Our main objective being clarification of the RHB for some specific zeta-functions, we fix the notation and short-hand and stick to it throughout. We write for the complex variable

$$s = \sigma + it$$

once and for all, with σ and t the real and imaginary parts, respectively.

The **Riemann zeta-function** $\zeta(s)$ is defined by

$$\zeta(s) = \prod_p \left(1 - \frac{1}{p^s}\right)^{-1} = \sum_{n=1}^{\infty} \frac{1}{n^s} \tag{1.1.1}$$

for $\sigma > 1$, where in the product, p runs through all the primes. We often refer to this as RZF and its additive form (Dirichlet series expression) is recalled in (2.1.8).

Some highlights of this book are in the spirit of Petkovšek et al. (1996), not exactly the contents but the title $A = B$, or more in the spirit of Ramanujan, whose discovery is often related to identification of remote-looking objects. We refer to this as $A = B$. Also $A = B = C$, whence $A = C$, etc. appears. Riemann's definition (1.1.1) of RZF already contains a lot of far-reaching inspirations. For example, the equality is between the product and the sum and it implies that $\mathbb{Z}$ is a unique factorization domain (UFD). This applies to many zeta-functions since they are defined by Euler products. The Dedekind zeta-function is one typical example and so are Dirichlet L-functions. The former is UFD w.r.t. ideal multiplication (historically this was an important step) while the latter is augmentation of RZF by characters. Examples of zeta-functions which do not have Euler products (save for some special cases) include Epstein zeta-functions, which are zeta-functions of a quadratic form and the Hurwitz and the Lerch zeta-function. One might feel persuaded if one observes that

K. Chakraborty et al., *Modular Relations and Parity in Number Theory*,
Infosys Science Foundation Series, https://doi.org/10.1007/978-981-96-6471-9_1

these are additive components of the zeta-function with the Euler product. Cf. $A = B$ for these: (1.2.1) (2.2.34).

The **Riemann functional equation (FER)** is stated in (2.1.9). Cf. Exercise 2.1.3. Since we use the short-hand "Ramified Functional Equation" (RFE) throughout, we refer to (2.1.9) as FER—Functional Equation of Riemann. Cf. Exercise 2.12.7 which views FER as a special case of HTFE, via Lemma 2.12.5.

We refer to (2.12.6) as the **Hecke (type) functional equation (HTFE).** For more details, cf. Sect. 2.12.

The perturbed Dirichlet series associated with RZF is the **Hurwitz zeta-function (HZF)** defined by (2.1.11). It is the zeta-function for the shifted sequence $\{n + x\}$.

Its counterpart is the boundary **Lerch zeta-function (LZF)** defined by (2.1.10), with the shift being accommodated in the exponential function. This is also called the polylogarithm function with complex exponential argument. Both of them reduce to the Riemann zeta-function when the shift is 1.

These are connected by the **Hurwitz formula** in (2.1.12) (i.e. the functional equation for the Hurwitz zeta-function) while its reciprocal is the **reciprocal Hurwitz formula (RHF)** in (2.1.13). These will be accommodated in Theorem 2.1.2, which gives the relation between RFES and the generalized Euler identity.

We remark that the link connecting the theory of HZF and the theory of the gamma function is the **Lerch formula (LF)** (2.1.46).

Definition 1.1.1 The following notation will be used throughout. Let

$$Q(m, n) = am^2 + bmn + cn^2 = a|m + n\tau|^2, \quad \tau = x + iy = \frac{b + i\sqrt{|\Delta|}}{2a} \tag{1.1.2}$$

be a positive definite binary quadratic form, where

$$a = \frac{\sqrt{|\Delta|}}{2 \operatorname{Im} \tau}, \quad \Delta = b^2 - 4ac < 0, \tag{1.1.3}$$

so that

$$c = a|\tau|^2, \quad b = 2a \operatorname{Re} \tau, \quad x = \frac{b}{2a}. \quad y = \frac{\sqrt{|\Delta|}}{2a}. \tag{1.1.4}$$

Then in conformity with (1.1.2), τ is the solution of the quadratic equation with imaginary part positive:

$$a\tau^2 - b\tau + c = 0. \tag{1.1.5}$$

We let $\mathcal{H}$ be the upper half-plane (UHP) once and for all:

$$\mathcal{H} = \{\tau = x + iy | y > 0\} = \{z = x + iy | y > 0\}. \tag{UHP}$$

Then the **Epstein zeta-function** (EZF) $Z(s, Q)$ associated to a positive definite quadratic form Q is defined by

$$Z(s, Q) := \sum_{m,n}{}' \frac{1}{Q(m,n)^s} = a^{-s}\zeta_{\mathbb{Z}^2}(s, \tau), \tag{1.1.6}$$

where here and in what follows, in such a sum in $m, n \in \mathbb{Z}$, the prime on the summation sign means the omission of the case $m = n = 0$ and the **Epstein-type Eisenstein series** (EES) is defined by

$$\zeta_{\mathbb{Z}^2}(s, \tau) := \sum_{m,n}{}' \frac{1}{|m + n\tau|^{2s}} = \sum_{m,n}{}' \frac{1}{\left((m + xn)^2 + (yn)^2\right)^s} \tag{1.1.7}$$

or

$$\zeta_{\mathbb{Z}^2}(s, \tau) = Z(s, Q_1), \quad Q_1(m, n) = (x^2 + y^2)n^2 + 2xmn + m^2. \tag{1.1.8}$$

Cf., e.g. Kuzumaki (2011), Zhang and Williams (1995), etc. Cf. the procedure in Sect. 2.5 of transforming a lattice $m\omega_1 + n\omega_2$ into $m + n\tau$.

EES (1.1.7) is a lattice function and as can be seen in Remark 3.1.1, the Hecke gamma transform shows its effect of separating the variables $m, n\tau$ resp. $(m + xn)^2, (yn)^2$.

The following expressions (and some more, especially (3.6.2)) will play a central role throughout the book.

$$Z(s, Q) = a^{-s}\zeta_{\mathbb{Z}^2}(s, \tau), \quad E(s, \tau) = \frac{y^s}{\zeta(2s)}\zeta_{\mathbb{Z}^2}(s, \tau). \tag{1.1.9}$$

We refer to the first one as (EZF≈EES) and the second one as (EMS≈EES). Here $E(s, z)$ is the **Eisenstein–Maass series** (EMS) defined by (3.2.6) as a non-holomorphic automorphic form. It has another expression (3.2.7).

Remark 1.1.1 As remarked on Davenport (1980, p. 43), the notation (1.1.2) is used by most of modern writers, while

$$Q(m, n) = am^2 + 2bmn + cn^2 = a|m + n\tau|^2, \tag{1.1.10}$$

$$\tau = x + iy = \frac{b + i\sqrt{|d|}}{a}, \quad d = \frac{1}{4}\Delta = b^2 - ac < 0$$

is the classical notation used by Dirichlet. We sometimes use the classical one. In the definition (1.1.7), we may also write

$$\zeta_{\mathbb{Z}^2}(s, \tau) = \sum_{m,n}{}' \frac{1}{|m\tau + n|^{2s}}.$$

We include the indefinite case by identifying it with a real quadratic field.

Exercise 1.1.1 Check the following. The special case of EES with $s = k \in \mathbb{Z}, k > 1$ is closely related to the classical Eisenstein series $G_k(\tau)$ which is a modular form of weight $2k$ and admits a Fourier expansion (Serre, 1973, pp. 83, 92).

Cf. Wang et al. (2025) in which a generalization $G_s(\tau)$ of $G_k(\tau)$ is considered. It is not precisely the same as $\zeta_{\mathbb{Z}^2}(s, \tau)$ but is more delicate, leading to the Kronecker limit formula without absolute value.

In view of this context, we sometimes refer to EES as a *holomorphic Eisenstein series by abuse of language*. This is the link that connects holomorphic and non-holomorphic automorphic forms by multiplication by the non-holomorphic factor y^s.

One of the penetrating ingredients in this book is the Chowla–Selberg integral formula (CSIF) (or the Fourier–Bessel expansion), cf. (2.5.26):

$$\Gamma(s)\Delta(s)\zeta_{\mathbb{Z}^2}(s,\tau) = 2\Delta(s)\Gamma(s)\zeta(2s) + 2\sqrt{\pi}\Delta(s)\Gamma\left(s-\frac{1}{2}\right)\zeta(2s-1)y^{-2s+1} \tag{1.1.11}$$

$$+ 8\Delta(s)\pi^s y^{\frac{1}{2}-s}\sum_{n=1}^{\infty} n^{s-\frac{1}{2}}\sigma_{1-2s}(n)\cos 2\pi nx\, K_{s-\frac{1}{2}}(2\pi yn),$$

where $\Delta(s)$ is a suitable integral function. Choosing it as is stated in (1.1.9) gives CSIF for EPZ and EMS. For all five zeta-functions that we treat, cf. Theorem 3.6.1.

Here $K_\nu(x)$ is the modified Bessel function of the third kind of index ν, which we refer to as the K-Bessel function, cf. Sects. 3.1 and 3.6 for its basics.

Exercise 1.1.2 Find a necessary condition on $\Delta(s)$ under which $\Gamma(s)\Delta(s)\zeta_{\mathbb{Z}^2}(s,\tau)$ is unchanged under $s \leftrightarrow 1-s$. Cf. Exercise 2.5.1.

Solution. By the asymmetric FER (2.5.37), we find that a necessary condition is

$$\left(\frac{y}{\pi}\right)^{2s-1}\Delta(1-s) = \Delta(s). \tag{1.1.12}$$

One of the choices is $\Delta(s) = \left(\frac{y}{\pi}\right)^s$. Other choices can be depicted from $y = \frac{\sqrt{|\Delta|}}{2a}$. Details are left to the reader. □

In our proof of Theorem 2.5.1, we use the beta transform, which procedure looks more involved than Chowla–Selberg's first method of dissection and Hardy–Hecke transform. There is a good reason to use it. Indeed, by the beta transform method, we may locate the origin of the second term on the right of (1.1.11). It arises from $2\Delta(s)\sum_{n=1}^{\infty}(ny)^{-2s}J_n(c)$, (2.5.31), with the residue of the integral $J_n(c)$, (2.5.32) at the pole $s = -\frac{1}{2}$.

It turns out that for deriving CSIF, EMS is simpler than EES but we cling to the latter. It will be a good exercise to check all transitions between five zeta-functions in Theorem 3.6.1.

The main object of study in Chap. 3 is the Hecke–Eisenstein series (HES) (3.4.1) which includes EMS (3.2.6).

This is the main object of study in Maass (1964), which is also studied in Miyake (1989).

1.2 Synopsis of Each Chapter and Illustrative Examples

In this section, we give synopsis of each chapter in a symbolic and qualitative way using the notation above.

Chapters 2 and 3 are intimately connected. In a fitting catchword, the transition from Chaps. 2 to 3 is a generalization of Hecke theory, Theorem 2.12.3 to Maass theory, Theorem 3.2.1 and its corollaries; from the teacher to the student. Or from the unramified HFE, symbolically 1 : 1 to the ramified RFEM 2 : 2; or from the FBE to the FWE, by the latter of which we may locate the genesis of the Maass forms as those satisfying FWE.

Chapter 2 starts from the ramified functional equation of Schnee-type RFES, symbolically 1 : 2.

Theorem 2.1.2 shows *the parity nature (generalized Euler identity)* of the Hurwitz and Lerch zeta-functions as well as ramified RHB correspondence. This is a precursor to RFEM 2 : 2 in Chap. 3.

In Chap. 2, we give two typical examples of this kind as expounded in Sect. 2.6.

The most famous one is the Dirichlet class number formula (CNF) which was the highlight in the beginning of analytic number theory due to Dirichlet. Let $\zeta_K(s)$ be the Dedekind zeta-function of a quadratic field. Then $A = B$ for this is

$$\zeta(s)L(s,\chi) = \zeta_K(s) = \sum_{\mathfrak{C}\in I/P} \zeta(s,\mathfrak{C}). \tag{1.2.1}$$

Here one side is the multiplicative decomposition (2.6.1) into the product of the Riemann zeta- and Dirichlet L-function, while the other side is the additive decomposition (2.5.5) into class zeta-functions (2.5.4). The latter decomposition is necessary since it is the class zeta-function which reduces to the Epstein zeta-function. For cyclotomic fields, multiplicative decomposition is replaced by (5.3.4).

Then $A = B$ for CNF is equating the two expressions for the residue of the Dedekind zeta-function resulting in (2.6.1).

On the other hand, $A = B$ for the **Lerch–Chowla–Selberg formula (LCSF)** is equating the two expressions for the Laurent constant (the one for the class zeta-function is obtained by the Kronecker limit formula). Details will be given in the case of imaginary resp. real quadratic field in Sect. 2.6 resp. in Sect. 2.9.

A qualitative version of the Lerch–Chowla–Selberg formula reads

Corollary 1.2.1 *The Lerch–Chowla–Selberg formula (2.6.4) for an imaginary quadratic field is* **a consequence of the functional equations** *for the Dedekind*

zeta-functions of the rational and the imaginary quadratic fields together with the decomposition theorem.

We state a quantitative formulation of the above passages. The residue of $\zeta_K(s, \mathfrak{C})$ in (2.5.42) is

$$\varkappa = \begin{cases} \frac{2\log\epsilon}{\sqrt{\Delta}} & \Delta > 0 \\ \frac{2\pi}{w\sqrt{-\Delta}} & \Delta < 0 \end{cases}. \tag{1.2.2}$$

In the imaginary quadratic case, this will be proved as Item (ii) of Theorem 2.5.3, which is one of crucial results on LCSF. $h\varkappa$ is the residue of $\zeta_K(s)$ and is a special case of (2.5.2).

The Laurent expansion (2.5.46) of $\zeta_K(s, \mathfrak{C})$ is of the form

$$\zeta_K(s, \mathfrak{C}) = \frac{\varkappa}{s-1} + c(\mathfrak{C}) + \cdots, \tag{1.2.3}$$

so that $A = B$ amounts to

$$\frac{L(1,\chi)}{s-1} + \gamma L(1,\chi) + L'(1,\chi) + \cdots = \frac{h\varkappa}{s-1} + \sum_{\mathfrak{C}} c(\mathfrak{C}) + \cdots, \tag{1.2.4}$$

whence the Dirichlet class number formula and the Lerch–Chowla–Selberg formula follow:

$$L(1,\chi) = h\varkappa, \quad \gamma L(1,\chi) + L'(1,\chi) = \sum_{\mathfrak{C}} c(\mathfrak{C}). \tag{1.2.5}$$

For CNF $L(1,\chi) = h\varkappa$, cf. Exercise 2.7.1. The most famous book that treats the Kronecker limit formula is Siegel (1964) in which the (absolute value of the) discriminant is assumed to be 1 and care must be taken when treating the general case. The most detailed treatment (in the case $\Delta < 0$) is Meyer (1981, pp. 174–205) which is concisely stated in Deninger (1984, pp. 184–185). Cf. Chakraborty et al. (2009b, pp. 119–130) and Chakraborty et al. (2009a) for higher derivatives.

As one can imagine, the results may be extended to the case of cyclotomic fields. Cf., e.g. Gut (1929), Deninger (1984).

The reason is intriguing: the Kronecker limit formula side is well known for a general Epstein zeta-function, Sect. 2.10 while the multiplicative decomposition side is assured by class field theory.

To reach LCSF we need to treat both sides of (1.2.5) as described by Corollary 1.2.1.

The left-hand side (LHS) is pertaining to an explicit expression for $L(1,\chi)$ and $L'(1,\chi)$, which we accommodate in the theory of periodic Dirichlet series, DFT (discrete Fourier transform) in Sect. 2.2.

LHS: Dirichlet L-function. Base change formula, Corollary 2.2.2, entails a number of important information including Gauss' formulas for the digamma function at rational arguments, Eisenstein formula, etc. These play an essential role in unifying Chowla's problem and Maillet–Demyanenko determinants (CPMD) in Chap. 5.

The right-hand side (RHS) is dealt with by the Kronecker limit formula. **RHS**: CSIF or Fourier–Whittaker expansion. Chapter 2 is in essence for concretizing the Hecke theory Sect. 2.12. For example, in the case of zeta-functions associated to holomorphic modular forms, CSIF is one of the equivalent conditions to HTFE and the q-expansion is more prevailing as the Hecke or Bochner modular relation (Theorem 2.12.3). One of the most well-known examples of the Bochner modular relation is Ramanujan's formula (2.12.7) for Lambert series. The key tool is the Hecke gamma transform, Proposition 2.1.2, which has played and will also play an important role in some contexts, esp. in the Plana summation formula (Theorem 4.4.1). It is a starting point for Sato's theory of PHV (prehomogeneous vector spaces), Remark 2.1.3.

In addition to extensive discussion on CSIF, we also state the Ewald expansion as the Riemann–Siegel formula in Sects. 2.14 and 2.15. And the Riesz sum as the character sum and Popov formula in Sect. 2.16, thus exhausting all Hecke modular relations in concrete examples.

Chapter 3 is devoted to the **ramified functional equation of Maass-type (RFEM)** 2 : 2, cf. Definition 3.2.1 and its applications, esp. unification of the theories of zeta-functions associated with Maass forms and Epstein-allied zeta-functions.

The fact that the q-expansion depends on holomorphy at infinity of the automorphic form in question led Maass to invent a new type of non-holomorphic automorphic forms known as Maass (wave) forms (Maass, 1949). They are characterized by the associated Fourier–Whittaker expansion. Cf. Bellman (1950).

As a direct continuation of the Fourier–Bessel expansion, we develop the theory of Fourier–Whittaker expansion which is an equivalent to the RFEM. The main result is Theorem 3.2.1. Symbolically, $G_{2,1}^{1,2} \leftrightarrow G_{2,2}^{2,2}$, where the left-hand side is the (modified) Whittaker function and the right-hand side is its Mellin transform (Lemma 3.3.1), cf. the G-function hierarchy (3.2.4). This is in contrast to the Fourier–Bessel expansion $G_{1,1}^{1,1} \leftrightarrow G_{0,2}^{2,0}$ in the Hecke modular relation, Theorem 2.12.3.

(EMS$\approx$EES) is the guiding link that connects holomorphic and non-holomorphic automorphic forms, and we understand that it gives CSIF for both $Z(s, Q)$ and $\zeta_{\mathbb{Z}^2}(s, \tau)$ once it is proved for $E(s, z)$ and vice versa. CSIF is no other than the Fourier–Bessel expansion but we keep the name CSIF because it is developed in Chowla and Selberg (1949) and Selberg and Chowla (1967), which are one of the clearest descriptions.

In addition to the proof of Theorem 2.5.1 by the beta transform method (Kuzumaki, 2011), cf. Remark 3.1.1, we give several other proofs, cf. Sect. 3.6. Our second proof uses the Fourier expansion in x to prove CSIF for $E(s, z)$ in Theorem 3.6.2.

Zhang and Williams (1995) uses (EZF$\approx$EES) to prove the Kronecker limit formula for $Z(s, Q)$. The third proof is given for Theorem 3.6.1 as contracted form of the ramified functional equation.

We also give CSIF and the Kronecker limit formula in a more general setting of Asai in Sect. 3.7. It is instructive to follow the proof in the light of Exercises 3.6.7 and 2.5.3. This will enhance the understanding of Chap. 2.

Example 3.4.1 states the Fourier–Whittaker expansion of HES (3.4.1) and illustrates the contents of Chap. 3. This reduces to CSIF for five Dirichlet series in Theorem 3.6.1. The chapter ends with the Petersson trace formula as a precursor to the Kuznetsov trace formula (Agarwal et al., 2023).

Chapter 4 is devoted to elucidation of a.a. number-theoretic summation formulas under the unifying *Principle*—Theorem 4.1.1—to be established in Sect. 4.1. Example 4.1.1 provides the case of the *processed Dedekind zeta-function* of the rational and quadratic field while Corollary 4.1.1 gives the unprocessed case. As in Definition 4.1.1 and Table 4.2, we accommodate FER (functional equation of Riemann), HTFE (Hecke-type functional equation), VFE (Voronoǐ functional equation) in the functional Eq. (4.1.9) as the cases $(r_1, r_2) = (1, 0)$, $(r_1, r_2) = (0, 1)$, $(r_1, r_2) = (2, 0)$, respectively, where r_1 resp. r_2 indicates the number of real resp. imaginary places of Ω.

Summation formulas covered in Chap. 4 are the general Lipschitz summation formula in Sect. 4.2, Plana's summation formula à la Koshlyakov in Sect. 4.4, the Poisson and Voronoǐ summation formula in Sect. 4.5. Although the Poisson summation formula (Theorem 4.5.3) is restricted to analytic functions, it is still remarkable since in many literature, it is applied as if it were outside of modular relations. Our restoration of Hamburger's work (Hamburger, 1922) is rather brief since the main ingredients are (variants of) Fourier–Bessel expansion, Riesz sums, etc., which have been known and are developed in preceding chapters. This chapter may be compared with Lindelöf (1905) whose whole Chap. 4 (35 pages) is devoted to the study of summation formulas.

Chapter 5 unifies CPMD and reveals the underlying principles. Let $f \in C(M)$ and suppose the DFT $\hat{f}(M) = \frac{1}{\sqrt{M}} \sum_{a=1}^{M} \varepsilon_a(M) f(a) = 0$. Then we have the expression (2.2.23) for $D(1, f)$ which reduces to the one for $L(1, \chi)$ for $f = \chi$ a non-principal Dirichlet character mod M.

Here the parity appears as an essential ingredient. For example, for odd character χ mod M, (2.2.23) amounts to (2.2.35), which in turn amounts to (5.2.17).

The (inverse) problem of Chowla refers to a proof of linear independence of cotangent values by the non-vanishing of $L(1, \chi)$. The determinant of the coefficient matrix of the system of linear equations

$$0 = \sum_{a,b \in T} c_a \cot\left(\pi \frac{ab^{-1}}{M}\right)$$

in unknowns c_a is transformed into the product of $L(1, \chi)$'s via the Dedekind determinant and Chowla's problem follows. Then by the second equality of (5.2.17), the determinant is expressed as the Maillet–Demyanenko determinant.

Some of the results in Chap. 5 may be summarized as

Theorem 1.2.1 (Theorems 5.2.2 and 5.4.1) *In the odd case, Chowla's problem and Maillet–Demyanenko determinants merge on the grounds of the Dedekind determinant and the product of (5.2.17). In the even case, MD part holds as in the odd* *case with Clausen functions in place of Bernoulli polynomials while the linear independence concerns about the Clausen function values which leads to Ramachandra units.*

Along with the main underlying principle of the present book is RHB correspondence or the modular relation, "Hecke's principle" (Bellman, 1961, p. 8) is ubiquitously (and implicitly) adopted: *Invariance under the group of transformations should always be made explicit, usually by means of an expansion in terms of group characters.* It is noteworthy that Hecke's principle is in conformity with M. Sato's intuition toward the PHV stated in Remark 2.1.3 with the Fourier integral as an expansion.

References

Agarwal, P., Kanemitsu, S., & Kuzumaki, T. (2023). Automorphic L-functions by way of the Kuznetsov sum formula. *Advances in Computational Mathematics and Modelling, 1*, 47–72.

Bellman, R. (1950). Generalized Eisenstein series and non-analytic automorphic functions. *Proceedings of the National Academy of Sciences of the United States of America, 36*, 356–359.

Bellman, R. (1961). *A brief introduction to theta functions*. New York: Holt Rinehart and Winston Inc.

Chakraborty, K., Kanemitsu, S., & Kuzumaki, T. (2009a). Finite expressions for higher derivatives of the Dirichlet L-function and the Deninger R-function. *Hardy-Ramanujan Journal, 32*, 38–53.

Chakraborty, K., Kanemitsu, S., & Tsukada, H. (2009b). *Vistas of special functions II*. Singapore etc: World Scientific.

Chowla, S., & Selberg, A. (1949). On Epstein's zeta-function (I). *Proceedings of the National Academy of Sciences of the United States of America,35*, 371–374; *Collected Papers of Atle Selberg I*, 367–370. Springer, 1989. *The Collected Papers of Sarvadaman Chowla II*, 719–722. CRM, 1999.

Davenport, H. (1980). *Multiplicative number theory*, 1st ed. Chicago: Markham. 1967, 2nd ed. New York etc: Springer.

Deninger, C. (1984). On the analogue of Chowla-Selberg formula for real quadratic fields. *Journal für die reine und angewandte Mathematik, 351*, 171–191.

Gut, M. (1929). Die Zetafunktion, die Klassenzahl und die Kronecker'shce Grenzformel eines beliebiges Kreiskörpers. *Commentarii Mathematici Helvetici, 1*, 160–226.

Hamburger, H. (1922). Über einige Beziehungen, die mit der Funktionalgleichung der Riemannschen ζ-Funktion äquivalent sind. *Mathematische Annalen, 85*, 129–140.

Kuzumaki, T. (2011). Asymptotic expansions for a class of zeta-functions. *The Ramanujan Journal, 24*, 331–343.

Lindelöf, E. (1905). *Calcul des résidus et ses applications a la théorie des fonctions*. Paris: Gauthier-Villars.

Maass, H. (1964). *Lectures on modular functions of one complex variable*. Bombay: Tata Institute. Revised in 1983.

Maass, H. (1949). Über eine neue Art von nichtanalytischen automorphen Funktionen und die Bestimmung Dirichletscher Reihen durch Funktionalgelichungen. *Mathematische Annalen, 121*, 141–183.

Meyer, C. (1981). *L-Funktionen quadratischer Zahlkörper*. Köln: Vorlesung.

Miyake, T. (1989). *Modular forms*. New York etc: Springer.

Petkovšek, M., Wilf, H. S., & Zeilberger, D. (1996). $A = B$. Wellesley: A. K. Peters.

Selberg, A., & Chowla, S. (1967). On Epstein's zeta-function. *Journal für die reine und angewandte Mathematik,227*, 86–110; Collected Papers of Atle Selberg I, Springer, 1989, 521–545; The Collected Papers of Sarvadaman Chowla II. *CRM,1999*, 1101–1125.

Serre, J.-P. (1973). *A course in arithmetic*. New York -Heidelberg-Berlin: Springer.

Siegel, C. L. (1964). *Analytische Zahlentheorie II*. Göttingen: University Göttingen.

Wang, N. -L., Kuzumkai, T. and Kanemitsu, S. (2025). A tapestry of ideas with Ramanujan's formula woven in. *Axioms,14*, 146. https://doi.org/10.3390/axioms14020146

Zhang, N.-X., & Williams, K. (1995). On the Epstein zeta function. *Tamkang Journal of Mathematics, 26*, 165–176.

Chapter 2
Ramified Functional Equations and (CSIF)

In this chapter the main figures are Epstein-type Eisenstein series (EES), Hecke functional equation (HFE), Chowla–Selberg integral formula (CSIF) = Fourier–Bessel expansion (FBE), K-Bessel function.

In this chapter we shall clarify the meaning of a ramified functional equation of Schnee's type (Definition 2.1.1) to its root as the union of the Riemann–Hecke–Bochner (RHB) correspondence and the parity, thus enabling to describe even and odd parts of an entity whenever it is possible to introduce parity.

Definition 2.0.1 We speak of the RHB correspondence or the modular relation to mean the genesis of the functional equation *under the Mellin transform pair* (2.1.2) as well as the equivalent assertions to the functional equations. Here the functional equation may be any type, ramified and unramified. Cf. Kanemitsu and Tsukada (2014).

In the specific case of the complex exponential function basis, we may express the ramified functional equation in the form of the generalized Euler identity and as with the Euler identity, we may deduce almost all results from one formula. This will be done in Sect. 2.0.

The ramified functional equation in (2.1.40) strongly motivates, as well as the Eisenstein formula (2.2.28), to shift to the study of Dirichlet series with periodic coefficients in Sect. 2.2. Already Schnee (1930) as formulated by Ishibashi and Kanemitsu (1999) established the same functional equation as the Hurwitz and Lerch zeta-functions. Supplementing the results in Wang et al. (2020) etc., we describe the Laurent coefficients of the relevant Dirichlet series including the Dirichlet L-functions, which in particular give ingredients of the one hand of the decomposition of the Dedekind zeta-function of an imaginary quadratic field.

Then as another direction of the RHB correspondence between the theta-function and the lattice zeta-function, we shall establish the Chowla–Selberg integral formula as the Fourier–Bessel expansion $H_{1,1}^{1,1} \leftrightarrow H_{0,2}^{2,0}$ in Sect. 2.5; this section is a link to the development in Chap. 3 on the real-analytic automorphic functions. CSIF gives the

K. Chakraborty et al., *Modular Relations and Parity in Number Theory*,
Infosys Science Foundation Series, https://doi.org/10.1007/978-981-96-6471-9_2

evaluation of the Laurent constant of the Dedekind zeta-function, the other part of the decomposition. Equating them, we deduce (genesis of) the Lerch–Chowla–Selberg formula in Sect. 2.7.

2.1 Preliminaries and RHB Correspondence for HZ and LZ

As a precursor, we shall describe the RHB correspondence between the theta-like functions and the Hurwitz and the Lerch-zeta-functions as a natural generalization of the Riemann zeta-function. Ehrenpreis (1987) contains another advanced treatment through Appel transform (p. 73). The odd part has been treated in Wang et al. (2020) to some extent and here we make clear the source of the transformation formula and the even and odd parts of the Hurwitz and the Lerch zeta-functions. The main results in this section are Theorems 2.1.1 and 2.1.2. The odd part alone in the former theorem entails the partial fraction expansion for the cotangent function, an equivalent to the functional equation for the Riemann zeta-function while the even part gives slightly restricted information, and the union of even and odd parts in the latter theorem describes the whole structure for an even function and includes the functional equation.

The **Mellin transform pair (MTP)** refers to the Mellin and inverse Mellin transform of the form

$$F(s) = \int_0^\infty \xi^s f(\xi) \frac{\mathrm{d}\xi}{\xi}, \quad a < \sigma < b, \tag{2.1.1}$$

$$f(x) = \frac{1}{2\pi i} \int_{(c)} x^{-s} F(s) \, \mathrm{d}s, \quad \operatorname{Re} x > 0, \quad a < c < b,$$

if

$$f(\xi) = O(\xi^{-b+\eta}), \quad \xi \to \infty. \quad f(\xi) = O(\xi^{-a-\eta}), \quad \xi \to +0$$

for every $\eta > 0$. Cf. Paris and Kaminski (2001, pp. 79–178), Oberhettinger (1974), Sneddon (1972, pp. 262–297) etc. Its special case, the Mellin transform pair for gamma function (MTPG)

$$\Gamma(s) = \int_0^\infty t^s e^{-t} \frac{\mathrm{d}t}{t}, \sigma > 0, \quad e^{-x} = \frac{1}{2\pi i} \int_{c-i\infty}^{c+i\infty} \Gamma(s) x^{-s} \, \mathrm{d}s, \quad c > 0 \tag{2.1.2}$$

is the genesis of the Hecke correspondence. As a memorial to Hecke, we call some specific cases allied to (2.1.2) the **Hecke gamma transform (HGT)** including

$$\Gamma(s) x^{-s} = \int_0^\infty t^s e^{-xt} \frac{\mathrm{d}t}{t} \tag{2.1.3}$$

for $x > 0$ and $\sigma > 0$. For the real merit of HGT, cf. Remark 3.1.1. For more details and extensions, cf. Proposition 2.1.2.

Let $\{\lambda_n\}$ be a strictly increasing sequence of real numbers with $\lambda_1 > 0$. Let

$$Z(s) = \sum_{n=1}^{\infty} \frac{a_n}{\lambda_n^s} \tag{2.1.4}$$

be absolutely convergent for $\sigma > \sigma_a^*$. Then (2.1.3) gives for $\sigma > \max\{\sigma_a^*, 0\}$

$$\Gamma(s)Z(s) = \int_0^{\infty} t^s L(t) \frac{\mathrm{d}t}{t}, \quad L(z) = \sum_{n=1}^{\infty} a_n e^{-\lambda_n z}, \tag{2.1.5}$$

which *enables to consider the rapidly convergent Lambert series $L(z)$ in place of $Z(s)$*. **We also refer to** (2.1.5) **as the (Hecke) gamma transform or (HGT) and we shall apply this as granted.** This is a *discrete version of HGT* and so is Exercise 2.1.11. For the *continuous version*, cf. Exercise 2.1.12, 3.6.4. Bearing Remark 3.1.1 in mind will support smooth reading.

In Hecke's case, the Lambert series is the q-expansion of a modular form and automorphy may be applied. Cf. Weil's comments above.

Exercise 2.1.1 Prove the following formulas for the gamma function which we very often use below without further notice.

Difference equation

$$\Gamma(s+1) = s\Gamma(s).$$

Residues of Γ

$$\operatorname*{Res}_{s=-m} \Gamma(s) = \frac{(-1)^m}{m!}, \quad m = 0, 1, 2, \cdots.$$

Duplication formula

$$\Gamma(2s) = \sqrt{\pi}^{-1} 2^{2s-1} \Gamma(s) \Gamma\left(s + \frac{1}{2}\right).$$

Reciprocity relation

$$\Gamma(s)\Gamma(1-s) = \frac{\pi}{\sin \pi s}.$$

The **Stirling formula** (Erdélyi et al., 1953, p. 47, (6))

$$\Gamma(s) \sim \sqrt{2\pi} e^{-\frac{\pi}{2}|t|} |t|^{\sigma - \frac{1}{2}}, \quad |t| \to \infty. \tag{2.1.6}$$

In addition to these, Kummer's Fourier series (2.1.47) is essential in the closed form (2.3.23) for $L'(1, \chi)$ for χ odd.

Proof can be found in all standard textbooks, e.g. Kanemitsu and Tsukada (2007), Sneddon (1972).

Exercise 2.1.2 Give a heuristic reasoning for the beta transform in Lemma 2.5.1.

Solution. We formally interpret the binomial expansion

$$(1+x)^{-s} = \sum_{r=0}^{\infty} \binom{-s}{r} x^r \approx \int_0^{\infty} \binom{-s}{r} x^r \, dr,$$

in which we substitute

$$\binom{-s}{r} = \frac{\Gamma(s+r)}{\Gamma(s)} \frac{(-1)^r}{\Gamma(r+1)}, \quad \frac{(-1)^r}{\Gamma(r+1)} \approx \Gamma(-r).$$

This gives a candidate for the required beta integral. Here we make the heuristic interpretation of the residue

$$\Gamma(-r) \approx \operatorname*{Res}_{s=-r} \Gamma(s) = \frac{(-1)^r}{r!}.$$

□

Example 2.1.1 We illustrate a typical procedure of applying the Cauchy residue theorem by proving the Mellin inversion formula (the second equality of) (2.1.2). It is, by definition, Meijer's G-function $G_{0,1}^{1,0}$, cf. Sect. 2.13. Thus we are to prove

$$G_{0,1}^{1,0}\left(x \,\middle|\, \begin{matrix} - \\ 0 \end{matrix}\right) = e^{-x}, \quad c > 0. \tag{2.1.7}$$

We integrate $\Gamma(s)x^{-s}$ along the rectangle with vertices at

$$c - iT, \; c + iT, \; -N + iT, \; -N - iT,$$

where we understand $T \to \infty, -T \to -\infty$ may be taken $iT \to i\infty, -iT \to -i\infty$ and $N \to \infty$ dose not coincide with the pole of $\Gamma(s)$. By the Stirling formula, the horizontal integrals tend to 0 as $T \to \infty$ and the integral in (2.1.7) may be replaced by the one along the vertical path $-N - i\infty \to -N + i\infty$ plus the sum of residues $\sum_{m=0}^{[N]} \frac{(-1)^m}{m!} x^m$. We call this "by **shifting the integration path** to the left up to $\sigma = -N$." The resulting vertical integral tends to 0 as $N \to \infty$ in view of difference equation. Hence the integral $\frac{1}{2\pi i} \int_{c-i\infty}^{c+i\infty} \Gamma(s)x^{-s} \, ds$ coincides with the sum of residues $\sum_{m=0}^{\infty} \frac{(-1)^m}{m!} x^m$, which is e^{-x}.

Hence **as long as the Stirling formula applies with exponential decay, the integral along $\sigma = c$ may be replaced by the integral along $\sigma = -N$ plus the sum of residues between $\sigma = c$ and $\sigma = -N$.** In the case only the polynomial order decay is available and there is a zeta-function factor, it sometimes happens that we may shift to $\sigma = -d, 0 < d$, then we may apply the functional equation and transform the integral along $\sigma = -d$ into a suitable form in some cases.

Cf. (2.12.7) as a typical example of the Bochner modular relation.

Recall the additive expression for the Riemann zeta-function (RZF)

$$\zeta(s) = \sum_{n=1}^{\infty} \frac{1}{n^s} \tag{2.1.8}$$

for $\sigma > 1$, where $n^{-s} = e^{-s \log n}$, with $\log n$ its principal value (once and for all). The series is absolutely convergent and continued analytically to a meromorphic function over the whole plane with a unique simple pole at $s = 1$ with residue 1. The analytic continuation is done by some expressions in $0 < \sigma < 1$ and by the functional equation (FER)

$$\xi(s) := \pi^{-\frac{s}{2}} \Gamma\left(\frac{s}{2}\right) \zeta(s) = \pi^{-\frac{1-s}{2}} \Gamma\left(\frac{1-s}{2}\right) \zeta(1-s) = \xi(1-s) \tag{2.1.9}$$

for $\sigma < 0$, where $\xi(s) = \xi_{\mathbb{Q}}(s)$ is the completed Dedekind zeta-function in Sect. 4.1. It turns out that for deeper theory on distribution of prime numbers, one needs the Euler product and Riemann himself defines by the Euler product first, cf. (1.1.1).

Sato (1992) gives a rather thorough table of functional equations which arise from the theta-transformation formula, cf. also Kanemitsu and Tsukada (2014, p. 34, Table 2.1). There, after the Riemann zeta-function, there comes the Dedekind zeta-function and Epstein zeta-functions associated to quadratic forms, etc. Riemann gave two proofs of the functional Eq. (2.1.9), one depending on a variant of the theta-functions ϑ_3 and the other, on the contour integral expression. For the former, cf. Exercise 2.1.7. Also cf. Davenport (1980, pp. 61–63), where the Dirichlet L-function case is also treated by the same technique. Cf. Remarks 2.1.2, 2.3.2.

Recall the boundary Lerch zeta-function (LZF) $\ell_s(x)$ defined by

$$\ell_s(x) = \sum_{n=1}^{\infty} \frac{e^{2\pi i x n}}{n^s}, \quad \sigma > 1 \quad \text{or} \quad \sigma > 0, x \notin \mathbb{Z}, \tag{2.1.10}$$

which has its counterpart, the Hurwitz zeta-function (HZF) defined by

$$\zeta(s, x) = \sum_{n=0}^{\infty} \frac{1}{(n+x)^s}, \quad \sigma > 1, \quad 0 < x \leq 1. \tag{2.1.11}$$

It is continued meromorphically over the whole plane with a simple pole at $s = 1$. These will be uplifted to more general zeta-functions in Sect. 4.1 and the range of x may be widened, cf. (4.1.15).

These are connected by the *Hurwitz formula* (HF) (i.e. the functional equation for the Hurwitz zeta-function): for $\sigma > 1, 0 < x \le 1$,

$$\zeta(1-s,x) = \frac{\Gamma(s)}{(2\pi)^s}\left(e^{-\frac{\pi i s}{2}}\ell_s(x) + e^{\frac{\pi i s}{2}}\ell_s(1-x)\right), \tag{2.1.12}$$

while its reciprocal (RHF) is

$$\ell_{1-s}(x) = \frac{\Gamma(s)}{(2\pi)^s}\left(e^{\frac{\pi i s}{2}}\zeta(s,x) + e^{-\frac{\pi i s}{2}}\zeta(s,1-x)\right), \quad 0 < x < 1. \tag{2.1.13}$$

By (2.1.13), the boundary Lerch zeta-function $\ell_{1-s}(x)$ is continued meromorphically over the whole plane with $s = 1$ a plausible singular point.

Exercise 2.1.3 Show that the limiting case with $x \to 1$ of (2.1.12) reads

$$\zeta(1-s) = \frac{2\Gamma(s)}{(2\pi)^s}\cos\left(\frac{\pi s}{2}\right)\zeta(s), \tag{2.1.14}$$

an asymmetric form of (FER) and that this is equivalent to (2.1.9).

Solution. This amounts to the proof of

$$\zeta(s) = \sqrt{\pi}^{2s-1}G(s)\zeta(1-s),$$

where $G(s)$ is the quotient of gamma factors

$$G(s) = \frac{\Gamma\left(\frac{1-s}{2}\right)}{\Gamma\left(\frac{s}{2}\right)}.$$

We assume the Laurent expansion (Proof will be given after Corollary 4.4.3)

$$\zeta(s,x) = \frac{1}{s-1} - \psi(x) + O(s-1), \quad \text{as} \quad s \to 1, \tag{2.1.15}$$

where

$$\psi(x) = \frac{\Gamma'}{\Gamma}(x) \tag{2.1.16}$$

is the **Euler digamma function** and γ indicates the **Euler constant**

$$\gamma = -\psi(1) = 0.5772156649\cdots. \tag{2.1.17}$$

Hence (2.1.15) reduces to

$$\zeta(s) = \frac{1}{s-1} + \gamma + O(s-1), \quad \text{as} \quad s \to 1. \tag{2.1.18}$$

$s = 0$ is a removable singularity of $\ell_s(x)$ if $x \in \mathbb{R} - \mathbb{Z}$ in view of (2.1.15). Indeed, comparison of both sides of (2.1.13) as $s \to 1$ leads to the relation

$$\psi(x) - \psi(1-x) = 2\pi i \ell_0(x) + \pi i, \tag{2.1.19}$$

which together with

$$\ell_0(x) = \frac{e^{2\pi i x}}{1 - e^{2\pi i x}} = \frac{1}{2}\left(-1 + i \cot \pi x\right) \tag{2.1.20}$$

for $x \in \mathbb{R} \setminus \mathbb{Z}$ entails the odd part formula for the digamma function

$$\psi(x) - \psi(1-x) = -\pi \cot \pi x \tag{2.1.21}$$

valid for non-integral values of x. Cf. odd part of Theorem 2.1.1 and the remark immediately after Theorem 2.3.3.

The Gaussian representation

$$\psi(x) = -\gamma - \frac{1}{x} + \sum_{n=1}^{\infty} \frac{x}{n(n+x)} \tag{2.1.22}$$

and (2.1.21) gives rise to the partial fraction expansion for the cotangent function

$$\begin{aligned} \cot \pi z &= \frac{1}{\pi z} + \frac{1}{\pi} \sum_{n-1}^{\infty} \left(\frac{1}{n+z} - \frac{1}{n-z} \right) \\ &= \frac{1}{\pi z} + \frac{2z}{\pi} \sum_{n=1}^{\infty} \frac{1}{z^2 - n^2} \end{aligned} \tag{2.1.23}$$

for non-integral values of z, where the last series is absolutely convergent. This is contained in the odd part assertion of Theorem 2.1.1 and will play an important role in some places.

We assemble the identities for $\ell_1(x)$, cf. e.g. Wang et al. (2020), Yamamoto (1977).

$$\begin{aligned} &\sum_{n=1}^{\infty} \frac{\cos(2\pi n x)}{n} + i \sum_{n=1}^{\infty} \frac{\sin(2\pi n x)}{n} = \ell_1(x) \\ &= -\log\left(1 - e^{2\pi i x}\right) = \sum_{n=1}^{\infty} \frac{e^{2\pi i n x}}{n} = A_1(x) - \pi i \overline{B}_1(x), \end{aligned} \tag{2.1.24}$$

$0 < x < 1$, where

$$A_1(x) = -\log 2|\sin \pi x| = \sum_{n=1}^{\infty} \frac{\cos(2\pi n x)}{n} \tag{2.1.25}$$

is its real part (even part), the first Clausen function (or the log sine function) and the imaginary part (odd part) is (2.1.36) with $\varkappa = 1$, which reads

$$x - [x] - \frac{1}{2} = \bar{B}_1(x) = -\frac{1}{\pi} \sum_{n=1}^{\infty} \frac{\sin(2\pi n x)}{n}, \tag{2.1.26}$$

where the second equality holds for $x \notin \mathbb{Z}$, and where $[x]$ indicates the greatest integer function. Cf. (2.2.25).

In a long-forgotten paper (Hutchinson, 1929) Hutchinson considers the zeta-function

$$Z_b(s) = Z(a, b, s) = \sum_{\nu=0}^{\infty} (a\nu + b)^{-s}, \quad 0 < b \le a,$$

which is $a^{-s}\zeta(s, x)$, where $x = \frac{b}{a}$, so that $0 < x \le 1$. Hence this is a slight generalization of the Hurwitz zeta-function. He considers the conjugate functions Z_b and Z_{a-b} and states the functional equation satisfied by the even and odd parts, (Hutchinson, 1929, (20), (21)):

$$\begin{aligned} Z_b + Z_{a-b} &= \frac{2}{\pi} \left(\frac{2\pi}{a}\right)^s \sin\left(\frac{\pi s}{2}\right) \Gamma(1-s) \sum_{n=1}^{\infty} \frac{\cos\left(2\pi \frac{b}{a} n\right)}{n^{1-s}} \\ &= \frac{1}{\pi} \left(\frac{2\pi}{a}\right)^s \sin\left(\frac{\pi s}{2}\right) \Gamma(1-s)(\ell_{1-s}(x) + \ell_{1-s}(1-x)) \end{aligned} \tag{2.1.27}$$

and

$$\begin{aligned} Z_b - Z_{a-b} &= \frac{2}{\pi} \left(\frac{2\pi}{a}\right)^s \cos\left(\frac{\pi s}{2}\right) \Gamma(1-s) \sum_{n=1}^{\infty} \frac{\sin\left(2\pi \frac{b}{a} n\right)}{n^{1-s}} \\ &= \frac{i}{\pi} \left(\frac{2\pi}{a}\right)^s \cos\left(\frac{\pi s}{2}\right) \Gamma(1-s)(\ell_{1-s}(x) - \ell_{1-s}(1-x)), \end{aligned} \tag{2.1.28}$$

where $x = \frac{b}{a}$.

He also proves the abundance principle (Knopp, 2000) for the functions that satisfy the functional equations (2.1.27) resp. (2.1.28), cf. Hutchinson (1929, Theorems 2, 3).

In what follows we use the notation

$$f^e(x) = \frac{1}{2}(f(x) + f(-x)), \quad f^o(x) = \frac{1}{2}(f(x) - f(-x)) \tag{2.1.29}$$
$$\text{or} \quad f^e(x) = \frac{1}{2}(f(x) + f(1-x)), \quad f^o(x) = \frac{1}{2}(f(x) - f(1-x))$$

as the case may be, so that

$$f = f^e + f^o. \tag{2.1.30}$$

It follows that f is even $\iff f = f^e$ resp. f is odd $\iff f = f^o$ (Table 2.1).

The following is a slightly modified table of Milnor (1983) (Table 2.2).

We say that a (continuous) function f satisfies the **Kubert identity** (or Kubert relation distribution property) $(*_s)$ if

$$\sum_{r=0}^{m-1} f\left(\frac{x+r}{m}\right) = m^{1-s} f(x) \tag{$*_s$}$$

holds for all $1 \le m \in \mathbb{Z}$ and $0 < x < 1$ and call the totality $\mathcal{K}_s$ of Kubert functions the **Kubert space**. Cf. Chap. 6, the first item. The complex vector space $\mathcal{K}_s$ is of dimension 2 and is spanned by $\ell_s(x)$ and $\ell_s(1-x)$ (s is not a negative integer) and similarly by the Hurwitz zeta basis (Milnor, 1983, p. 287).

The importance of the Hurwitz and the Lerch zeta-functions is reinforced by the fact that they form bases of two important vector spaces $\mathcal{K}_s$ and $D(M)$. Cf. Theorem 2.2.1 for the latter space.

Table 2.1 Parity and cosmic dual force

Objects	Odd	Even
Character	Odd	Even
Quadr. field	Imaginary	Real
Function	Bernoulli polyn.	Clausen ftn.
Class no.	Relative class no.	Class no. of max. real subfld
Determinant	Maillet	Maillet with Clausen ftn.

Table 2.2 Even and odd functions in Kubert space

$s =$	Odd	Even
-2	$\frac{\cos \pi x}{\sin^3 \pi x} = \frac{2}{\pi^3}\zeta^o(3, x) = 4i\ell_{-2}(x)$	$2\zeta^e(3, x)$
-1	$2\zeta^o(2, x)$	$\frac{1}{\sin^2 \pi x} = \frac{2}{\pi^2}\zeta^e(2, x)$
0	$-\pi \cot \pi x = 2\pi i \ell_0(x) + \pi i = 2\psi^o(x)$	$\bar{B}_0(x) = 1$
1	$\bar{B}_1(x) = -\frac{2}{2\pi i}\ell_1^o(x)$	$A_1(x) = \ell_1^e(x)$
2	$A_2(x) = \frac{1}{\pi i}\ell_2^o(x)$	$\bar{B}_2(x) = \frac{1}{\pi^2}\ell_2^e(x)$

Table 2.3 Transform and its inversion

Objects	Easier	Wanted
Process	Transform	Inversion
RHB corr.	Mellin transf.	Inverse Mellin
Fourier–Bessel	Boundary Lerch	Perturbed Dirichlet ser.
Diff., int.	Diff.	Int.

Here (2.1.20) and

$$\frac{\partial}{\partial x}\ell_s(x) = 2\pi i \ell_{s-1}(x), \quad x \in \mathcal{H}$$

give

$$\frac{1}{\sin^2 \pi x} = -4\ell_{-1}(x), \quad \frac{\cos \pi x}{\sin^3 \pi x} = 4i\ell_{-2}(x),$$

etc. In general

$$\begin{aligned} \ell_{-k}(x) &= \frac{1}{(2\pi i)^k}\frac{\mathrm{d}^k}{\mathrm{d}x^k}\ell_0(x) = \frac{1}{2(2\pi i)^k}\frac{\mathrm{d}^k}{\mathrm{d}x^k}(-1 + i\cot \pi x) \\ &= \frac{i}{2(2\pi i)^k}\frac{\mathrm{d}^k}{\mathrm{d}x^k}\cot \pi x, \end{aligned} \tag{2.1.31}$$

the last equality being true for $k \geq 1$, which may be thought of as $A_{1-k}(x)$. In a wide sense, a.a. processes are to make the object easier and then return to the original (Table 2.3).

We state the special case of (2.1.27) and (2.1.28) as a theorem.

Theorem 2.1.1 *The even part*

$$\begin{aligned} &\frac{1}{2}(\zeta(s,x) + \zeta(s,1-x)) = \zeta^e(s,x) \\ &= \frac{1}{\pi}(2\pi)^s \sin\left(\frac{\pi s}{2}\right)\Gamma(1-s)\ell_{1-s}^e(x) \\ &= (2\pi)^{s-1}\sin\left(\frac{\pi s}{2}\right)\Gamma(1-s)(\ell_{1-s}(x) + \ell_{1-s}(1-x)) \end{aligned} \tag{2.1.32}$$

is in RHB correspondence with the theta-transformation formula (2.1.64). *But it does not lead to the functional equation since it does not hold for* $x = 1$.

On the other hand, the odd part

$$\begin{aligned}\frac{1}{2}(\zeta(s,x)-\zeta(s,1-x)) &= \zeta^o(s,x) \\ &= -\frac{i}{\pi}(2\pi)^s\cos\left(\frac{\pi s}{2}\right)\Gamma(1-s)\ell^o_{1-s}(x) \\ &= -i(2\pi)^{s-1}\cos\left(\frac{\pi s}{2}\right)\Gamma(1-s)(\ell_{1-s}(x)-\ell_{1-s}(1-x))\end{aligned} \tag{2.1.33}$$

is in RHB correspondence (in the form of (2.1.78)*) with the transformation formula* (2.1.79) *for the conjugate theta-function* (2.1.77).

In the limit as $s \to 1$, *the odd part* (2.1.33) *entails the odd part formula* (2.1.21) *for* ψ *and in the long run entails the partial fraction expansion for the cotangent function* (2.1.23).

It is known (Kanemitsu & Tsukada, 2007, Chap. 5) *that* (2.1.14) *is equivalent to the functional equation* (2.1.9).

Proof of the even part is given in Lemma 2.1.1 while the odd part will be settled in Corollary 2.1.1 as well as toward the end of this section.

Proposition 2.1.1 *For* $1 \le \varkappa \in \mathbb{Z}$ *and* $0 < x < 1$ *we have*

$$\ell_\varkappa(x) = \frac{(2\pi i)^{\varkappa-1}}{\varkappa!}\left(A_\varkappa(x) - \pi i \bar{B}_\varkappa(x)\right), \tag{2.1.34}$$

where the even part $A_\varkappa(x)$ *is the Clausen function of order* $\varkappa$

$$\begin{aligned}A_\varkappa(x) &= \frac{\varkappa!}{2(2\pi i)^{\varkappa-1}}\sum_{n=-\infty}^{\infty}{}' \frac{\mathrm{sgn}(n)e^{2\pi i n x}}{n^\varkappa} \\ &= \frac{\varkappa!}{2(2\pi i)^{\varkappa-1}}\left(\ell_\varkappa(x) + (-1)^{\varkappa-1}\ell_\varkappa(1-x)\right)\end{aligned} \tag{2.1.35}$$

while the odd part $\bar{B}_\varkappa(x)$ *is the periodic Bernoulli polynomial of order* $\varkappa$:

$$\bar{B}_\varkappa(x) = -\frac{\varkappa!}{(2\pi i)^\varkappa}\sum_{n=-\infty}^{\infty}{}' \frac{e^{2\pi i n x}}{n^\varkappa} = -\frac{\varkappa!}{(2\pi i)^\varkappa}\left(\ell_\varkappa(x) - (-1)^{\varkappa-1}\ell_\varkappa(1-x)\right) \tag{2.1.36}$$

These are complementary to each other and satisfy the parity relation

$$A_\varkappa(x) = (-1)^{\varkappa-1}A_\varkappa(1-x), \quad \bar{B}_\varkappa(x) = (-1)^\varkappa \bar{B}_\varkappa(1-x). \tag{2.1.37}$$

Proof Equation (2.1.24) is the key, which may be proved by Dirichlet's test. Higher order case (2.1.34) follows by integration, implying (2.1.35) and (2.1.36). □

Exercise 2.1.4 Check that first a few Bernoulli polynomials are

$$B_1(x) = x - \frac{1}{2}, \quad B_2(x) = x^2 - x + \frac{1}{6}, \tag{2.1.38}$$

$$B_3(x) = x^3 - \frac{3}{2}x^2 + \frac{1}{2}x, \quad B_4(x) = x^4 - 2x^2 + x^3 - \frac{1}{30}, \cdots \tag{2.1.39}$$

Definition 2.1.1 Specifying the general case of the ramified functional equation (Kanemitsu & Tsukada, 2014, Chap. 7), we say that the Dirichlet series $\varphi(s)$ and $\psi_1(s)$, $\psi_2(s)$ satisfy the **ramified functional equation of Schnee type (RFES)** if the following holds:

$$\begin{aligned} &\Gamma\left(\frac{s}{2}\right)\Gamma\left(\frac{1+s}{2}\right)\varphi(s) \\ &= \Gamma\left(\frac{1-s}{2}\right)\Gamma\left(\frac{1+s}{2}\right)\psi_1(1-s) + \Gamma\left(\frac{s}{2}\right)\Gamma\left(1-\frac{s}{2}\right)\psi_2(1-s). \end{aligned} \tag{2.1.40}$$

For a slightly more general case with a more general gamma factor and only one Dirichlet series, cf. e.g. Li et al. (2020, Sect. 6).

Theorem 2.1.2 *The genesis of the functional equation* (2.1.9) *for the Riemann zeta-function lies in the union of the even part* (2.1.32) *and the odd part* (2.1.33)*:*

$$\begin{aligned} &\zeta(s, x) \\ &= \frac{1}{\pi}(2\pi)^s\Gamma(1-s)\left(\sin\left(\frac{\pi s}{2}\right)\ell^e_{1-s}(x) - i\cos\left(\frac{\pi s}{2}\right)\ell^o_{1-s}(x)\right) \\ &= (2\pi)^{s-1}\Gamma(1-s)\times \\ &\left(\sin\left(\frac{\pi s}{2}\right)(\ell_{1-s}(x) + \ell_{1-s}(1-x)) - i\cos\left(\frac{\pi s}{2}\right)(\ell_{1-s}(x) - \ell_{1-s}(1-x))\right), \end{aligned} \tag{2.1.41}$$

a reminiscence of Euler's identity, where the even resp. odd part is in RHB correspondence with the theta-function resp. the conjugate theta-function. This leads to the functional equation (2.1.14) *in view of vanishing of the odd part, i.e. the odd part is discarded in the functional equation. This amounts to RFES* (2.1.40) *with*

$$\begin{aligned} \varphi(s) &= \pi^{-\frac{s}{2}}\zeta(s, x), \\ \psi_1(s) &= \pi^{-\frac{s}{2}}\ell^e_s(x) = \frac{1}{2}(\pi^{-\frac{s}{2}}\ell_s(x) + \pi^{-\frac{s}{2}}\ell_s(1-x)) \\ \psi_2(s) &= -i\pi^{-\frac{s}{2}}\ell^o_s(x) = -\frac{i}{2}(\pi^{-\frac{s}{2}}\ell_s(x) - \pi^{-\frac{s}{2}}\ell_s(1-x)). \end{aligned} \tag{2.1.42}$$

This is the reciprocal of what is enunciated in Li et al. (2020)*. We may claim that the zeta-functions φ, ψ_1, ψ_2 are in* **ramified RHB correspondence** *with the theta-functions ϑ, $\tilde{\vartheta}$ in the sense that φ resp. ψ_i is in RHB correspondence with ϑ resp. $\tilde{\vartheta}$ with respective gamma factors.*

Proof Equation (2.1.41) follows from Theorem 2.1.1. Equation (2.1.41) with $x = 1$ leads to (2.1.14). To prove that (2.1.42) satisfies (2.1.40), we use the well-known results for the gamma function in Exercise 2.1.1. Substituting

$$\Gamma(1-s) = \sqrt{\pi}^{-1} 2^{-s} \Gamma\left(\frac{1-s}{2}\right) \Gamma\left(1-\frac{s}{2}\right), \quad \sin\left(\frac{\pi s}{2}\right) = \frac{\pi}{\Gamma\left(\frac{s}{2}\right) \Gamma\left(1-\frac{s}{2}\right)},$$
$$\cos\left(\frac{\pi s}{2}\right) = \sin\left(\frac{\pi(1-s)}{2}\right) = \frac{\pi}{\Gamma\left(\frac{1-s}{2}\right) \Gamma\left(\frac{1+s}{2}\right)}$$

in the right-hand side of (2.1.41) and multiplying by $\Gamma\left(\frac{s}{2}\right)\Gamma\left(\frac{1+s}{2}\right)$, we deduce that

$$\Gamma\left(\frac{s}{2}\right) \Gamma\left(\frac{1+s}{2}\right) \zeta(s,x) = \frac{1}{\sqrt{\pi}} \pi^s \tag{2.1.43}$$
$$\times \left(\Gamma\left(\frac{1-s}{2}\right) \Gamma\left(\frac{1+s}{2}\right) \ell^e_{1-s}(x) - \Gamma\left(\frac{s}{2}\right) \Gamma\left(1-\frac{s}{2}\right) i\ell^o_{1-s}(x)\right).$$

Finally, multiplying by $\frac{1}{2}\pi^{-\frac{s}{2}}$, we conclude (2.1.40).

The modification (2.1.42) is common and has the effect of taking the basic sequence as $\{\sqrt{\pi}n\}$ in place of $\{n\}$. □

For another proof, cf. Exercise 2.1.6, which shows that **RFES is a consequence of the Euler identity**. Ramified functional equation of Schnee type is an example of the ramified functional equation and more general case will be treated in Sect. 3.2.

Corollary 2.1.1 *The derivative of* (2.1.41) *at* $s = 0$ *amounts to*

$$\zeta'(0,x) = -(\log 2\pi + \gamma)\, \bar{B}_1(x) + \frac{1}{2} A_1(x) - \frac{i}{\pi} (\ell_1')^o(x), \tag{2.1.44}$$

where

$$(\ell_1')^o(x) = \frac{1}{2}(\ell_1'(x) - \ell_1'(1-x)) = i \sum_{n=1}^{\infty} \frac{\log n}{n} \sin 2\pi n x. \tag{2.1.45}$$

Under the assumption of the **Lerch formula**

$$\zeta'(0,x) = \log \frac{\Gamma(x)}{\sqrt{2\pi}}, \tag{2.1.46}$$

Equation (2.1.44) *leads to* **Kummer's Fourier series**

$$\log \frac{\Gamma(x)}{\sqrt{2\pi}} = \sum_{n=1}^{\infty} \left(\frac{1}{2n} \cos 2\pi n x + \frac{\gamma + \log 2\pi n}{\pi n} \sin 2\pi n x \right). \tag{2.1.47}$$

Cf. Corollary 2.16.2.

From (2.1.36) *and* (2.1.41) *the formula*

$$\zeta(1-\varkappa, 1-\{x\}) = -\frac{1}{\varkappa}\bar{B}_{\varkappa}(x) \tag{2.1.48}$$

follows.

The even part (2.1.32) *leads to the parity relation* $B_1(x) + B_1(1-x) = 0$, *whence the second formula in* (2.1.37).

Proof The derivative of (2.1.41) at $s = 0$ amounts to

$$\begin{aligned} 2\pi i\zeta'(0,x) &= \left(\log 2\pi - \Gamma'(1)\right)(\ell_1(x) - \ell_1(1-x)) \\ &\quad + \frac{\pi i}{2}(\ell_1(x) + \ell_1(1-x)) + \ell_1'(x) - \ell_1'(1-x) \end{aligned} \tag{2.1.49}$$

on correcting Wang et al. (2020, (4.38)). Substituting (2.1.35), (2.1.36) and

$$\ell_1'(x) := \frac{\partial}{\partial s}\ell_{1-s}(x)\Big|_{s=0} = \sum_{n=1}^{\infty} \frac{\log n}{n} e^{2\pi i n x}, \tag{2.1.50}$$

and invoking (2.1.46), we deduce (2.1.44) and (2.1.45). Adding (2.1.49) with x replaced by $1-x$ to (2.1.49), we find the Fourier series for the Clausen function $A_1(x)$ in (2.1.35) with $\varkappa = 1$.

Equation (2.1.41) with $s = 1-\varkappa$ implies that the Fourier series (2.1.36) is equal to $\zeta(1-\varkappa, x)$, whence (2.1.48).

The limiting case as $s \to 1$ of (2.1.33) immediately leads to (2.1.19) in view of (2.1.15).

Then as above, we conclude (2.1.21). □

Definition 2.1.2 We define the $\varkappa$th Bernoulli number by (2.1.48) with $x = 0$:

$$B_{\varkappa} = -\varkappa\zeta(1-\varkappa), \quad \varkappa \geq 2, \quad B_0 = 1, \quad B_1 = -\frac{1}{2} \tag{2.1.51}$$

and assume their basic properties, cf. e.g. Kanemitsu and Tsukada (2007, Chap. 1). Cf. (2.3.18).

Exercise 2.1.5 Generalize (2.1.48) to the zeta-regularization formula

$$\bar{B}_s(x) = -s\zeta(1-s, 1-\{x\}) \tag{2.1.52}$$

and confirm that it entails (2.1.51).

Kurokawa (1994) and Weil (1975) emphasize the importance of the Lerch formula (2.1.46) as the connecting link of the gamma function and the Hurwitz zeta-function, $A = B$. This appears as an essential ingredient in the finite form of LCSF in Sect. 2.7; cf. especially Table 2.5.

Exercise 2.1.6 Prove (RFES) in the form of Li et al. (2020, Lemma 1). The Dirichlet series (corresponding to (2.1.42))

$$\varphi(s) = \frac{1}{\pi^{\frac{s}{2}}} \ell_s(x), \tag{2.1.53}$$

$$\psi_1(s) = \frac{1}{2\pi^{\frac{s}{2}}} \zeta(s, x) + \frac{1}{2\pi^{\frac{s}{2}}} \zeta(s, 1-x) \tag{2.1.54}$$

and

$$\psi_2(s) = \frac{i}{2\pi^{\frac{s}{2}}} \zeta(s, x) - \frac{i}{2\pi^{\frac{s}{2}}} \zeta(s, 1-x)$$

satisfy the ramified functional equation (2.1.40) as a consequence of Euler's identity.

Solution. Euler's identity is equivalent to

$$\begin{cases} e^{\frac{\pi}{2}i(1-s)} & = \sin\frac{\pi}{2}s + i\cos\frac{\pi}{2}s \\ e^{-\frac{\pi}{2}i(1-s)} & = \sin\frac{\pi}{2}s - i\cos\frac{\pi}{2}s \end{cases}. \tag{2.1.55}$$

We rewrite the reciprocal Hurwitz formula (RHF) as $(0 < x < 1)$

$$\ell_s(x) = f_1(s) f_2(s), \tag{2.1.56}$$

say, where $f_1(s) = \frac{\Gamma(1-s)}{(2\pi)^{1-s}}$ and $f_2(s) = e^{\frac{1-s}{2}\pi i}\zeta(1-s, x) + e^{-\frac{1-s}{2}\pi i}\zeta(1-s, 1-x)$. Substituting (2.1.55), we find that

$$\begin{aligned} f_2(s) &= \sin\frac{\pi}{2}s\left(\zeta(1-s, x) + \zeta(1-s, 1-x)\right) \\ &+ i\cos\frac{\pi}{2}s\left(\zeta(1-s, x) - \zeta(1-s, 1-x)\right). \end{aligned} \tag{2.1.57}$$

We transform the first factor accordingly

$$f_1(s) = \frac{\pi^{s-\frac{3}{2}}}{2} \frac{\Gamma\left(\frac{1-s}{2}\right)}{\Gamma\left(\frac{s}{2}\right)} \frac{\pi}{\sin\frac{\pi}{2}s} \quad \text{or} \quad = \frac{\pi^{s-\frac{3}{2}}}{2} \frac{\Gamma\left(1-\frac{s}{2}\right)}{\Gamma\left(\frac{1+s}{2}\right)} \frac{\pi}{\cos\frac{\pi}{2}s}. \tag{2.1.58}$$

Substituting (2.1.57) and (2.1.58), we transform (2.1.56) into

$$\ell_s(x) = \frac{\pi^{s-\frac{1}{2}}\Gamma\left(\frac{1}{2}-\frac{s}{2}\right)}{2\Gamma\left(\frac{s}{2}\right)}\Big(\zeta(1-s,x)+\zeta(1-s,1-x)\Big) \tag{2.1.59}$$
$$+i\,\frac{\pi^{s-\frac{1}{2}}\Gamma\left(1-\frac{s}{2}\right)}{2\Gamma\left(\frac{s}{2}+\frac{1}{2}\right)}\Big(\zeta(1-s,x)-\zeta(1-s,1-x)\Big)$$
$$= \frac{\pi^{\frac{s}{2}}\Gamma\left(\frac{1}{2}-\frac{s}{2}\right)}{\Gamma\left(\frac{s}{2}\right)}\psi_1(1-s)+\frac{\pi^{\frac{s}{2}}\Gamma\left(1-\frac{s}{2}\right)}{\Gamma\left(\frac{s}{2}+\frac{1}{2}\right)}\psi_2(1-s).$$

Clearing the denominators (and multiplying by $\pi^{-\frac{s}{2}}$ in (2.1.59), we arrive at (2.1.40). □

Remark 2.1.1 Milnor (1983, Appendix 1) incorporates the parity relation in (2.1.12) and Theorem 2.1.1 in its ultimate form.

Equation (2.1.41) is valid for all x in the form

$$\zeta(s,1-\{x\}) = \frac{1}{\pi}(2\pi)^s\Gamma(1-s)\left(\sin\left(\frac{\pi s}{2}\right)\ell_{1-s}^{e}(x)+i\cos\left(\frac{\pi s}{2}\right)\ell_{1-s}^{o}(x)\right),$$

where

$$\{x\} = x-[x] = \bar{B}_1(x)+\frac{1}{2} \tag{2.1.60}$$

is the fractional part of x, $[x]$ being the integral part of x defined below (2.1.26). Cf. the passage after (2.2.12).

We note that we need the full force (2.1.41) or (2.1.11) to derive Kummer's Fourier series (2.1.47). With (the derivative of) the even part (2.1.32) alone, we can obtain only the Fourier series for the Clausen function $A_1(x)$ in (2.1.35), not Kummer's Fourier series. It can be shown that Kummer's Fourier series is, in the long run, equivalent to the functional equation for the Riemann zeta-function, (Kanemitsu & Tsukada, 2007, p. 108; Chakraborty et al., 2009a; Chakraborty, Kanemitsu and Tsukada, pp. 168–175).

Under (2.1.22), all Eqs. (2.1.19), (2.1.23), (2.1.21), (2.1.20) are equivalent:

$$\psi(x)-\psi(1-x) = 2\pi i\ell_0(x)+\pi i = 2\pi i\frac{e^{2\pi i x}}{1-e^{2\pi i x}}+\pi i \tag{2.1.61}$$
$$= -\pi\cot\pi x = -\frac{1}{x}-2x\sum_{n=1}^{\infty}\frac{1}{x^2-n^2},\quad 0<x<1.$$

Equation (2.1.36) resp. (2.1.21) being the case $s=0$ resp. (the limiting case as) $s\to 1$ of (2.1.41) are connected by the generalized (inverse) Eisenstein formula (2.2.28). Cf. Corollary 2.2.3.

Ogg (1969a) is of particular interest since the Jacobi triple product (Sect. 2.11) of the theta-function is deduced by comparing two transformation formulas, one from the definition as a series (2.1.62) and the other as the logarithm of the product and these are handled by the corresponding zeta-functions. We make use of Ogg's result, stated in Lemma 2.1.1, which establishes the RHB correspondence with the theta-transformation formula as well as the case of the even part of Theorem 2.1.1. Hence this is an RHB proof of the product formula with reference to uniqueness of an elliptic function.

Ogg (1969a, (1)) should read (n should read n^2)

$$\vartheta(\tau, x) = \vartheta_3(\tau, \pi x) = \sum_{n=-\infty}^{\infty} e^{2\pi i n x + \pi i \tau n^2} = \sum_{n=-\infty}^{\infty} z^n q^{n^2} \tag{2.1.62}$$

$$= 1 + \sum_{n=1}^{\infty} 2\cos(2\pi n x) q^{n^2},$$

where $z = e^{2\pi i x}$ and we write

$$q = e^{\pi i \tau}, \quad \tau \in \mathcal{H} \tag{2.1.63}$$

once and for all.

$\vartheta_3(\tau, x)$ is one of 4 Jacobi theta-functions. Cf. e.g. Chan (2020), Whittaker and Watson (1927). Throughout in what follows $\vartheta(\tau, x)$ *always means* $\vartheta_3(\tau, \pi x)$. $\vartheta(it, 0) = \sum_{n=-\infty}^{\infty} e^{-\pi t n^2}$ is the theta-function used by Riemann to deduce the functional equation for $\zeta(s)$, cf. Remark 2.1.2. It is clear that $\vartheta(\tau, x)$ is a doubly periodic function in τ and x with period 2 and 1, respectively:

$$\vartheta(\tau, x) = \vartheta(\tau + 2, x + 1),$$

Its characteristic property is the **theta-transformation formula** (cf. (3.6.30))

$$e^{\pi i \tau x^2} \vartheta(\tau, \tau x) = \sqrt{\frac{i}{\tau}} \vartheta\left(-\frac{1}{\tau}, x\right), \tag{2.1.64}$$

which we may express as

$$f_j(\tau, x) = \sqrt{\frac{i}{\tau}} f_k\left(-\frac{1}{\tau}, x\right), \quad \{j, k\} = \{1, 2\},$$

on putting

$$f_1(\tau, x) = e^{\pi i \tau x^2} \vartheta(\tau, \tau x), \quad f_2(\tau, x) = \vartheta(\tau, x).$$

Indeed, for $j = 2, k = 1$, this is (2.1.64). For $j = 1, k = 2$, we have

$$f_1\left(-\frac{1}{\tau}, x\right) = e^{\pi i (-\frac{1}{\tau}) x^2} \vartheta\left(-\frac{1}{\tau}, -\frac{1}{\tau} x\right) = e^{-\pi i \tau y^2} \vartheta\left(-\frac{1}{\tau}, y\right)$$

on writing $x = -\tau y$. Substituting from (2.1.64), we conclude (2.1.65).

We use the case of $\tau = it, t > 0$

$$f_j(it, x) = \sqrt{\frac{1}{t}} f_k\left(\frac{i}{t}, x\right), \quad \{j, k\} = \{1, 2\}. \tag{2.1.65}$$

Let

$$f_j(\infty) = \lim_{\operatorname{Im}\tau \to \infty} f_j(\tau, x), \quad j = 1, 2.$$

Then

$$f_1(\infty) = \begin{cases} 1 & x \in \mathbb{Z} \\ 0 & x \notin \mathbb{Z} \end{cases}, \quad f_2(\infty) = 1.$$

For $x \in \mathbb{R}$ consider the Mellin transform

$$\Phi_j(2s, x) = \int_0^\infty t^s (f_j(it, x) - f_j(\infty)) \frac{dt}{t}, \quad j = 1, 2, \tag{2.1.66}$$

where $\sigma > 0$. Subtracting $f_j(\infty)$ is to assure the convergence of the integral at the lower limit (cf. Example 2.12.1).

Lemma 2.1.1 (Ogg) *The functional equation*

$$\Phi_j(s, x) = \Phi_k(1 - s, x), \quad \{j, k\} = \{1, 2\} \tag{2.1.67}$$

is in RHB correspondence with (2.1.64) *and constitutes the even part in Theorem 2.1.1.*

$\Phi_j(s, x)$ with correction terms is EBV (entire and bounded in every vertical strip, Ogg, 1969b, p. xiii).

Proof Note that

$$f_1(it, x) = e^{-\pi t x^2} \vartheta(it, itx) = \sum_{n=-\infty}^{\infty} e^{-\pi t (n+x)^2}, \tag{2.1.68}$$

$$f_2(it, x) = \vartheta(it, x) = \sum_{n=-\infty}^{\infty} e^{2\pi i n x - \pi t n^2}.$$

We use (HGT) (2.1.3) with $y = (n + x)^2$ resp. $y = n^2$ to find that

$$\Phi_j(2s, x) = \pi^{-s} \Gamma(s) \phi_j(2s, x), \quad j = 1, 2, \tag{2.1.69}$$

where

$$\phi_1(s,x) = \sum_{n=-\infty, n+x\neq 0}^{\infty} |n+x|^{-s}, \quad |n+x|^{-s} = e^{-s\log|n+x|} \tag{2.1.70}$$

is the lattice zeta-function (2.5.27) and

$$\phi_2(s,x) = \sum_{n=-\infty, n\neq 0}^{\infty} e^{2\pi i n x}|n|^{-s} = \ell_s(x) + \ell_s(1-x) = \sum_{n=1}^{\infty} 2\cos(2\pi n x) n^{-s}. \tag{2.1.71}$$

Equation (2.1.69) leads to (2.1.67) since the functional equation in RHB correspondence with (2.1.65) is

$$\pi^{-\frac{s}{2}}\Gamma\left(\frac{s}{2}\right)\phi_j(s,x) = \pi^{-\frac{1-s}{2}}\Gamma\left(\frac{1-s}{2}\right)\phi_k(1-s,x). \tag{2.1.72}$$

By the same argument as in the proof of Theorem 2.1.2, we may transform (2.1.72) into the even part of Theorem 2.1.1 since by convention $0^{-s} = 0$, (2.1.69) may be expressed as

$$\phi_1(s,x) = \zeta(s,x) + \zeta(s,1-x).$$

□

Remark 2.1.2 $\phi_1(s,x)$ in (2.1.70) appears again as the lattice function in (2.5.27). This being a perturbed Dirichlet series, which are associated with the Fourier–Bessel expansion (Kanemitsu & Tsukada, 2014, Chap. 4), we will see two ample examples of the Fourier–Bessel expansion in subsequent sections.

In the above argument, the variable $2s$ is used instead of s which means that one considers the sequence $\{n^2\}$ rather than $\{n\}$ to the effect that one considers the Epstein zeta-function in a unary quadratic form. Cf. Exercise 2.12.7.

It would be inspirational to scrutinize Riemann's way of using the theta-function to prove FER. This will be treated again in Exercise 2.12.7 as the Epstein zeta-function of the unary quadratic form. As in Exercise 2.1.7, **Riemann's method leads to the Ewald expansion**, (2.12.45), i.e. the expansion with incomplete gamma function coefficients. It is developed by Lavrik and later by Terras, cf. Kanemitsu and Tsukada (2014, Chap. 5). The use of the full series amounts to Theorem 2.1.2.

Exercise 2.1.7 Analyze Riemann's way of using the theta-function to prove the functional equation (2.1.9).

Solution. Let

$$\omega(\tau,x) = \sum_{n=1}^{\infty} e^{2\pi i n x} q^{n^2}$$

be the half of the series. Then

$$\vartheta(\tau, x) = \vartheta(\tau, -x) = 1 + 2\omega^e(\tau, x) = 1 + \omega(\tau, x) + \omega(\tau, -x),$$

whence we have

$$2e^{\pi i\tau x^2}\omega^e(\tau, \tau x) = e^{\pi i\tau x^2}\vartheta(\tau, \tau x) - e^{\pi i\tau x^2} = \sqrt{\frac{i}{\tau^3}}\vartheta\left(-\frac{1}{\tau}, x\right) - e^{\pi i\tau x^2}. \tag{2.1.73}$$

Let

$$\omega(t) = \frac{1}{2}\vartheta(it, 0) - \frac{1}{2} = \sum_{n=1}^{\infty} e^{-\pi n^2 t}, \quad \mathrm{Re}\, t > 0.$$

Then (2.1.73) reads for $\omega(t)$

$$\omega(t^{-1}) = -\frac{1}{2} + \frac{1}{2}\sqrt{t} + \sqrt{t}\omega(t). \tag{2.1.74}$$

By HGT with s replaced by $\frac{s}{2}$

$$\pi^{-\frac{s}{2}}\Gamma\left(\frac{s}{2}\right)\zeta(s) = \int_0^{\infty} t^{\frac{s}{2}}\omega(t)\,\frac{\mathrm{d}t}{t}. \tag{2.1.75}$$

Riemann divides the integral in (2.1.75) into two parts, $\int_0^1$ and $\int_1^{\infty}$, make the change of variable $t \leftrightarrow \frac{1}{t}$ in the first integral and apply (2.1.74) to obtain (a symmetric form of) the functional equation (FER) or (2.1.9). Note that in the Hardy–Hecke transform (3.1.22), both t and $\frac{1}{t}$ are involved. □

Exercise 2.1.8 Confirm that the solution to Exercise 2.1.7 is another manifestation of the rasing property of the gamma transform in the spirit of Remark 3.1.1. Another example is Exercise 2.1.11.

Solution. By the gamma transform, we have (2.1.75) or (according to Hecke)

$$\pi^{-s}\Gamma(s)\zeta(2s) = \int_0^{\infty} t^{s}\omega(t)\,\frac{\mathrm{d}t}{t}, \tag{2.1.76}$$

where $\omega(t)$ is (a half of) the theta series. We apply the theta-transformation formula, but the resulting series is not the Hardy–Hecke transform as in Exercise 2.5.3, and we are led to the Ewald expansion. Here we perceive the difference between the shifted and non-shifted sequences. □

The difference between the gamma factor of RZF and HLF is explained in Remark 2.1.2, Sect. 2.12 and several other places.

We settle the case of the odd part in Theorem 2.1.1. It looks natural to introduce the *conjugate* theta-function as the theta-function-like (modular) function corresponding to (2.1.28). This was known to de la Vallée Poussin, cf. Remark 2.3.2. For $z = e^{2\pi ix}$ and $q = e^{\pi i\tau}$,

$$\tilde{\vartheta}(\tau, x) = \sum_{n=-\infty}^{\infty} ne^{2\pi inx+\pi i\tau n^2} = \sum_{n=-\infty}^{\infty} nz^n q^{n^2} = z\frac{\partial}{\partial z}\vartheta(\tau, x) \tag{2.1.77}$$
$$= \sum_{n=1}^{\infty} 2in\sin(2\pi nx)q^{n^2}.$$

Equation (2.1.40) necessitates the validity of

$$\Gamma\left(\frac{1+s}{2}\right)\varphi(s) = \Gamma\left(1-\frac{s}{2}\right)\psi_2(s). \tag{2.1.78}$$

It follows from RHB correspondence (i.e. from the Mellin inversion (2.1.2)) that for (2.1.78) to hold it is necessary and sufficient that the transformation formula is to be of the form

$$e^{\pi i\tau x^2}\tilde{\vartheta}(\tau, \tau x) = \sqrt{\frac{i}{\tau^3}}\tilde{\vartheta}\left(-\frac{1}{\tau}, x\right). \tag{2.1.79}$$

Summarizing, we have

Theorem 2.1.3 *The conjugate theta-function* (2.1.77) *satisfies the transformation formula* (2.1.79) *which is in RHB correspondence with the functional equation* (2.1.78). *It is odd w.r.t. x and $\tilde{\vartheta}(\tau, 0) = 0$.*

For product expressions for theta-allied functions, cf. Sect. 2.11.

We quote results of Zhu and Mehta (2021) which forms the basis for the odd part in Theorem 2.1.1.

Lemma 2.1.2 (i) *The partial fraction expansion for the cotangent function and the functional equation* (2.1.9) *for the Riemann zeta-function are equivalent.*
(ii) *The infinite product expansion for the sine function and the partial fraction expansion for the cotangent function are equivalent.*

The gamma function is defined as the Mellin transform of $e^{-\xi}$ as in (2.1.2). The extension of its validity plays an important role in our discussion and is presented as Proposition 2.1.2.

Exercise 2.1.9 Prove the following.
(i) Suppose $f(z)$ is analytic inside and on the arc $C : \theta \in [\theta_1, \theta_2]$, $\theta_1 < \theta_2$, of the circle $|z - a| = r$ save for $z = a$, at which it has at most a simple pole with residue k, i.e.

$$\lim_{r\to+0}(z-a)f(z) = k,$$

then

$$\lim_{r\to+0}\int_C f(z)\,\mathrm{d}z = ik(\theta_2-\theta_1).$$

(ii) Suppose $f(z)$ is analytic on the arc $C: z = Re^{i\theta}$, $\theta\in[\theta_1,\theta_2]$, $\theta_1<\theta_2$, that

$$f(Re^{i\theta}) = o(1), \quad R\to\infty.$$

that $x>0$, and that C is a part of the upper semi-circle. Then

$$\int_C e^{ixz} f(z)\,\mathrm{d}z = o(1).$$

Solution. Proof is given on Chakraborty et al. (2016, p. 51, Lemma 1.2) in the case of finite a. (ii) may be thought of as the case of $a=\infty$ and the proof is given on Chakraborty et al. (2016, pp. 42–54, Lemma 1.3) as Jordan's lemma. The finite case follows from the following.

$$\int_C f(z)\,\mathrm{d}z = \int_{\theta_1}^{\theta_2} f(a+e^{i\theta}) ire^{i\theta}\,\mathrm{d}\theta, \quad f(a+re^{i\theta}) = \frac{k}{re^{i\theta}} + O(1), \quad r\to 0.$$

The case of $a=\infty$ with $\theta\in[\theta_1,\theta_2]\subset\left[0,\frac{\pi}{2}\right]$ follows from the estimate

$$\int_C f(z)\,\mathrm{d}z = o\left(\int_{\theta_1}^{\theta_2} Re^{-xR^\sigma\sin\theta}\,\mathrm{d}\theta\right) = o\left(\int_0^{\frac{\pi}{2}} R^\sigma e^{-xR\frac{2}{\pi}\theta}\,\mathrm{d}\theta\right)$$
$$= o(R^{\sigma-1}) = o(1)$$

by Jordan's inequality. □

Proposition 2.1.2 *The Mellin transform pair* (2.1.2) *valid for* $\sigma>0$ *resp.* $\mathrm{Re}\,x>0$, $0<c$ *extends to the right-half plane (RHP)*

$$\mathrm{Re}\,x\geq 0, \quad x\neq 0: \quad 0\neq x = |x|e^{i\theta}, \quad -\frac{\pi}{2}\leq\theta\leq\frac{\pi}{2} \tag{2.1.80}$$

for $0<\sigma<1$ *resp.* $0<c<1$ *as the Hecke gamma transform and its inversion: The Hecke gamma transform* (2.1.3), *which holds for* $\mathrm{Re}\,x>0, \sigma>0$ *extends to* $\mathrm{Re}\,x=0, x\neq 0, 0<\sigma<1$ *and the purely imaginary case* $x=iy$, $0<y$ *amounts to the one-sided Fourier transform*

$$e^{-\frac{\pi i}{2}s}y^{-s}\Gamma(s) = \int_0^\infty \xi^s e^{-iy\xi}\frac{\mathrm{d}\xi}{\xi} = \sqrt{2\pi}F_+[\cdot^{s-1}](y). \tag{2.1.81}$$

Or as cosine and sine transforms,

$$C[\cdot^{s-1}](y) = \sqrt{\frac{2}{\pi}} y^{-s} \cos \frac{\pi}{2} s\Gamma(s), \quad S[\cdot^{s-1}](y) = \sqrt{\frac{2}{\pi}} y^{-s} \sin \frac{\pi}{2} s\Gamma(s). \quad (2.1.82)$$

Proof Kimura (1998, pp. 169–171) gives a detailed proof of HGT including the case $y < 0$. We modify (Sneddon, 1972, pp. 64–66) to make clear the use of Exercise 2.1.9. We integrate the function $f(z) = z^{s-1}e^{-xz}$, $0 < \sigma < 1$, $x > 0$ along the contour

$$C = c_r \cup [r, R] \cup C_R \cup \{\mathrm{Re}\, z = 0, \mathrm{Im}\, z : R \to r\},$$

where $c_r : z = re^{i\theta}, \theta : \frac{\pi}{2} \to 0$, $C_R : z = Re^{i\theta}, \theta : 0 \to \frac{\pi}{2}$, where $0 < r < R$. Then on C_R, we have $|f(z)| = o(1)$ for $0 < \sigma < 1$, so that $\int_{C_R} = o(1)$ by Exercise 2.1.9, (ii). On the vertical line $\{\mathrm{Im}\, z : R \to r\}$, we have

$$\int_{iR}^{ir} f(z)\, \mathrm{d}z = -\int_r^R \left(e^{\frac{\pi i}{2}} y\right)^s e^{-ixy} \frac{\mathrm{d}y}{y} = -e^{\frac{\pi i}{2}s} I,$$

where $I = \int_r^R \xi^{s-1} e^{-ix\xi}\, \mathrm{d}\xi$. Hence by Cauchy's theorem, we have

$$0 = \int_C f(z)\, \mathrm{d}z = \int_r^R \xi^{s-1} e^{-x\xi}\, \mathrm{d}\xi - e^{\frac{\pi i}{2}s} I + o(1)$$

as $r \to 0$, $R \to \infty$. Hence this leads to (2.1.81). □

Remark 2.1.3 (i) Sneddon (1972, p. 55, Lemma 2) states the counterpart $a = \infty$ case. It reads as the $a = 0$ and $r \to \infty$ case of Exercise 2.1.9, (i): If

$$\lim_{z \to \infty} z f(z) = k,$$

uniformly, then

$$\lim_{r \to \infty} \int_C f(z)\, \mathrm{d}z = ik(\theta_2 - \theta_1).$$

This follows from

$$f(Re^{i\theta}) = \frac{k}{Re^{i\theta}} + o\left(\frac{1}{R}\right), \quad R \to \infty.$$

But this does not work for $f(z)$ as is stated on Sneddon (1972, p. 56, ll. 1-2) and we need to appeal to Jordan's lemma.
(ii) Kimura (1998, p. 169, Lemma 4.20) reads: For $0 < \sigma < 1$ and $\mathbb{R} \ni x \neq 0$,

$$\int_0^\infty \xi^s e^{2\pi i \xi x} \frac{\mathrm{d}\xi}{\xi} = \left(2\pi |x| e^{\frac{\pi i}{2} \operatorname{sgn} x}\right)^{-s} \Gamma(s), \quad (2.1.83)$$

which is the purely imaginary case $\mathrm{Re}\, x = 0$ of HGT:

$$\int_0^\infty \xi^s e^{-\xi(-i|x|\,\mathrm{sgn}\,x)} \frac{\mathrm{d}\xi}{\xi} = (ix)^{-s}\Gamma(s).$$

Kimura (1998, pp. vii–x) views (2.1.83) as the **Fourier transform**

$$\int_0^\infty e^{2\pi i\xi x}\xi^{s-1}\,\mathrm{d}\xi = \left(2\pi e^{\frac{\pi i}{2}\,\mathrm{sgn}\,x}\right)^{-s}\Gamma(s)|x|^{-s}, \tag{2.1.84}$$

which sends the power ξ^{s-1} to the power $|x|^{-s}$. This led M. Sato to the search of systematically finding polynomials $f(x)$ over a vector space V whose complex power is sent to a complex power of another polynomial over the dual space V^* under the Fourier transform. This leads to systematic discovery of zeta-functions that satisfy the functional equation. Sato in 1961 found that there is a big-scale action of groups and was led to establishing the theory of Prehomogeneous vector spaces (PHV).

Exercise 2.1.10 Examine the case $y < 0$ of (2.1.81).

Exercise 2.1.11 Prove the integral representation for RZF

$$\zeta(s) = \frac{1}{\Gamma(s)}\int_0^\infty \frac{x^s}{e^x - 1}\frac{\mathrm{d}x}{x} \tag{2.1.85}$$

Solution. Summing (2.1.3) with $x = n$ over $n = 1, \cdots$ for $\sigma > 1$ and the geometric series, we deduce (2.1.85). □

Exercise 2.1.12 Use the gamma transform to prove the integral representation (Novikov, 1980)

$$\frac{\Gamma^2\left(\frac{s}{2}\right)}{\Gamma(s)} = 4(zz')^{s/2}\int_0^\infty \frac{1}{\left(zx^2 + z'x^{-2}\right)^s}\frac{\mathrm{d}x}{x}, \tag{2.1.86}$$

where $z > 0$, $z' > 0$ and for $\sigma > 0$ first and then for all s (save for non-positive integers) by analytic continuation.

This is Hecke's trick in deriving the Kronecker limit formula for a real quadratic field (cf. Sect. 2.9). A slight variation is the following.

$$\frac{\Gamma^2\left(\frac{s}{2}\right)}{\Gamma(s)} = 2\int_0^\infty \frac{1}{(e^v + e^{-v})^s}\,\mathrm{d}v, \tag{2.1.87}$$

which is $c(s)$ in Zagier's notation (Zagier, 1975).
Solution. Mutiplying (2.1.3) with s replaced by $\frac{s}{2}$ and with $x = z$ resp. $x = z'$, we have

$$\Gamma^2\left(\frac{s}{2}\right) = (zz')^{\frac{s}{2}}\int_0^\infty\int_0^\infty e^{-(zu+z'v)}(uv)^{\frac{s}{2}-1}\,\mathrm{d}u\mathrm{d}v.$$

Now by the change of variable $u = x^2y$, $v = x^{-2}y$, the last integral becomes

$$4(zz')^{-\frac{s}{2}}\int_0^\infty \frac{\mathrm{d}x}{x}\int_0^\infty e^{-(zx^2+z'x^{-2})y}y^s\,\frac{\mathrm{d}y}{y},$$

the inner integral of the right-hand side of which is $\frac{1}{(zx^2+z'x^{-2})^s}\Gamma(s)$, we conclude (2.1.86).

To prove (2.1.87), we make the change of variable $\sqrt{\frac{z}{z'}}x^2 = e^v$. Then

$$\int_0^\infty \frac{1}{\left(zx^2+z'x^{-2}\right)^s}\frac{\mathrm{d}x}{x} = \frac{1}{2}(zz')^{-s/2}\int_0^\infty \frac{1}{(e^v+e^{-v})^s}\,\mathrm{d}v. \tag{2.1.88}$$

Comparing (2.1.86) and (2.1.88) we conclude (2.1.87). □

2.2 Periodic Dirichlet Series

In connection with Remark 2.1.1, we summarize the results on periodic Dirichlet series as developed by Schnee (1930), Yamamoto (1977), Funakura (1990), Ishibashi and Kanemitsu (1999), etc. This includes Dirichlet L-function as a special case of *the Dirichlet series $D(s, f)$ with a periodic coefficient f of period M with orthogonality* $\sum_{a=1}^{M} f(a) = 0$.

As will be seen, it is the theory of DFT (discrete Fourier transform) as developed, e.g. in Frazier (1999), Chen (2010, Chap. 4) (under the name of Arithmetical Fourier transform—AFT) (cf. also Feng et al., 2021).

The theory of DFT for arithmetic functions has been developed in Li et al. (2013, Chap. 3) resp. in Li et al. (2017, pp. 109–114) in the case of periodic functions resp. in the case of a finite group. Cf. also Kanemitsu and Tsukada (2007, Sect. 8.1) and Chakraborty et al. (2009a, Sects. 4.1, 4.3).

Definition 2.2.1 Let $C(M)$ be the vector space of all periodic arithmetic functions f with period M:

$$C(M) = \{f : \mathbb{Z} \to \mathbb{C} | f(n+M) = f(n)\}. \tag{2.2.1}$$

Let

$$\varepsilon_j(a) = e^{2\pi i j a/M}, \quad 1 \le j \le M, \tag{2.2.2}$$

where a is an integer variable. Let

$$\mathcal{E} := \{\frac{1}{\sqrt{M}}\varepsilon_j | 1 \le j \le M\}.$$

Let χ_a be the characteristic function χ_a mod M (Chakraborty et al., 2009c, p. 73).

$$\chi_a(n) = \begin{cases} 1 & n \equiv a \bmod M \\ 0 & n \not\equiv a \bmod M \end{cases} \tag{2.2.3}$$

and let

$$\mathcal{X} = \{\chi_a | 1 \leq a \leq M\}.$$

For $f \in C(M)$ let

$$D(s, f) = \sum_{n=1}^{\infty} \frac{f(n)}{n^s} \tag{2.2.4}$$

and let $D(M)$ denote the set of all Dirichlet series of the form (2.2.4):

$$D(M) = \{D(s, f) | f \in C(M)\}$$

for $\sigma > 1$ in the first instance.

Correspondingly to $\mathcal{E}$ resp. $\mathcal{X}$ we let

$$\mathcal{E}(D) = \{\ell_s\left(\frac{a}{M}\right)\}, \quad \mathcal{X}(D) = \{\zeta\left(s, 1 - \left\{\frac{a}{M}\right\}\right)\}, \tag{2.2.5}$$

where $\ell_s(x)$ resp. $\zeta(s, x)$ is the Lerch zeta-function in (2.1.10) resp. the Hurwitz zeta-function in (2.1.11).

The following is a basic but far-reaching theorem which is used in Chap. 5 and will be elucidated in the light of characters in Sect. 5.6, cf. Exercise 5.6.1. Here we give a computational proof, providing some additional material.

Theorem 2.2.1 *$C(M)$ is an inner product space of dimension M isomorphic to $D(M)$ of all Dirichlet series $D(s, f)$ with periodic coefficients $f \in C(M)$. $C(M)$ has ONBs $\mathcal{E}$ consisting of (normalized) additive characters* (2.2.2) *and $\mathcal{X}$ consisting of characteristic functions* (2.2.3) *corresponding to orthogonality of characters (5.2.6). (5.2.7) amounts to the base change formula*

$$\sum_{a=1}^{M} f(a)\chi_a(n) = f(n) = \hat{\hat{f}}(-n) = \frac{1}{\sqrt{M}} \sum_{a=1}^{M} \hat{f}(a)\varepsilon_a(n), \tag{2.2.6}$$

where $\hat{f}$ is the DFT of $f \in C(M)$ defined by (2.2.8) *(a consequence of (5.2.11)).*

$D(M)$ has bases $\{\ell_s\left(\frac{a}{M}\right)\}$ resp. $\{\zeta\left(s, 1 - \{\frac{a}{M}\}\right)\}$ corresponding to (2.2.2) *resp.* (2.2.3). *The base change formula* (2.2.17), *which is a basis of merging (CPMD) in Theorem 5.2.2 is in the long run a consequence of (5.2.7) via* (2.2.6).

Proof The inner product of $f_1, f_2 \in C(M)$ is defined naturally by

$$(f_1, f_2) = \sum_{a \bmod M} f_1(a)\overline{f_2(a)},$$

where the bar $\bar{\cdot}$ means the complex conjugation of $\cdot$. $C(M)$ becomes an inner product space.

We have

$$(\varepsilon_j, \varepsilon_k) = \delta_{jk} M, \quad 1 \le j, k \le M, \tag{2.2.7}$$

so that $\mathcal{E}$ is an orthonormal system (ONS). Suppose it is an orthonormal basis (ONB). Then the orthogonal expansion reads

$$f = \sum_{b=1}^{M} \left(f, \frac{1}{\sqrt{M}} \varepsilon_b \right) \frac{1}{\sqrt{M}} \varepsilon_b = \sum_{b=1}^{M} \hat{f}(b) \frac{1}{\sqrt{M}} \varepsilon_b$$

for every $f \in C(M)$. Thus the **DFT** $\hat{f}$ (or the bth Fourier coefficient) of $f \in C(M)$ is to be defined by

$$\hat{f}(b) = \left(f, \frac{1}{\sqrt{M}} \varepsilon_b \right) = \frac{1}{\sqrt{M}} \sum_{a=1}^{M} \varepsilon_b(-a) f(a). \tag{2.2.8}$$

Once defined by (2.2.8), we find by orthogonality that

$$\hat{\hat{f}}(-a) = \frac{1}{\sqrt{M}} \sum_{b=1}^{M} \hat{f}(b) \varepsilon_b(a) = f(a), \tag{2.2.9}$$

the **Fourier inversion** or Fourier expansion formula. This is the right-hand side of equality. This also shows that $C(M)$ is generated by $\mathcal{E}$, so that the set $\mathcal{E}$ forms a basis (i.e. ONB) of $C(M)$ and *a fortiori* $C(M)$ is a vector space of dimension M. By (2.2.7), (2.2.8) reads for $f = \varepsilon_j$

$$\hat{\varepsilon}_j(b) = \sqrt{M} \delta_{bj} = \sqrt{M} \chi_j(b), \tag{2.2.10}$$

where χ_j is the characteristic function (2.2.3). $\mathcal{X}$ is another basis and provides the left-hand side of (2.2.6) and the inverse of (2.2.10):

$$\sqrt{M} \hat{\chi}_a(n) = \sum_{j=1}^{M} \varepsilon_n(-j) \chi_a(j) = \varepsilon_a(-n). \tag{2.2.11}$$

Since

$$\sum_{n=1}^{\infty} \frac{|f(n)|}{n^\sigma} \ll \zeta(\sigma),$$

the series in (2.2.4) is absolutely convergent for $\sigma > 1$. Then $D(M)$ forms a vector space of dimension M canonically isomorphic to $C(M)$. Hence the sets $D(s, \varepsilon_a)$

and $D(s, \chi_a)$ $(1 \le a \le M)$ are bases of $D(M)$. The first one coincides with $\mathcal{E}(D)$ in (2.2.5). The second one is $\{D(s, \chi_a) \mid 1 \le a \le M\}$, where

$$D(s, \chi_a) = \sum_{n=1}^{\infty} \frac{\chi_a(n)}{n^s} = \sum_{\substack{n=1 \\ n \equiv a \bmod M}}^{\infty} \frac{1}{n^s} = \zeta(s, a, M) = M^{-s}\zeta\left(s, \frac{a}{M}\right), \quad (2.2.12)$$

where $\zeta(s, a, M)$ indicates the partial zeta-function. Hence $\mathcal{X}(D)$ is another basis. Note that it is $\zeta(s, 1 - \{\frac{a}{M}\})$ that belongs to $D(M)$ rather than $\zeta(s, \{\frac{a}{M}\})$, cf. Remark 2.1.1.

Hence the right-hand side equality of (2.2.9) implies that

$$D(s, f) = \frac{1}{\sqrt{M}} \sum_{a=1}^{M} \hat{f}(a)\ell_s\left(\frac{a}{M}\right) = \frac{1}{\sqrt{M}} \sum_{a=1}^{M-1} \hat{f}(a)\ell_s\left(\frac{a}{M}\right) + \frac{\hat{f}(M)}{\sqrt{M}}\zeta(s). \quad (2.2.13)$$

It follows that $D(s, f)$ can be continued meromorphically over the whole plane and that it is entire if and only if $\hat{f}(M) = 0$ which reads by (2.2.8)

$$\hat{f}(M) = \frac{1}{\sqrt{M}} \sum_{a=1}^{M} f(a) = 0. \quad (2.2.14)$$

In parallel to (2.2.13), we have another expression

$$D(s, f) = \sum_{a=1}^{M} f(a)\zeta(s, a, M) = \frac{1}{M^s} \sum_{a=1}^{M} f(a)\zeta\left(s, \frac{a}{M}\right).$$

□

As in (2.1.29) we let

$$\begin{aligned} f^e &= \frac{1}{2}\left(f(n \bmod M) + f(-n \bmod M)\right), \\ f^o &= \frac{1}{2}\left(f(n \bmod M) - f(-n \bmod M)\right) \end{aligned} \quad (2.2.15)$$

be the even resp. odd part of f: $f = f^e + f^o$. Then the parity inherits to the DFT and we have

$$\begin{aligned} \hat{f^e}(b) &= \frac{1}{2}(\hat{f}(b) + \hat{f}(-b)) = \frac{1}{\sqrt{M}} \sum_{a=1}^{M} \cos\left(2\pi \frac{b}{M} a\right) f(a), \\ \hat{f^o}(b) &= \frac{1}{2}(\hat{f}(b) - \hat{f}(-b)) = i\frac{1}{\sqrt{M}} \sum_{a=1}^{M-1} \sin\left(2\pi \frac{b}{M} a\right) f(a). \end{aligned} \quad (2.2.16)$$

Hence we may express evenness resp. oddness of f by writing $f = f^e$ resp. $f = f^o$.

Table 2.4 Vector spaces and their bases

Vector space	$C(M)$	$D(M)$
Add. character basis	ε_a	$\ell_s\left(\frac{a}{M}\right)$
Characteristic. ftn. basis	χ_a	$\zeta\left(s, 1-\{\frac{a}{M}\}\right)$

Corollary 2.2.1 *$D(s, \hat{f}^o)$ is an entire function and so is $L(s, \chi)$ for non-principal Dirichlet character χ.*

Proof Condition (2.2.14) is satisfied for odd f and for non-principal χ by orthogonality of characters (Table 2.4). □

The following is a slightly modified version of Wang et al. (2020, Proposition 1).

Corollary 2.2.2 (base change formula) *For $f \in C(M)$*

$$\frac{1}{M^s}\sum_{a=1}^{M} f(a)\zeta\left(s, \frac{a}{M}\right) = \sum_{a=1}^{M} f(a)\zeta(s, a, M) = D(s, f) \tag{2.2.17}$$

$$= \frac{1}{\sqrt{M}}\sum_{a=1}^{M} \hat{f}(a)\ell_s\left(\frac{a}{M}\right) = \frac{1}{\sqrt{M}}\sum_{a=1}^{M-1} \hat{f}(a)\ell_s\left(\frac{a}{M}\right) + \frac{\hat{f}(M)}{\sqrt{M}}\zeta(s),$$

which entails the formula for the Laurent constant (cf. (2.2.22)*)*

$$-\frac{\hat{f}(M)}{\sqrt{M}}(\log M + \gamma) - \frac{1}{M}\sum_{a=1}^{M} f(a)\psi\left(\frac{a}{M}\right) = \lim_{s\to 1}\left(D(s, f) - \frac{\hat{f}(M)}{\sqrt{M}}\zeta(s)\right) \tag{2.2.18}$$

$$= \frac{1}{\sqrt{M}}\sum_{a=1}^{M-1} \hat{f}(a)\ell_1\left(\frac{a}{M}\right)$$

as well as the odd part corresponding to (2.1.33)

$$\frac{1}{M^s}\sum_{a=1}^{M-1} f(a)\zeta^o\left(s, \frac{a}{M}\right) = \frac{1}{M^s}\sum_{a=1}^{M-1} f^o(a)\zeta\left(s, \frac{a}{M}\right) \tag{2.2.19}$$

$$= D(s, f^o)$$

$$= \frac{1}{\sqrt{M}}\sum_{a=1}^{M-1} \hat{f^o}(a)\ell_s\left(\frac{a}{M}\right) = \frac{1}{\sqrt{M}}\sum_{a=1}^{M-1} \hat{f}(a)\ell_s^o\left(\frac{a}{M}\right)$$

by (2.2.16)*. Equation* (2.2.18) *with $\hat{f}(M) = 0$ reads*

$$-\frac{1}{M}\sum_{k=1}^{M} f(k)\psi\left(\frac{k}{M}\right) = D(1, f) = \frac{1}{\sqrt{M}}\sum_{k=1}^{M-1} \hat{f}(k)\ell_1\left(\frac{k}{M}\right). \tag{2.2.20}$$

The Laurent constant of (2.2.19) *is*

$$\frac{\pi}{2M}\sum_{k=1}^{M-1} f(k)\cot\left(\frac{\pi k}{M}\right) = -\frac{1}{M}\sum_{k=1}^{M-1} f(k)\psi^o\left(\frac{k}{M}\right) = D(1, f^o) \tag{2.2.21}$$
$$= \frac{1}{\sqrt{M}}\sum_{k=1}^{M-1} \hat{f}(k)\ell_1^o\left(\frac{k}{M}\right) = -\frac{\pi i}{\sqrt{M}}\sum_{k=1}^{M-1} \hat{f}(k)\bar{B}_1\left(\frac{k}{M}\right).$$

Proof Equation (2.2.18) is the equality between the Laurent constants of both sides of (2.2.17), where (2.1.15) and (2.1.18) are used. Equation (2.2.19) follows since $\hat{f}^o(M) = 0$. □

Remark 2.2.1 We note that the base change formula for $C(M)$ resp. $D(M)$ is another manifestation of $A = B$ in which the right-hand side A resp. the left-hand side B is the expression in terms of the orthogonality of characters resp. of characters of a finite Abelian group. Cf. Table 2.4.

Exercise 2.2.1 Prove that Wang et al. (2020, (47)) should read

$$\gamma_0(f) = \lim_{s\to 1}\left(D(s, f) - \frac{\hat{f}(M)}{\sqrt{M}}\frac{1}{s-1}\right) \tag{2.2.22}$$
$$= -\frac{\hat{f}(M)}{\sqrt{M}}\log M - \frac{1}{M}\sum_{k=1}^{M} f(k)\psi\left(\frac{k}{M}\right)$$
$$= \frac{1}{\sqrt{M}}\sum_{k=1}^{M-1} \hat{f}(k)\ell_1\left(\frac{k}{M}\right) + \frac{\hat{f}(M)}{\sqrt{M}}\gamma.$$

Let $f \in C(M)$ and suppose (2.2.14), Then prove the following expressions for $D(1, f)$.

$$-\frac{1}{M}\sum_{k=1}^{M} f(k)\psi\left(\frac{k}{M}\right) = D(1, f) = \frac{1}{\sqrt{M}}\sum_{k=1}^{M-1} \hat{f}(k)\ell_1\left(\frac{k}{M}\right). \tag{2.2.23}$$

Theorem 2.2.2 *Suppose $a \not\equiv 0 \bmod M$ by which we understand either we extend ψ to a periodic function or $1 \le a \le M-1$. Then Gauss' first formula for the digamma function* (2.1.21) *at rational argument*

$$\psi\left(\frac{a}{M}\right) = -\gamma - \log M + \pi i\sum_{b=1}^{M-1} \varepsilon_a(-b)\bar{B}_1\left(\frac{b}{M}\right) - \sum_{b=1}^{M-1} \varepsilon_a(-b)A_1\left(\frac{b}{M}\right) \tag{2.2.24}$$

(cf. Exercise 2.2.2) resp. Gauss' second formula, cf. Srivastava and Choi (2001, p. 19, (49)) *and* (2.1.24)

$$\sum_{b=1}^{M} e^{2\pi i \frac{a}{M} b} \psi\left(\frac{b}{M}\right) = \ell_1\left(\frac{a}{M}\right) = -\log\left(1 - e^{2\pi i \frac{a}{M}}\right) \tag{2.2.25}$$

is a consequence of (2.2.18) *with* $f = \chi_a$ *resp.* $f = \varepsilon_a$.

It is known Hashimoto et al. (2008) that (2.2.24) is equivalent to finite expressions for $L(1, \chi)$. We state this as Theorem 2.3.3. If $a \equiv 0 \bmod M$, then (2.2.24) reduces to (2.1.17) and (2.2.25) to the Kubert identity with $s = 1$

$$M^{-1} \sum_{r=0}^{M-1} \psi\left(\frac{x+r}{M}\right) = \psi(x) - \log M. \tag{$*_1$}$$

Exercise 2.2.2 Check that (2.2.24) has other expressions:

$$\psi\left(\frac{a}{M}\right) = -\gamma - \log M + \pi i \sum_{b=1}^{M-1} \varepsilon_a(-b) \bar{B}_1\left(\frac{b}{M}\right) - \sum_{j=1}^{M-1} \cos\frac{2\pi a j}{M} A_1\left(\frac{j}{M}\right) \tag{2.2.26}$$

$$= -\gamma - \log M - \frac{\pi}{2}\cot\frac{a}{M}\pi + \sum_{j-1}^{M-1} \cos\frac{2\pi a j}{M} \log 2 \sin\frac{\pi j}{M}$$

$$= -\gamma - \log 2M - \frac{\pi}{2}\cot\frac{a}{M}\pi + 2 \sum_{0<j<M/2} \cos\frac{2\pi a j}{M} \log \sin\frac{\pi j}{M}.$$

Here the first and the second equalities follow from (2.2.27) and (2.1.25).

Substituting (2.2.8) in the far-right side of (2.2.21) and using the inverse Eisenstein formula (2.2.28), we just return to the left-hand side. Cf. Remark 5.2.2.

Corollary 2.2.3 (i) *The pair of the Eisenstein formula*

$$\sqrt{M}\widehat{\cot}\left(\frac{\pi a}{M}\right) = \sum_{b=1}^{M-1} \varepsilon_a(-b) \cot\left(\frac{b}{M}\pi\right) = 2iM\bar{B}_1\left(\frac{a}{M}\right) \tag{2.2.27}$$

and the (inverse) Eisenstein formula

$$\sqrt{M}\hat{\bar{B}}_1\left(\frac{a}{M}\right) = \sum_{b=1}^{M-1} \varepsilon_a(-b) \bar{B}_1\left(\frac{b}{M}\right) = -\frac{1}{2i}\cot\left(\frac{\pi a}{M}\right) \tag{2.2.28}$$

is a consequence of

$$\frac{\pi}{M}\sum_{b=1}^{M-1} f(b)\cot\left(\frac{\pi b}{M}\right) = \frac{-2\pi i}{\sqrt{M}}\sum_{b=1}^{M-1} \hat{f}(b)\bar{B}_1\left(\frac{b}{M}\right) \tag{2.2.29}$$

with $f = \varepsilon_a$ *resp.* $f = \chi_a$ *for which* (2.2.10) *resp.* (2.2.11) *holds.*
(ii) *We have*

$$\begin{aligned} D'(1, f^o) &= \frac{1}{\sqrt{M}}\sum_{a=1}^{M-1} \hat{f}(a)(\ell_1')^o\left(\frac{a}{M}\right) \\ &= \frac{i}{\sqrt{M}}\sum_{a=1}^{M-1} \hat{f}(a)\sum_{n=1}^{\infty}\frac{\log n}{n}\sin\left(2\pi\frac{a}{M}n\right) \end{aligned} \tag{2.2.30}$$

which leads, under the assumption of the Lerch formula (2.1.46) *and the functional equation* (2.1.12), *to*

$$\begin{aligned} D'(1, f^o) = &\frac{i}{\sqrt{M}}\frac{\pi}{2}\sum_{a=1}^{M-1} \hat{f}(a)\left(\log\Gamma\left(\frac{a}{M}\right) - \log\Gamma\left(1-\frac{a}{M}\right)\right) \\ &+ \frac{i}{\sqrt{M}}\pi(\gamma + \log 2\pi)\sum_{a=1}^{M-1}\hat{f}(a)\bar{B}_1\left(\frac{a}{M}\right). \end{aligned} \tag{2.2.31}$$

Proof (i) (2.2.33) follows from (2.2.17) with $f = \chi_a$ in view of (2.2.11). The odd part of (2.2.18) reads

$$\begin{aligned} &-\frac{1}{M}\sum_{k=1}^{M-1} f(k)\left(\psi\left(\frac{k}{M}\right) - \psi\left(1-\frac{k}{M}\right)\right) \\ &= \frac{1}{\sqrt{M}}\sum_{k=1}^{M-1}\hat{f}(k)\left(\ell_1\left(\frac{k}{M}\right) - \ell_1\left(-\frac{k}{M}\right)\right). \end{aligned} \tag{2.2.32}$$

By (2.2.11), (2.2.32) with $f = \chi_b$ amounts to

$$\begin{aligned} &-\frac{1}{M}\left(\psi\left(\frac{b}{M}\right) - \psi\left(1-\frac{b}{M}\right)\right) \\ &= \frac{1}{M}\sum_{k=1}^{M-1}\varepsilon_b(-k)\left(\ell_1\left(\frac{k}{M}\right) - \ell_1\left(-\frac{k}{M}\right)\right). \end{aligned}$$

Substituting (2.1.36) with $\varkappa = 1$ and (2.1.21), we conclude (2.2.28).

Equation (2.2.30) follows by differentiating (2.2.17) whereby we appeal to (2.1.50). For (2.2.31), the Hurwitz formula (2.1.12) and the Kummer Fourier series (2.1.47) are used. □

In Li et al. (2010a) and references therein, a generalization of the Eisenstein formula

$$\sum_{j=1}^{M-1} e^{-2\pi i \frac{a}{M} j} \ell_s\left(\frac{j}{M}\right) = M^{1-s}\zeta\left(s, \frac{a}{M}\right) - \zeta(s) \tag{2.2.33}$$

has been used, it being understood that for $s = 1$ the right-hand side means the limit as $s \to 1$. This follows from (2.2.17) with $f = \chi_a$, cf. Eisenstein (1844), and also follows from the odd part (2.2.19) of (2.2.17).

Corollary 2.2.4 *Let χ be a non-principal Dirichlet character mod M. Then*

$$\frac{1}{M^s}\sum_{a=1}^{M-1} \chi(a)\zeta\left(s, \frac{a}{M}\right) = L(s, \chi) = \frac{1}{\sqrt{M}}\sum_{a=1}^{M-1} \hat{\chi}(a)\ell_s\left(\frac{a}{M}\right), \tag{2.2.34}$$

which entails

$$-\frac{1}{M}\sum_{a=1}^{M-1} \chi(a)\psi\left(\frac{a}{M}\right) = L(1, \chi) = \frac{1}{\sqrt{M}}\sum_{a=1}^{M-1} \hat{\chi}(a)\ell_1\left(\frac{a}{M}\right). \tag{2.2.35}$$

The above is about the Laurent constant of $D(s, f)$. Higher order Laurent coefficients will be studied in Sect. 2.4.

As another instance of RFES in Definition 2.1.1, we state the classical case:

Theorem 2.2.3 (Schnee, 1930; Ishibashi & Kanemitsu, 1999) *Suppose f is a periodic arithmetic function with period M. Then*

$$D(1-s, f) = \left(\frac{\pi}{M}\right)^{\frac{1}{2}-s}\left(\frac{\Gamma\left(\frac{s}{2}\right)}{\Gamma\left(\frac{1-s}{2}\right)}D(s, \hat{f}^e) + i\frac{\Gamma\left(\frac{1+s}{2}\right)}{\Gamma\left(1-\frac{s}{2}\right)}D(s, \hat{f}^o)\right), \tag{2.2.36}$$

or (RFES)

$$\begin{aligned} &\Gamma\left(\frac{s}{2}\right)\Gamma\left(\frac{1+s}{2}\right)\left(\frac{\pi}{M}\right)^{-\frac{s}{2}} D(s, f) \\ &= \Gamma\left(\frac{1-s}{2}\right)\Gamma\left(\frac{1+s}{2}\right)\left(\frac{\pi}{M}\right)^{-\frac{1-s}{2}} D(1-s, \hat{f}^e) \\ &+ i\Gamma\left(\frac{s}{2}\right)\Gamma\left(1-\frac{s}{2}\right)\left(\frac{\pi}{M}\right)^{-\frac{1-s}{2}} D(1-s, \hat{f}^o). \end{aligned} \tag{2.2.37}$$

This is equivalent to the generalized Euler identity

$$\tilde{D}(1-s, f) = \sqrt{\pi}^{-1} 2^{1-s} \Gamma(s) \left(\cos\left(\frac{\pi s}{2}\right) \tilde{D}(s, \hat{f}^e) + i \sin\left(\frac{\pi s}{2}\right) \tilde{D}(s, \hat{f}^o) \right), \tag{2.2.38}$$

where

$$\tilde{D}(s, f) = \left(\frac{\pi}{M}\right)^{-\frac{s}{2}} D(s, f).$$

Proof Equation (2.2.36) amounts to (2.2.37), i.e. the ramified functional equation (2.1.40), on clearing the denominators, multiplying by $\left(\frac{\pi}{M}\right)^{\frac{s-1}{2}}$ and replacing s by $1-s$. □

For the special case of Dirichlet L-function, cf. Theorem 2.3.2. The main theorem of Salerno et al. (1991) is a slightly different expression for the general Euler identity (2.2.38). We are thankful to Professor Atul Dixit for informing this paper.

It follows that

$$\begin{aligned} D(1, f^o) &= i\sqrt{\pi}^{-1} 2\left(\frac{\pi}{M}\right)^{\frac{1}{2}} D(0, \hat{f}^o) \lim_{s\to 0} \Gamma(s) \sin\left(\frac{\pi s}{2}\right) \\ &= i\frac{1}{\sqrt{M}} 2D(0, \hat{f}^o)\frac{\pi}{2} \end{aligned}$$

whence

$$D(1, f^o) = i\frac{\pi}{\sqrt{M}} D^o(0, \hat{f}^o).$$

Let χ be a Dirichlet character mod u, then $\chi \in C(ku)$ for $k = 1, 2, \ldots$. For a positive integer d we define $\chi^{(d)} \in C(du)$ by

$$\chi^{(d)}(n) = \begin{cases} \chi\left(\frac{n}{d}\right), & \text{if } d \mid n, \\ 0, & \text{if } d \nmid n. \end{cases}$$

Let

$$X(M) = \{\chi^{(d)} | \chi \text{ Dirichlet character } \bmod u \text{ and } du = M\}.$$

In Yamamoto (1977, Proposition 1.1) Yamamoto showed the following result:

Lemma 2.2.1 $\mathrm{X}(M)$ *is an orthogonal system (OS) of* $C(M)$.

It is immediate to deduce the following result from Lemma 2.2.1.

Lemma 2.2.2 *Let* $M > 2$. *Then*

$$\mathrm{X}_0(M) = \left\{ \frac{1}{\sqrt{\varphi(\frac{M}{d})}} \chi^{(d)} \,\middle|\, d|M \right\}$$

is an ONB of $C(M)$.

We note that the normality of the system in Lemma 2.2.2 entails the fact that the Euler function $\varphi(n)$ is the number of natural numbers $\leq n$ coprime to n. The system leads to an interesting Parseval type formula, cf. Chakraborty et al. (2009b).

This subsection continues on to Sect. 5.6.

2.3 Dirichlet Characters and L-Functions

In this section we introduce Dirichlet characters χ resp. associated Dirichlet L functions $L(s,\chi) = D(s,\chi)$ as important members of $C(M)$ resp. $D(M)$, cf. Apostol (1976a), Davenport (1980), Li et al. (2013, Chap. 3), Yamamoto (1977), etc.

Definition 2.3.1 The additive characters ε_j in (2.2.2) are characters of the residue classes $\mathbb{Z}/M\mathbb{Z}$. The reduced residue classes $G_M := (\mathbb{Z}/M\mathbb{Z})^\times$ is an Abelian group of order $\varphi(M)$, the Euler function counting the number of integers $1, \cdots, M$ relatively prime to M. Let $\hat{G}_M$ be the (multiplicative) character group of G_M. Then the values of $\chi \in \hat{G}_M$ are necessarily the roots of unity. From each reduced residue class character $\chi \in \hat{G}_M$ there arises its 0-extension denoted by the same symbol:

$$\chi(n) = \begin{cases} \chi(n + M\mathbb{Z}) & (n, M) = 1 \\ 0 & (n, M) > 1, \end{cases}$$

where (n, M) indicates the greatest common divisor of n and M. Then $\chi \in C(M)$ and is called a Dirichlet character mod M. The 0-extension of the trivial character is called the principal character and denoted χ_0. For a non-principal character χ there exists the least period $f|M$ (f is a divisor of M) and the unique Dirichlet character χ^* to the modulus f such that

$$\chi(n) = \chi_0\chi^*(n) = \begin{cases} \chi^*(n) & (n, M) = 1 \\ 0 & (n, M) > 1, \end{cases}$$

where χ_0 is the principal character mod M. f is called the conductor of χ and χ is said to be induced by χ^*. χ^* is not induced by any other character mod d with $d|f$. Such a character is called **primitive**, i.e. a Dirichlet character mod M is primitive if and only if its conductor is M. The Dirichlet series (2.2.4) with Dirichlet character coefficients is called the Dirichlet L-functions and denoted

$$L(s,\chi) = D(s,\chi) = \sum_{n=1}^{\infty} \frac{\chi(n)}{n^s} = \prod_p \left(1 - \frac{\chi(p)}{p^s}\right)^{-1}$$

for $\sigma > 1$ in the first instance. It is continued meromorphically over the whole plane with a possible simple pole at $s = 1$. Only for primitive characters, many formulas for L-functions take their simplest forms.

Theorem 2.3.1 (i) *For Dirichlet characters χ, ψ mod M, the orthogonality relation*

$$\sum_{a=1}^{M} \chi(a)\bar{\psi}(a) = \begin{cases} 0, & \chi \neq \psi \\ \varphi(M), & \chi = \psi \end{cases} \tag{2.3.1}$$

holds, and in particular,

$$\sum_{a=1}^{M} \chi(a) = \begin{cases} 0, & \chi \neq \chi_0 \\ \varphi(M), & \chi = \chi_0 \end{cases}.$$

(ii) *If χ is a non-principal character mod M induced by the primitive character χ^*, then*

$$L(s,\chi) = L(s,\chi^*)P(s,\chi^*), \quad P(s,\chi^*) = \prod_{p|M}\left(1 - \frac{\chi^*(p)}{p^s}\right). \tag{2.3.2}$$

If χ is a primitive character mod M, then $P(s,\chi^*) = 1$.
(iii) $0 \neq f \in C(M)$ *is a Dirichlet character mod M if it is multiplicative and $f(n) = 0$ for $(n, M) > 1$.*

Proof of (ii) given on Li et al. (2013, p. 101) depends on the relative primality principle on Li et al. (2013, p. 83) which in turn is based on the Möbius inversion formula. Chen (2010) makes full use of this formula in the context of arithmetical Fourier series.

Another proof of (ii) may be given by the multiplicative Eratosthenes sieve on Li et al. (2013, pp. 93–94).

$$\chi(p) = \begin{cases} 0, & p \mid M \\ \chi^*(p), & p \nmid M \end{cases}.$$

Then

$$\begin{aligned} L(s,\chi^*) &= \prod_{p}\left(1 - \frac{\chi^*(p)}{p^s}\right)^{-1} \\ &= \prod_{p \mid M}\left(1 - \frac{\chi^*(p)}{p^s}\right)^{-1} \prod_{p \nmid M}\left(1 - \frac{\chi^*(p)}{p^s}\right)^{-1} \\ &= \prod_{p \mid M}\left(1 - \frac{\chi^*(p)}{p^s}\right)^{-1} L(s,\chi) \end{aligned}$$

whence (2.3.2).

(iii) is a characterization by Prachar (1957, p. 192). Cf. also Spira (1970).

Remark 2.3.1 That a Dirichlet character χ mod M is a 0-extension of the residue class character $\chi \in \hat{G_M}$, where $G_M = (\mathbb{Z}/M\mathbb{Z})^*$ implies that any sum over the full period may be reduced to the sum over reduced residue classes:

$$\sum_{a=1}^{M-1} \chi(a) f(a) = \sum_{\substack{1 \le a \le M-1 \\ (a,M)=1}} \chi(a) f(a) = \sum_{\bar{a} \in G_M} \chi(\bar{a}) f(\bar{a}),$$

where we extend f to a periodic function with period M and χ in the third term is the residue class character.

Exercise 2.3.1 (Iwasawa, 1975, p. 4)

Let χ_1 and χ_2 be primitive characters with conductors f_1 and f_2, respectively. Then there is a unique primitive character with conductor f, $f \mid f_1 f_2$ such that

$$\chi(a) = \chi_1(a)\chi_2(a) \tag{2.3.3}$$

for $(a, f_1 f_2) = 1$. χ is called the product of χ_1 and χ_2 and denoted by: $\chi = \chi_1\chi_2$. Note that (2.3.3) does not necessarily hold for $(a, f_1 f_2) > 1$.

Prove that the set of all primitive characters forms a group under the multiplication (2.3.3).

Solution. The identity is given by the principal character χ_0^* mod 1: $\chi_0^*(n) = 1$ for all n. The inverse is given by the complex conjugate $\bar{\chi}$: $\bar{\chi}(a) = \overline{\chi(a)}$. □

Definition 2.3.2 Define the **general Gauss sum**

$$\tau(\chi, n) = \sum_{a=1}^{M} \chi(a) e^{\frac{2\pi i n a}{M}} = \sqrt{M}\hat{\chi}(-n), \tag{2.3.4}$$

where $\hat{\chi}$ is the DFT of χ (2.2.8) and the (normalized) **Gauss sum**

$$\tau(\chi) = \tau(\chi, 1) = \sum_{a=1}^{M} \chi(a) e^{\frac{2\pi i a}{M}}.$$

Lemma 2.3.1 (separability of the general Gauss sum) (i) *Suppose χ is a non-principal character mod M. Then the general Gauss sum* (2.3.4) *is separable, i.e.*

$$\tau(\chi, n) = \bar{\chi}(n)\tau(\chi) \tag{2.3.5}$$

if and only if χ is a primitive character mod M.
(ii) *For a Kronecker character, which is real primitive,*

$$\tau(\chi_d) = \begin{cases} i\sqrt{|d|} & d < 0 \\ \sqrt{d} & d > 0 \end{cases}. \tag{2.3.6}$$

(i) is due to Apostol (1976a, p.171, Theorem 8.19), Li et al. (2013, p. 131, Theorem 5.4).

It follows from (2.3.4) and (2.3.5) that for a primitive χ

$$\hat{\chi}(n) = \frac{\tau(\chi)}{\sqrt{M}}\chi(-1)\bar{\chi}(n) \tag{2.3.7}$$

and

$$\tau(\chi)\tau(\bar{\chi}) = |\tau(\chi)|^2 = M. \tag{2.3.8}$$

Correspondingly to Theorem 2.2.3 (cf. Theorem 2.1.2), we have

Theorem 2.3.2 *A non-principal character χ mod M is primitive if and only if the functional equation* (2.2.36) *holds true, which reads*

$$\left(\frac{\pi}{M}\right)^{-\frac{1-s}{2}} L(1-s,\chi) = \frac{\tau(\chi)}{\sqrt{M}}\left(\frac{\pi}{M}\right)^{-\frac{s}{2}} \left(\frac{\Gamma\left(\frac{s}{2}\right)}{\Gamma\left(\frac{1-s}{2}\right)}L^e(s,\bar{\chi}) + \frac{\Gamma\left(\frac{1+s}{2}\right)}{\Gamma\left(1-\frac{s}{2}\right)}iL^o(s,\bar{\chi})\right) \tag{2.3.9}$$

or

$$\begin{aligned}\left(\frac{\pi}{M}\right)^{-\frac{1-s}{2}} L(1-s,\chi) &= \frac{\tau(\chi)}{\sqrt{M}}\left(\frac{\pi}{M}\right)^{-\frac{s}{2}} \\ &\times \left(\frac{\Gamma\left(\frac{s}{2}\right)}{\Gamma\left(\frac{1-s}{2}\right)}\frac{1+\chi(-1)}{2}L(s,\bar{\chi}) + \frac{\Gamma\left(\frac{1+s}{2}\right)}{\Gamma\left(1-\frac{s}{2}\right)}\frac{1-\chi(-1)}{2}iL(s,\bar{\chi})\right).\end{aligned} \tag{2.3.10}$$

In the form of a generalized Euler identity

$$\begin{aligned}L(s,\chi) &= \frac{\tau(\chi)}{M^s}\frac{1}{\pi}(2\pi)^s\Gamma(1-s)\times \\ &\left(\sin\left(\frac{\pi s}{2}\right)L^e(1-s,\bar{\chi}) - i\cos\left(\frac{\pi s}{2}\right)L^o(1-s,\bar{\chi})\right) \\ &= \frac{\tau(\chi)}{M^s}\frac{1}{\pi}(2\pi)^s\Gamma(1-s)\times \\ &\left(\sin\left(\frac{\pi s}{2}\right)\frac{1+\chi(-1)}{2}L(1-s,\bar{\chi}) - i\cos\left(\frac{\pi s}{2}\right)\frac{1-\chi(-1)}{2}L(1-s,\bar{\chi})\right) \\ &= \frac{\tau(\chi)}{M^s}\frac{1}{\pi}(2\pi)^s\Gamma(1-s)\times \\ &\left(\sin\left(\frac{\pi s}{2}\right)(1-\mathfrak{a}(\chi))L(1-s,\bar{\chi}) - i\cos\left(\frac{\pi s}{2}\right)\mathfrak{a}(\chi)L(1-s,\bar{\chi})\right),\end{aligned} \tag{2.3.11}$$

where $\mathfrak{a}(\chi)$ is the parity symbol

$$\mathfrak{a} = \mathfrak{a}(\chi) = \frac{1-\chi(-1)}{2} = \begin{cases} 0 & \chi(-1) = 1 \\ 1 & \chi(-1) = -1 \end{cases}. \tag{2.3.12}$$

This is the genesis of the functional equation for the Dirichlet L-function as described by de La Vallée Poussin (as in Davenport).

Proof Equation (2.3.10) is a direct application of Theorem 2.2.3 together with (2.3.7), the expression of the DFT of χ in view of separability of the general Gauss sum, and (2.3.11) follows from (2.3.10) as in the proof of Theorem 2.1.2. Reverse assertion is due to Li et al. (2013, Exercise 72, p. 132). □

Then the common expression for the functional equation for the Dirichlet L-function reads

$$\xi(s,\chi) := \left(\frac{\pi}{M}\right)^{-\frac{s}{2}} \Gamma\left(\frac{s+\mathfrak{a}}{2}\right) L(s,\chi) = \frac{i^{\mathfrak{a}}\tau(\chi)}{\sqrt{M}} \xi(1-s,\bar{\chi})$$

can be read out from (2.3.10).

In the case of Dirichlet characters, their parity is separated, i.e. they are either even or odd and the parity symbol (2.3.12) describes their parity. χ is even resp. odd if and only if $\mathfrak{a} = 0$ $(\chi(-1) = 1)$ resp. $\mathfrak{a} = 1$ $(\chi(-1) = -1)$ and the associated functional equation reduces to an unramified one.

As a special case of the functional equation, Theorem 2.3.2, we have for a primitive odd χ

$$L(1,\chi) = -\pi i \frac{\tau(\chi)}{f} L(0,\bar{\chi}). \tag{2.3.13}$$

Lemma 2.3.2 (i) *Suppose χ is an odd character with conductor f. Then we have*

$$L(1,\chi) = \pi i \frac{\tau(\chi)}{f} B_{1,\bar{\chi}}, \tag{2.3.14}$$

where

$$B_{1,\bar{\chi}} = \sum_{a=1}^{f-1} \bar{\chi}(a) \bar{B}_1\left(\frac{a}{f}\right) \tag{2.3.15}$$

is the first generalized Bernoulli number.

(ii) *Let $\chi \neq \chi_0$ be a Dirichlet character* mod M *with conductor $f = f_\chi$ and let $P(s,\chi^*)$ be the product in (2.3.2). Then for $k \in \mathbb{N}$, the kth generalized Bernoulli number is defined by*

$$B_{k,\chi} = M^{k-1} \sum_{a=1}^{M-1} \chi(a) B_k\left(\frac{a}{M}\right) = B_{k,\chi^*} P(1-k,\chi^*), \tag{2.3.16}$$

where B_{k,χ^*} is defined by the first equality with M replaced by f. In particular,

$$B_{1,\chi} = \sum_{a=1}^{M-1} \chi(a)\bar{B}_1\left(\frac{a}{M}\right) = B_{1,\chi^*}P(0,\chi^*) = -i\frac{f}{\pi\tau(\chi^*)}L(1,\chi^*)P(0,\chi^*). \tag{2.3.17}$$

(iii) *The proper form of the functional equation of* $L(s,\chi)$ *for an odd imprimitive* χ *induced by the primitive* χ^* *seems to be*

$$P(1-s,\bar{\chi}^*)L(s,\chi) = -\frac{i\tau(\chi)}{\pi M^s}(2\pi)^s\Gamma(1-s)\cos\left(\frac{\pi s}{2}\right)P(s,\chi^*)L(1-s,\bar{\chi}).$$

Proof (i) follows from Corollary 2.2.4 in view of (2.1.36). Equation (2.3.17) follows from (2.3.13) and (2.3.14). Equation (2.3.16) is contained in Kanemitsu and Kuzumaki (2001, Lemma 2) and can be proved by the same Eratosthenes sieve as is used to prove (2.3.2). □

The generalized Bernoulli numbers (2.3.16) have been introduced to express the value $L(1-k,\chi)$:

$$L(1-k,\chi) = -\frac{B_{k,\chi}}{k}. \tag{2.3.18}$$

Cf. Washington (1983, pp. 30–31, 37). Cf. (2.1.48). The first generalized Bernoulli number (2.3.15) appears again as the sum S_1 in (2.16.1). Note that (2.3.14), in the long run, is a finite expression for the class number formula as with those in Theorem 2.3.3, in the imaginary quadratic case. These are connected by the base change formula (5.2.17). Cf. Table 2.9.

It remains to specify the theta-like functions and their transformation formulas which are in RHB correspondence with $D(s, f^e)$ resp. Theorem 2.1.1 as in Theorem 2.1.1.

$$\psi(t, f^e) = \vartheta(it, 0, f^e) = \sum_{n=-\infty}^{\infty} f^e(n)e^{-\frac{\pi}{M}tn^2}, \tag{2.3.19}$$

$$\tilde{\psi}(t, f^o) = \tilde{\vartheta}(it, 0, f^o) = \sum_{n=-\infty}^{\infty} nf^o(n)e^{-\frac{\pi}{M}tn^2},$$

which satisfies the transformation formula

$$\sqrt{\frac{1}{t}}\psi\left(\frac{1}{t}, f^e\right) = \psi(t, f^e), \quad \sqrt{\frac{1}{t^3}}\psi\left(\frac{1}{t}, f^o\right) = \tilde{\psi}(t, f^o).$$

These follow from the Mellin inversion (2.1.2), i.e. RHB correspondence.

Remark 2.3.2 In the case of Dirichlet L-functions, (2.3.19) has been used by de la Vallée Poussin (1896, p. 302), cf. e.g. Davenport (1980, pp. 68–71) . Similarly the conjugate theta-function (2.1.77) above.

Since our main concern above was in the odd part, only the Lerch zeta expression of Corollary 2.2.2 was used and Corollary 2.2.4 contains the unused part, i.e. (2.2.35) for a non-principal character χ. It is Kanemitsu and Tsukada (2007, p. 173, (8.28)) and is a basis of the main result of Hashimoto et al. (2008), Kanemitsu and Tsukada (2007, pp. 174–175, Theorem 8.2), which we state as Theorem 2.3.3. Cf. Sect. 5.6 for other applications.

Theorem 2.3.3 *Suppose χ is a Dirichlet character* mod M. *Then Gauss' first formula* (2.2.24) *for the digamma function at rational argument is equivalent to finite expressions for $L(1,\chi)$ (χ* mod M*)*

$$L(1,\chi) = \frac{\pi}{2M}\sum_{a=1}^{M-1} \chi(a)\cot\frac{a}{M}\pi \tag{2.3.20}$$

for χ odd and

$$L(1,\chi) = \frac{1}{\sqrt{M}}\sum_{a=1}^{M-1} \widehat{\chi}(a)A_1\left(\frac{a}{M}\right) = -\frac{1}{\sqrt{M}}\sum_{a=1}^{M-1} \widehat{\chi}(a)\log\left(2\sin\frac{a}{M}\pi\right) \tag{2.3.21}$$

for $\chi \neq \chi_0$ even, where $\widehat{\chi}$ is the DFT of χ or the general Gauss sum (2.3.4).

The finite expression (2.3.20) resp. (2.3.21) leads to a **finite form of the class number formula for an imaginary resp. real quadratic field**. In the imaginary quadratic case, it is here that the odd part formula for the digamma function (2.1.21) shows its effect. These will be applied in Chap. 5. For the χ_0 case, cf. Sect. 5.6.

Equation (2.2.30) reads for a primitive odd character χ

$$L'(1,\chi) = -\frac{\tau(\chi)}{M}\sum_{a=1}^{M-1} \bar{\chi}(a)(\ell_1')^o\left(\frac{a}{M}\right) \tag{2.3.22}$$

and under the Lerch formula (2.1.46)

$$L'(1,\chi) = \pi i\frac{\tau(\chi)}{M}\left(\sum_{a=1}^{M-1} \bar{\chi}(a)\log\Gamma\left(\frac{a}{M}\right) + (\gamma+\log 2\pi)B_{1,\bar{\chi}}\right), \tag{2.3.23}$$

where $B_{1,\bar{\chi}}$ is the first generalized Bernoulli number (2.3.15).

Corollary 2.3.1 *For the Kronecker character with $d < 0$,* (2.3.22) *and* (2.3.6) *lead to*

$$\gamma L(1,\chi_d) + L'(1,\chi_d) = -\frac{i}{\sqrt{|d|}}\sum_{a=1}^{|d|-1} \bar{\chi}_d(a)(\ell_1')^o\left(\frac{a}{|d|}\right) + \gamma L(1,\chi_d). \tag{2.3.24}$$

Equations (2.3.23) *and* (2.3.14) *lead to*

$$\gamma L(1,\chi)+L'(1,\chi)=\pi i\frac{\tau(\chi)}{M}\sum_{a=1}^{M-1}\bar{\chi}(a)\log\Gamma\left(\frac{a}{M}\right)+(2\gamma+\log 2\pi)L(1,\chi), \tag{2.3.25}$$

and

$$\gamma L(1,\chi_d)+L'(1,\chi_d)=-\frac{\pi}{\sqrt{|d|}}\sum_{a=1}^{|d|-1}\bar{\chi}_d(a)\log\Gamma\left(\frac{a}{|d|}\right)+(2\gamma+\log 2\pi)L(1,\chi_d). \tag{2.3.26}$$

2.4 Laurent Coefficients

We consider higher order Laurent coefficients of the periodic Dirichlet series $D(s, f)$ and mainly state results of Wang et al. (2020). Let

$$D(s,f)=\frac{\frac{\hat{f}(M)}{\sqrt{M}}}{s-1}+\frac{1}{\sqrt{M}}\sum_{k=0}^{\infty}\frac{\gamma_k(f)}{k!}(s-1)^k$$

resp.

$$\zeta(s,a,M)=\frac{\frac{1}{M}}{s-1}+\sum_{k=0}^{\infty}\frac{\gamma_k(a,M)}{k!}(s-1)^k$$

be the Laurent expansion of $D(s.f)$ resp. the partial zeta-function (2.2.12) around $s=1$, where

$$\gamma_k(a,M)=\lim_{x\to\infty}\sum_{\substack{n\le x\\ n\equiv a \pmod M}}\frac{\log^{k+1}n}{M(k+1)}.$$

$\gamma_k(a, M)$ or rather $(-1)^k\gamma_k(a, M)$ is called the kth **generalized Euler constant** and is extensively studied by Lehmer (1975), Briggs (1961), Kanemitsu (1987), Shirasaka (2002), et al. and culminated by Wang et al. (2020). We also use the Laurent expansion generalizing (2.1.15):

$$\zeta(s,x)=\frac{1}{s-1}-\psi(x)+\sum_{k=1}^{\infty}\frac{\gamma_k(x)}{k!}(s-1)^k.$$

The above two expansions reduce to the Laurent expansion for the Riemann zeta-function, which is often written as

$$\zeta(s)=\frac{1}{s-1}+\sum_{k=0}^{\infty}\frac{(-1)^k}{k!}\gamma_k(s-1)^k$$

Then we have the following theorems generalizing Ishibashi and Kanemitsu (1999, Theorem 2) and Shirasaka (2002, Theorem).

Theorem 2.4.1 (base change formula)

$$\frac{\frac{\hat{f}(M)}{\sqrt{M}}}{s-1}+\sum_{k=0}^{\infty}\frac{1}{k!}\left(\sum_{a=1}^{M} f(a)\gamma_k(a,M)\right)(s-1)^k = D(s,f) \tag{2.4.1}$$

$$=\frac{\frac{\hat{f}(M)}{\sqrt{M}}}{s-1}+\frac{1}{\sqrt{M}}\sum_{k=0}^{\infty}\frac{1}{k!}\left(\sum_{a=1}^{M-1}\hat{f}(a)(-1)^k\sigma_k^a+\hat{f}(M)(-1)^k\gamma_k\right)(s-1)^k$$

where

$$\sigma_k^a=\sum_{n=1}^{\infty}\frac{e^{2\pi i n\frac{a}{M}}}{n}\log^k n=(-1)^k\frac{\partial^k}{\partial^k s}\ell_1\left(\frac{a}{M}\right),$$

i.e. $\ell_1\left(\frac{a}{M}\right)$ *is a generating function of* σ_k^a *and a fortiori,*

$$\frac{1}{M}\sum_{a=1}^{M} f(a)\sum_{j=0}^{k}\binom{k}{j}(\log M)^{k-j}\gamma_j\left(\frac{a}{M}\right)=\sum_{a=1}^{M} f(a)\gamma_k(a,M) \tag{2.4.2}$$

$$=\frac{\gamma_k(f)}{\sqrt{M}}=\frac{1}{\sqrt{M}}\sum_{a=1}^{M-1}\hat{f}(a)\frac{\partial^k}{\partial^k s}\ell_1\left(\frac{a}{M}\right)+\frac{\hat{f}(M)}{\sqrt{M}}(-1)^k\gamma_k,$$

the case $k=0$ *leading to* (2.2.18).

Proof Equation (2.4.1) follows from (2.2.17) of Corollary 2.2.2.

Equation (2.4.2) follows from (2.4.1) save for the first equality which is a consequence of (2.2.12). □

The Laurent constant ($k=0$ case) has been elucidated in Corollary 2.2.2 with $\gamma_0(x)=-\psi(x)$ etc. We restrict to the $k=0$ case. The odd and even parts of (2.4.2) is a slight generalization of the Dirichlet class number formula in the finite form, e.g. through

$$L^{(k)}(1,\chi)=\sum_{a=1}^{M}\chi(a)\gamma_k(a,M).$$

Cf. Hashimoto et al. (2008), Kanemitsu and Tsukada (2007, pp. 176–182), etc. Gauss' first formula (2.2.24) which has been used in Lehmer (1975, p. 135), Funakura (1990) etc. is a consequence of (2.4.2) and so is Gauss' second formula (2.2.25).

Funakura (1990) deduces as his main theorem, the Kronecker limit formula for $D(s,f)$ which is compactified as Corollary 2.4.1. His proof depends on the first equality of Corollary 2.2.2, and therein he uses Gauss' first formula to transform the result into the required form. For its elucidation, we refer to Wang et al. (2020).

Corollary 2.4.1 (Funakura, 1990, Theorem 5)

$$\frac{1}{\sqrt{M}}\gamma_0(f) = \lim_{s\to 1}\left(D(s,f) - \frac{\frac{\hat{f}(M)}{\sqrt{M}}}{s-1}\right) = \frac{1}{\sqrt{M}}\sum_{a=1}^{M-1}\hat{f}(a)\ell_1\left(\frac{a}{M}\right) + \frac{\hat{f}(M)}{\sqrt{M}}\gamma, \tag{2.4.3}$$

which may be expressed as

$$= \frac{1}{\sqrt{M}}\sum_{a=1}^{M-1}\hat{f}(a)A_1\left(\frac{a}{M}\right) - \frac{\pi i}{\sqrt{M}}\sum_{a=1}^{M-1}\hat{f}(a)\bar{B}_1\left(\frac{a}{M}\right) + \frac{\hat{f}(M)}{\sqrt{M}}\gamma \tag{2.4.4}$$

and as

$$= -\frac{1}{\sqrt{M}}\sum_{a=1}^{M-1}\hat{f}(a)\log 2\sin\frac{\pi a}{M} + \frac{\pi}{2M}\sum_{a=1}^{M-1}f(a)\cot\frac{\pi a}{M} + \frac{\hat{f}(M)}{\sqrt{M}}\gamma. \tag{2.4.5}$$

Equations (2.4.4) and (2.4.5) follow from (2.1.24). This is the Lerch zeta part of Corollary 2.2.2.

As another example of generalized Euler constants, we state Stankus' theorem (Stankus, 1983) on the first generalized Euler constant $\gamma_1 = \gamma_0(1)$ as elucidated by Li (2005). It asserts that for $x \geq 2$ and $K \geq 2$ an integer, we have

$$\begin{aligned}\gamma_1 = \frac{1}{2}\Big(&\sum_{k=1}^{\infty}\frac{d(k)}{k}e^{-\frac{k^2}{x}} - \frac{1}{8}\log^2 x - \frac{3}{4}\gamma\log x - \frac{1}{8}\big(\gamma^2 + \zeta(2)\big) \\ &+ \sum_{m=1}^{K-1}\frac{(-1)^{m+1}B_{2m}^2}{m!(2m)^2}\frac{1}{x^m} + R_1(x,K)\Big),\end{aligned} \tag{2.4.6}$$

where

$$|R_1(x,K)| \leq C_1(K,\varepsilon)x^{-K+\varepsilon},\ \varepsilon > 0, \tag{2.4.7}$$

and where $C_1(K,\varepsilon)$ is a constant depending only on K and $\varepsilon > 0$. The numerical value is

$$\gamma_1 = -0,14563168\cdots.$$

As can be seen by (2.4.8), the modular relation (2.4.8) involves both x and $\frac{1}{x}$, so that in (2.4.6) x cannot be taken too large. Indeed Stankus chooses $x = 2.5$, which is rather small and the situation remains the same as Ramanujan's formula for zeta-values (at positive odd argument) (2.12.7). Cf. Kanemitsu et al. (2002), Example 3.4.1, etc.

We shall locate (2.4.6) as the modular relation

$$\sum_{k=1}^{\infty} \frac{d(k)}{k} G_{0,1}^{1,0}\left(\frac{k^2}{x}\,\middle|\,\begin{matrix}-\\0\end{matrix}\right) = \mathrm{P}(x) + \pi \sum_{k=1}^{\infty} d(k) G_{1,4}^{2,1}\left(\pi^4 x k^2\,\middle|\,\begin{matrix}1\\0,0,\frac{1}{2},\frac{1}{2}\end{matrix}\right), \tag{2.4.8}$$

a consequence of the functioned equation

$$\pi^{-s-1}\Gamma\left(\frac{s+1}{2}\right)^2 \zeta(s+1)^2 = \pi^s \Gamma\left(-\frac{s}{2}\right)^2 \zeta(-s)^2, \tag{2.4.9}$$

where $G_{p,q}^{m,n}$ signifies the Meijer G-function in Sect. 2.13 and $d(k)G_{1,4}^{2,1}$ depends on K.

$$\begin{aligned}\mathrm{P}(x) &= \sum\nolimits_{s=0,-2,\cdots,-2(K-1)} \mathrm{Res}\, \frac{\zeta^2(s+1)}{s}\Gamma\left(\frac{s}{2}+1\right)x^{\frac{s}{2}} \\ &= \frac{1}{8}\log^2 x + \frac{3}{4}\gamma \log x + \frac{1}{8}(\gamma^2 + \zeta(2)) + 2\gamma_1 - \sum_{m=1}^{K-1} \frac{(-1)^{m+1} B_{2m}^2}{m!(2m)^2}\frac{1}{x^m}\end{aligned}$$

in conformity with (2.4.6).

Stankus' argument is as follows. By (2.1.7), the left side of (2.4.8) is $\sum_{k=1}^{\infty} \frac{d(k)}{k} e^{-\frac{k^2}{x}}$ and at the same time is equal to

$$\frac{1}{2\pi i}\int_{(c)} \frac{\zeta^2(s+1)}{s}\Gamma\left(\frac{s}{2}+1\right)x^{\frac{s}{2}}\,\mathrm{d}s, \quad c > 1.$$

As in Example 2.1.1 we move the line of integration to $\sigma = -2K + \varepsilon, \varepsilon > 0$, encountering the poles at $s = -2m,\ m = 1, \cdots,\ K-1$ with residues $\frac{(-1)^m B_{2m}^2}{m!(2m)^2 x^m}$ and with the resulting integral

$$I_1 = \frac{1}{2\pi i}\int_{(-2K+\varepsilon)} \frac{\zeta^2(s+1)}{s}\Gamma\left(\frac{s}{2}+1\right)x^{\frac{s}{2}}\,\mathrm{d}s.$$

Then he applies the functional equation (2.4.9) to rewrite I_1 as

$$\begin{aligned}&\frac{1}{2\pi i}\int_{(2K-2\varepsilon)}\frac{\Gamma\left(-\frac{s}{2}\right)\Gamma\left(\frac{s}{2}\right)^2}{\Gamma\left(\frac{1}{2}-\frac{s}{2}\right)^2}\zeta(s)^2x^{-\frac{s}{2}}\,\mathrm{d}s\\&=2\frac{1}{2\pi i}\int_{(K-\varepsilon)}\frac{\Gamma\left(-s\right)\Gamma\left(s\right)^2}{\Gamma\left(\frac{1}{2}-s\right)^2}\zeta(2s)^2x^{-s}\,\mathrm{d}s\end{aligned}$$

which is the second term of the right-hand side of (2.4.8) and this is estimated as in (2.4.7). Thus (2.4.6) is a slight deformation of the modular relation (2.4.9).

Exercise 2.4.1 Simplify the above argument by considering

$$\frac{1}{2\pi i}\int_{(c)}\Gamma(s)\zeta^2(2s+1)x^s\,\mathrm{d}s\left(=\sum_{k=1}^{\infty}\frac{d(k)}{k}e^{-\frac{k^2}{x}}\right)$$

and

$$\zeta(2s+1)^2=\pi^{4s+1}\frac{\Gamma\left(-s\right)^2}{\Gamma\left(s+\frac{1}{2}\right)^2}\zeta(-s)^2.$$

Solution. The argument is as in Example 2.1.1 with the residual function being the same. □

Higher order generalized Euler constants of various kinds have been extensively studied. Cf. e.g. Kanemitsu (1987), Chakraborty et al. (2009a).

2.5 (CSIF) and Kronecker Limit Formula

In this section, we shall prove the Chowla–Selberg integral formula (CSIF) and deduce the Kronecker limit formula as well as the half of the Dirichlet class number formula for a quadratic field. The K-Bessel function plays a central role for which we refer to Sects. 3.1 and 3.6. In the latter, we shall continue the study of CSIF based on the Fourier expansion and DEs.

We assume some basics from algebraic number theory and state some results which will be used in what follows. Useful references are Goldstein (1971), Kitaoka (2025), Takagi (1971), etc. In elucidating Koshlyakov's results in later chapters, we treat the rational and quadratic fields, denoting the field Ω (some more material on this will be stated in Sect. 4.1). However, for the moment, we consider a general algebraic number field, denoted K, up to cyclotomic fields with degree

$$n=[K:\mathbb{Q}]=r_1+2r_2,$$

with r_1 resp. r_2 real resp. imaginary conjugates.

Let

$$\zeta_K(s) = \sum_{0 \neq \mathfrak{a} \in \mathfrak{O}} (N\mathfrak{a})^{-s}, \ \sigma > 1 \tag{2.5.1}$$

be the **Dedekind zeta-function (DZF)** of K, where $N\mathfrak{a} = [\mathfrak{O} : \mathfrak{a}]$ is the norm of the ideal $\mathfrak{a}$ and where $\mathfrak{O} = \mathfrak{O}_K$ is the **ring of algebraic integers** in K. Then on one hand, it amounts to finding a closed form for the residue and the Laurent constant of $\zeta_K(s)$ corresponding to Sect. 2.4.

The class number formula follows by equating the found residue and the one arising from decomposition theorem, Lemma 2.6.1.

The Kronecker limit formula refers to a closed form for the Laurent constant of the Epstein zeta-function of a binary quadratic form. The method is well known and is due to Kronecker and Hecke in the case of imaginary and real quadratic fields, respectively. There are many references, e.g. Siegel (1961) for imaginary quadratic fields and Hecke (1917), Zagier (1975), Novikov (1980), Bump and Goldfeld (1987), etc. for real quadratic fields, and references therein. For further developments, cf. e.g. Yoshida (1999).

Let I/P denote the **absolute ideal class group**, where I and P signify the group of all fractional ideals and the principal fractional ideals of K, respectively. The order $|I/P|$ is finite and is called the **class number** of K, denoted $h = h_K$. The Dedekind zeta-function has a simple pole at $s = 1$ with residue

$$\rho = h\varkappa = \frac{2^{r+1}\pi^{r_2} R h}{w\sqrt{|\Delta|}}, \tag{2.5.2}$$

where Δ is the **discriminant**, R is the **regulator**, w is the order of the group W_K of roots of unity in K and $\varkappa$ is the residue of the class zeta-function, cf. (1.2.2). Let $\mathfrak{O}_K^\times$ denote the **unit group** of K. r is the rank of the free Abelian group $\mathfrak{O}_K^\times / W_K$ and $r = r_1 + r_2 - 1$ (Dirichlet unit theorem). This may be expressed if the extension degree $n \leq 2$ as

$$\rho = -\frac{2^{r+1}\pi^{r_2}\zeta_\Omega^{(r)}(0)}{\sqrt{|\Delta|}}. \tag{2.5.3}$$

These will be proved in Theorem 2.5.1 for an imaginary quadratic field on account of

$$\zeta_K(s) = \frac{1}{w}\sum_Q Z(s, Q)$$

and in Sect. 2.6 for a real quadratic field.

For an ideal class $\mathfrak{C} \in I/P$ let

$$\zeta(s, \mathfrak{C}) = \zeta_K(s, \mathfrak{C}) = \sum_{0 \neq \mathfrak{a} \in \mathfrak{C}} (N\mathfrak{a})^{-s}, \ \sigma > 1 \tag{2.5.4}$$

be the class zeta-function. Then we have the additive decomposition

$$\zeta_K(s) = \sum_{\mathfrak{C} \in I/P} \zeta(s, \mathfrak{C}). \tag{2.5.5}$$

Therefore, it suffices to consider the class zeta-function and the approach for studying it (for a real or an imaginary quadratic field) remains the same as described, e.g. in Meyer (1981), Zagier (1975), etc. For a quadratic field $K = \mathbb{Q}(\sqrt{D})$, its unit group $\mathfrak{O}_K^\times$ is the finite group W_K of roots of unity of order w for an imaginary quadratic field and the infinite group $\{\pm\varepsilon^n | n \in \mathbb{Z}\}$ for a real quadratic field, respectively. Here ε signifies the **fundamental unit** (the smallest unit > 1) of K (Dirichlet unit theorem).

Now fix $\mathfrak{b} \in \mathfrak{C}^{-1}$. Then for $\mathfrak{a} \in \mathfrak{C}$ in (2.5.4), $\mathfrak{ab} = (\lambda) \in P$, where $\lambda \in \mathfrak{b}$. Hence $(N\mathfrak{a})^{-1} = N\mathfrak{b}(N\mathfrak{ab})^{-1} = N\mathfrak{b}|N(\lambda)|^{-1}$. Hence

$$\zeta(s, \mathfrak{C}) = \sum_{\lambda \in \mathfrak{b}/\mathfrak{O}_K^\times}{}' \frac{1}{\left(\frac{|N(\lambda)|}{N\mathfrak{b}}\right)^s}, \tag{2.5.6}$$

where the prime on the summation sign means that the value 0 is to be omitted.

We make a passing remark about the primitive element D and the discriminant Δ of a quadratic field $\mathbb{Q}(\sqrt{D})$. D is any integer $\neq 0, \pm 1$ and Δ is the determinant in (2.5.8) with $\mathfrak{b} = \mathfrak{O}$ and they may be different up to a square factor. But by abuse of language, one often writes $\mathbb{Q}(\sqrt{\Delta})$ and says that $\mathbb{Q}(\sqrt{D})$ is a quadratic field of discriminant Δ.

From now on we restrict to an imaginary quadratic case: $D < 0$. Then the absolute value sign may be suppressed in view of $N\lambda = N(\lambda) = |\lambda|^2 > 0$ and (2.5.6) reads

$$\zeta(s, \mathfrak{C}) = \frac{1}{w} \sum_{\lambda \in \mathfrak{b}}{}' \frac{1}{\left(\frac{N(\lambda)}{N\mathfrak{b}}\right)^s} = \frac{1}{w} Z(s, Q), \tag{2.5.7}$$

say, where w indicates the number of roots of unity contained in K, $Q = Q(m, n) = \frac{|N(\lambda)|}{N\mathfrak{b}}$ is a positive definite quadratic form with discriminant $\Delta < 0$ and $Z(s, Q)$ is an Epstein zeta-function in (1.1.6). Cf. e.g. Li et al. (2022, Sect. 2.1). Cf. Table 3.3 and (3.6.2).

The customary procedure is the following. Cf. e.g. Davenport (1980, pp. 43–53), Hecke (1923), Landau (1921), Serre (1973, pp. 81–82). If

$$\mathfrak{b} = \mathbb{Z}\omega_1 \oplus \mathbb{Z}\omega_2, \quad \omega := \frac{\omega_2}{\omega_1}, \quad \operatorname{Im}\omega > 0, \tag{2.5.8}$$

then expressing the well-known formula

$$\det \begin{pmatrix} \omega_1 & \bar{\omega}_1 \\ \omega_2 & \bar{\omega}_2 \end{pmatrix} = N\mathfrak{b}\sqrt{\Delta}$$

in the form

$$\delta(\mathfrak{b}) := \det \begin{pmatrix} \omega_1 & \bar{\omega}_1 \\ \omega_2 & \bar{\omega}_2 \end{pmatrix}^2 = N\mathfrak{b}^2 \Delta \tag{2.5.9}$$

we may rewrite (2.5.7) as

$$\zeta(s, \mathfrak{C}) = \frac{1}{w} |\Delta|^{-\frac{s}{2}} \sum_{\lambda \in \mathfrak{b}}{}' \frac{1}{\left(\frac{|N(\lambda)|}{\sqrt{|\delta(\mathfrak{b})|}}\right)^s} = \frac{1}{w} |\Delta|^{-\frac{s}{2}} Z(s, Q), \tag{2.5.10}$$

say, where $Q = \frac{|N(\lambda)|}{\sqrt{|\delta(\mathfrak{b})|}}$ is a positive definite quadratic form of discriminant -1.

Noting that (2.5.7) is unchanged if we replace $\mathfrak{b}$ by $\alpha\mathfrak{b}$, $K \ni \alpha \neq 0$, we may choose the basis of $\mathfrak{b}$ as the simpler one $\{1, \tau\}$ with $\tau \in \mathcal{H}$, where $\mathcal{H}$ means the upper half-plane (UHP):

$$\mathfrak{b} = \mathbb{Z} \oplus \mathbb{Z}\tau, \quad \lambda = m + n\tau, \quad m, n \in \mathbb{Z}, \tag{2.5.11}$$

which leads to the expression (1.1.7) with $a = \frac{\sqrt{|\Delta|}}{2y}$. For (2.5.9) reads

$$\left(\sqrt{\delta(\mathfrak{b})} =\right) \det \begin{pmatrix} 1 & 1 \\ \tau & \bar{\tau} \end{pmatrix} = N\mathfrak{b}(-i)\sqrt{|\Delta|}, \quad \text{or} \quad N\mathfrak{b} = \frac{\tau - \bar{\tau}}{i\sqrt{|\Delta|}} = \frac{2\,\mathrm{Im}\,\tau}{\sqrt{|\Delta|}}. \tag{2.5.12}$$

(This is a standard procedure, cf. e.g. Serre, 1973, pp. 82–83 or Rademacher, 1973, pp. 117–137, Chap. 8.)

Also correspondingly to (2.5.10),

$$\zeta(s, \mathfrak{C}) = \frac{1}{w} |\Delta|^{-\frac{s}{2}} \sum_{\lambda \in \mathfrak{b}}{}' \frac{1}{\tilde{Q}(m, n)^s}, \quad \tilde{Q}(m, n) = \frac{N(\lambda)}{2\,\mathrm{Im}\,\tau} \tag{2.5.13}$$

we see that $\frac{|N(\lambda)|}{2\,\mathrm{Im}\,\tau} = \frac{|m+n\tau|^2}{2\,\mathrm{Im}\,\tau}$ is a positive definite quadratic form of discriminant -1.

Although (2.5.10) or (2.5.13) is often used in deriving the Kronecker limit formula for an imaginary quadratic field, we use the setting as in Kumagai (2002), Selberg and Chowla (1967), Zhang and Williams (1995) etc. to state the general Chowla–Selberg integral formula with discriminant $\Delta < 0$, i.e. through (2.5.7), we view the class zeta-function as the Epstein zeta-function with a positive definite quadratic form of discriminant Δ. This corresponds to the choice of basis (2.5.11) in (2.5.7). Below we treat the Epstein zeta-function independently (through the Epstein-type Eisenstein series).

We consider $\vartheta(\tau, x) = \vartheta_3(\tau, \imath x)$ of (2.1.62) with $x = 0$.

$$\theta(\tau) := \vartheta(\tau, 0) = \sum_{n=-\infty}^{\infty} q^{n^2} = 1 + \sum_{n=1}^{\infty} 2q^{n^2} = \prod_{n=1}^{\infty} \left(1 - q^{2n}\right) \left(1 + q^{2n-1}\right)^2, \tag{2.5.14}$$

the last equality because of the Jacobi triple product (2.11.1). The Dedekind eta-function $\eta(\tau)$ is essentially the (multiplicative) even part of the theta-function defined by

$$\eta(\tau) = q^{\frac{1}{12}} \prod_{n=1}^{\infty} (1 - q^{2n}) = q^{\frac{1}{12}} \theta^e(\tau) \tag{2.5.15}$$

or

$$\log \eta(\tau) = \frac{1}{12} \log q + \sum_{n=1}^{\infty} \log(1 - q^{2n}) = \frac{1}{12} \pi i \tau - \sum_{n=1}^{\infty} \sigma_{-1}(n) q^{2n} \tag{2.5.16}$$

$$\log \overline{\eta(\tau)} = \log \eta(-\bar{\tau}) = -\frac{1}{12} \pi i \bar{\tau} - \sum_{n=1}^{\infty} \sigma_{-1}(n) \bar{q}^{2n}.$$

Also recall (2.1.77) in the form

$$\tilde{\theta}(\tau) = \tilde{\vartheta}(\tau, 0) = \sum_{n=-\infty}^{\infty} n q^{n^2}. \tag{2.5.17}$$

Equation (2.5.17) entails

$$\theta(\tau) = \int_0^1 \frac{1}{z} \tilde{\vartheta}(\tau, z) \, \mathrm{d}z.$$

The following lemma is a basis of the beta transform. In previous literature it used to be called the Mellin-Barnes integral, but this term refers to a much wider class of functions, cf. Paris and Kaminski (2001) and Remark 2.5.2. In view of Theorem 3.1.1 to the effect that the beta transform and the Hardy transform lead to the same Fourier–Bessel expansion, we may claim that this was first used by Hardy (1908) in the form of the Hardy transform. In the form of the beta integral, this was used by Koshlyakov (1954) and extensively used by Berndt (1969) later. Cf. Table 3.1.

Lemma 2.5.1 *The* **beta transform** *reads*

$$(1 + x)^{-s} = \frac{1}{2\pi i} \int_{(c)} \frac{\Gamma(z + s)\,\Gamma(-z)}{\Gamma(s)} x^z \mathrm{d}z = \frac{1}{\Gamma(s)} G_{1,1}^{1,1}\left(x^{-1} \middle| \begin{matrix} 1 \\ s \end{matrix}\right) \tag{2.5.18}$$

for $x > 0$, $-\operatorname{Re} s < c < 0$, where (c) signifies the vertical Bromwich path $z = c + it$, $-\infty < t < \infty$.

Theorem 2.5.1 (i) *Under the notation (1.1.2)–(1.1.7), in particular, $\tau = x + iy$, $y = \frac{\sqrt{|\Delta|}}{2a}$, we have the* **Chowla–Selberg integral formula**

$$\Gamma(s)Z(s,Q) = \Gamma(s)a^{-s}\zeta_{\mathbb{Z}^2}(s,\tau) \tag{2.5.19}$$
$$= 2\Gamma(s)\zeta(2s)a^{-s} + \frac{2^{2s}a^{s-1}\sqrt{\pi}}{|\Delta|^{s-\frac{1}{2}}}\Gamma\left(s-\frac{1}{2}\right)\zeta(2s-1) + R(s),$$

where

$$R(s) = \frac{8\pi^s 2^{s-\frac{1}{2}}}{\sqrt{a}|\Delta|^{\frac{s}{2}-\frac{1}{4}}}\sum_{n=1}^{\infty} n^{s-\frac{1}{2}}\sigma_{1-2s}(n)\cos 2\pi nx K_{s-\frac{1}{2}}(2\pi yn), \tag{2.5.20}$$

and where

$$\sigma_\alpha(n) = \sum_{d|n} d^\alpha \tag{2.5.21}$$

is the sum-of-divisors function and K_ν is the K-Bessel function introduced in Sect. 3.1. (2.5.19) *with* (2.5.20) *is equivalent to the functional equation* ($Z(s) = Z(s,Q)$)

$$\left(\frac{2\pi}{\sqrt{|\Delta|}}\right)^{-s}\Gamma(s)Z(s) = \left(\frac{2\pi}{\sqrt{|\Delta|}}\right)^{-(1-s)}\Gamma(1-s)Z(1-s). \tag{2.5.22}$$

(ii) *The Laurent expansion at $s = 1$ reads*

$$Z(s) = \frac{\frac{2\pi}{\sqrt{|\Delta|}}}{s-1} + \frac{2\pi}{\sqrt{|\Delta|}}\left(2\gamma + \log\frac{a}{|\Delta|}\right) - \frac{4\pi}{\sqrt{|\Delta|}}\log|\eta(\tau)|^2 + O(s-1). \tag{2.5.23}$$

Proof (i) Equation (2.5.19) (with (2.5.20)) has been proved in many places and elucidated as the Fourier–Bessel expansion (Kanemitsu & Tsukada, 2014, Chap. 4), i.e. as the correspondence $G_{1,1}^{1,1} \leftrightarrow G_{0,2}^{2,0}$. However, we give a slightly modified proof of Kuzumaki (2011) to exhibit the use of the beta transform as opposed to the Hecke gamma transform (2.1.3), elucidating the computations in Zhang and Williams (1995).

It suffices to prove (2.5.19) with (2.5.20) in the form

$$R(s) = \frac{4a^{-s}}{\Gamma(s)}\sum_{m,n=1}^{\infty}(ny)^{-2s}\frac{\cos(2\pi mnx)}{m}I_{m,n}(s,\tau), \tag{2.5.24}$$

where $I(m,n)$ has the following expressions, of which we mainly use the first K-Bessel function expression. For the other two, the confluent hypergeometric function and the Whittaker function expression, cf. Sect. 3.1.

$$I_{m,n}(s,\tau) = \frac{1}{\sqrt{\pi}} G_{0,2}^{2,0}\left((\pi mny)^2 \middle| \begin{matrix} - \\ s, \ \frac{1}{2} \end{matrix} \right) = 2\pi^s (mny)^{s+\frac{1}{2}} K_{s-\frac{1}{2}}(2\pi mny), \tag{2.5.25}$$

$$I_{m,n}(s,\tau) = e^{-2\pi mny}(2\pi mny)^{2s}\Psi(s, 2z; 4\pi mny),$$
$$I_{m,n}(s,\tau) = (\pi mny)^s W_{0,s-\frac{1}{2}}(4\pi mny).$$

We shall prove (2.5.19) with $R(s)$ in (2.5.24) in a slightly more general form

$$\Delta(s)\zeta_{\mathbb{Z}^2}(s,\tau) = 2\Delta(s)\zeta(2s) + \frac{2\sqrt{\pi}}{\Gamma(s)}\Delta(s)\Gamma(s-\frac{1}{2})\zeta(2s-1)y^{1-2s} \tag{2.5.26}$$
$$+ \frac{4\Delta(s)}{\Gamma(s)}\sum_{m,n=1}^{\infty}(ny)^{-2s}\frac{\cos(2\pi mnx)}{m} I_{m,n}(s,\tau),$$

where $\Delta(s) = a^{-s}$ leads to (2.5.19). For other choices, cf. Solution to Exercise 2.5.4. It suffices to prove the case $\Delta(s) = 1$: (2.5.33), (2.5.34) and (2.5.25).

Recall the lattice zeta-function (2.1.70)

$$\phi_1(s,\{x\}) = \zeta_{\mathbb{Z}}(s,x) = \sum_{n=-\infty}^{\infty} |n+\{x\}|^{-s} \tag{2.5.27}$$

for $\{x\} > 0$, where $\{x\}$ is the fractional part of x, (2.1.60), then ϕ_1 is the even part

$$\phi_1(s,\{x\}) = 2\zeta^e(s,\{x\}) = \zeta(s,\{x\}) + \zeta(s,1-\{x\}). \tag{2.5.28}$$

In (2.5.28) we use the functional equation (2.1.12) for the Hurwitz zeta-function: for $\sigma > 1$, $\{nx\} > 0$.

Then we obtain

$$\phi_1(-2s,\{nx\}) = \frac{2\Gamma(1+2s)}{(2\pi)^{1+2s}}\sin(-\pi s)\left\{\ell_{1+2s}(nx) + \ell_{1+2s}(1-nx)\right\}, \tag{2.5.29}$$

i.e. the even part of $\ell_{1+2s}(nx)$, valid for $\{nx\} = 0$, too by interpreting the left-hand side by the right.

By Exercise 2.5.2, we have

$$\zeta_{\mathbb{Z}^2}(s,\tau) = 2\zeta(2s) + \frac{2}{\Gamma(s)}\sum_{n=1}^{\infty}(ny)^{-2s} \tag{2.5.30}$$
$$\times \frac{1}{2\pi i}\int_{(c)} \Gamma(z+s)\Gamma(-z)\phi_1(-2z,\{nx\})(ny)^{-2z}\,dz.$$

Using the formulas $\sin(-\pi s) = \pi\Gamma(-s)^{-1}\Gamma(1+s)^{-1}$ and $\Gamma(1+2s) = 2^{2s}\pi^{-\frac{1}{2}}\Gamma(s+\frac{1}{2})\Gamma(s+1)$ and (2.5.29), the integrand on the right of (2.5.30) reduces to

$$\Gamma(z+s)\Gamma(-z)\zeta_{\mathbb{Z}}(-2z, nx)(ny)^{-2z}$$
$$= \frac{\Gamma(z+\frac{1}{2})\Gamma(z+s)}{\pi^{2z+\frac{1}{2}}} \{\ell_{2z+1}(nx) + \ell_{2z+1}(1-nx)\} (ny)^{-2z}.$$

Hence

$$\zeta_{\mathbb{Z}^2}(s,\tau) = 2\zeta(2s) + \frac{2}{\Gamma(s)} \sum_{n=1}^{\infty} (ny)^{-2s} J_n(c), \tag{2.5.31}$$

where

$$\begin{aligned} J_n(c) &= J_n(c,s) \qquad\qquad (2.5.32)\\ &= \frac{1}{2\pi i} \int_{(c)} \frac{\Gamma(z+s)\Gamma(z+\frac{1}{2})}{\pi^{2z+\frac{1}{2}}} \{\ell_{2z+1}(nx) + \ell_{2z+1}(1-nx)\} (ny)^{-2z} \mathrm{d}z. \end{aligned}$$

Moving the line of integration on the right of (2.5.32) to the right up to $\mathrm{Re}\, z = c_1$, where $-\frac{1}{2} < c_1 < 0$, thereby noting the pole at $z = -\frac{1}{2}$, we deduce that

$$\begin{aligned} &\frac{2}{\Gamma(s)} \sum_{n=1}^{\infty} (ny)^{-2s} J_n(c)\\ &= 2\zeta(2s) - \frac{2}{\Gamma(s)} \sum_{n=1}^{\infty} (ny)^{-2s+1} \frac{\Gamma(s-\frac{1}{2})}{\pi^{-\frac{1}{2}}} \{\ell_0(nx) + \ell_0(1-nx)\}\\ &\quad + \frac{2}{\Gamma(s)} \sum_{n=1}^{\infty} (ny)^{-2s} J_n(c_1), \end{aligned}$$

whence that

$$\zeta_{\mathbb{Z}^2}(s,\tau) = 2\zeta(2s) + \frac{2\sqrt{\pi}\Gamma(s-\frac{1}{2})}{\Gamma(s)} \zeta(2s-1) y^{-2s+1} + \frac{2}{\Gamma(s)} \sum_{n=1}^{\infty} (ny)^{-2s} J_n(c_1) \tag{2.5.33}$$

on using $\ell_0(nx) + \ell_0(1-nx) = -1$ which follows from $\ell_0(x) = \frac{e^{2\pi i x}}{1-e^{2\pi i x}}$.

Now

$$\begin{aligned} J_n(c_1) &= \frac{1}{2\pi i\sqrt{\pi}} \sum_{m=1}^{\infty} \left(e^{2\pi i mnx} + e^{-2\pi i mnx}\right) \int_{(c_1)} \frac{\Gamma(z+s)\Gamma(z+\frac{1}{2})}{m(\pi mny)^{2z}} \mathrm{d}s \qquad (2.5.34)\\ &= \frac{1}{2\pi i\sqrt{\pi}} \sum_{m=1}^{\infty} \frac{2\cos(2\pi mnx)}{m} \int_{(-c_1)} \Gamma(s-z)\Gamma\left(\frac{1}{2}-z\right) (\pi mny)^{2z} \mathrm{d}z\\ &= \sum_{m=1}^{\infty} \frac{2\cos(2\pi mnx)}{m} I_{m,n}(s,\tau), \end{aligned}$$

where $I_{m,n}(s,\tau)$ is the G-function given as the first equality in (2.5.25).

The equivalence of CSIF and (2.5.22) is a Hecke modular relation, Theorem 2.12.3. Cf. Exercise 2.5.1. This completes the proof of (i).

(ii) To deduce (2.5.23), we use (2.5.19) in the form

$$Z(s) = 2\zeta(2s)a^{-s} + \frac{2^{2s}a^{s-1}\sqrt{\pi}}{|\Delta|^{s-\frac{1}{2}}}\frac{\Gamma\left(s-\frac{1}{2}\right)\zeta(2s-1)}{\Gamma(s)} + \frac{1}{\Gamma(s)}R(s). \quad (2.5.35)$$

Near $s=1$, the second term on the right of (2.5.35) has the expansion

$$\frac{\frac{2\pi}{\sqrt{|\Delta|}}}{s-1} + \frac{2\pi}{\sqrt{|\Delta|}}\left(2\gamma + \log\frac{a}{|\Delta|}\right)$$

while the first term becomes $2\zeta(2)a^{-1} = \frac{\pi^2}{3a}$, whence it follows that

$$Z(s) = \frac{\frac{2\pi}{\sqrt{|\Delta|}}}{s-1} + \frac{2\pi}{\sqrt{|\Delta|}}\left(2\gamma + \log\frac{a}{|\Delta|}\right) + \frac{\pi^2}{3a} + R(1) + O(s-1),$$

where

$$\begin{aligned} R(1) &= \frac{8\sqrt{2}\pi}{\sqrt{a}|\Delta|^{\frac{1}{4}}}\sum_{n=1}^{\infty} n^{\frac{1}{2}}\sigma_{-1}(n)\cos 2\pi xn\sqrt{\frac{1}{4yn}}e^{-2\pi yn} \\ &= \frac{8\pi}{\sqrt{|\Delta|}}\sum_{n=1}^{\infty}\sigma_{-1}(n)(\cos 2\pi xn)e^{-2\pi yn} \end{aligned}$$

in view of (3.1.7).

Recalling $q = e^{\pi i\tau}$ in (2.1.63), we may write

$$2e^{-2\pi yn}\cos 2\pi xn = q^{2n} + \bar{q}^{2n},$$

whence by (2.5.16)

$$\begin{aligned} R(1) &= \frac{4\pi}{\sqrt{|\Delta|}}\sum_{n=1}^{\infty}\sigma_{-1}(n)(q^{2n}+\bar{q}^{2n}) \\ &= -\frac{4\pi}{\sqrt{|\Delta|}}\left(\log\eta(\tau) + \log\overline{\eta(\tau)}\right) + \frac{4\pi}{\sqrt{|\Delta|}}\left(\frac{\pi i}{12}(\tau-\bar{\tau})\right). \end{aligned}$$

Here the last term is

$$\frac{4\pi}{\sqrt{|\Delta|}}\frac{\pi i}{12}2i\,\mathrm{Im}\,\tau = -\frac{\frac{2}{3}\pi^2}{\sqrt{|\Delta|}}y, \quad (2.5.36)$$

which cancels $\frac{\pi^2}{3a}$, amounting to (2.5.23). □

Remark 2.5.1 This theorem will be uplifted to Theorem 3.6.2. The proof above depending on the beta transform and the proof in Exercise 3.6.7 depending on the gamma and Hardy–Hecke transform (by the first method of Chowla and Selberg) is a manifestation of Theorem 3.1.1.

Exercise 2.5.1 Prove that CSIF implies (2.5.22). Cf. Exercise 1.1.2.

Solution. The easiest to check is CSIF (3.6.5) for $E^*(s, z)$. The asymmetric form

$$\Gamma\left(\frac{1-2s}{2}\right)\zeta(1-2s) = \pi^{\frac{1}{2}-2a}\Gamma(2s)\,\zeta(2s) \tag{2.5.37}$$

of (FER), the invariance of $n^{s-\frac{1}{2}}\sigma_{1-2s}(n)$ and (3.1.5). □

Exercise 2.5.2 Confirm (2.5.30) in the light of Remark 3.1.1.

Solution. Collecting the terms with $n = 0$ in the definition of $\zeta_{\mathbb{Z}^2}(s, \tau)$ (1.1.7), (3.6.2) gives $2\sum_{m=1}^{\infty} m^{-2s} = 2\zeta(2s)$. In the remaining sum $2\sum_{n=1}^{\infty}\sum_{m=-\infty}^{\infty}\{(m+nx)^2 + n^2y^2\}^{-s}$, we transform the summand into $(ny)^{-2s}\left(1+\left(\frac{m+nx}{ny}\right)^2\right)^{-s}$ and apply the beta transform in Lemma 2.5.1, separating 1 and $x \leftrightarrow \left(\frac{m+nx}{ny}\right)^2$. We choose c such that $-\sigma < c < -\frac{1}{2}$, and sum over m to deduce (2.5.30). □

Exercise 2.5.3 Confirm that the first method of Chowla and Selberg in Exercise 3.6.7 is the decomposing effect of the gamma transform and the Hardy–Hecke transform in the light of Remark 3.1.1.

Also confirm the continuous version of (HGT) in Exercise 3.6.4 has the same effect.

Solution. As in the above exercise, we isolate the term $2\zeta(2s)$. In the remaining sum $2\sum_{n=1}^{\infty}\sum_{m=-\infty}^{\infty}\{(m+nx)^2 + (ny)^2\}^{-s}$, we apply the Hecke gamma transform, separating $(m+nx)^2$ and $(ny)^2$. The sum over m is the theta series for which we apply the theta-transformation formula. Then the resulting integral is the Hardy–Hecke transform. □

Exercise 2.5.4 Choosing $\Delta(s)$ in (2.5.26) suitably, deduce Theorem 3.6.1. Cf. Exercise 3.6.7.

Solution. Choosing

$$\Delta(s) = \frac{y^s}{\zeta(s)},\quad \frac{1}{2}\left(\frac{y}{\pi}\right)^s,$$

etc. we can cover the 4th and 5th cases in Table 3.3. □

Exercise 2.5.5 Let ϑ_1 be one of 4 theta-functions, cf. (2.1.62) and Chan (2020, pp. 30–31, 60–61). In place of (2.5.23), prove the following Landau's form of Kronecker's limit formula. Use

$$\vartheta_1'(\tau, 0) = 2q^{\frac{1}{4}} \prod_{j=1}^{\infty} (1 - q^{2j})^3 \tag{2.5.38}$$

instead of (2.5.16) to derive

$$\begin{aligned} Z(s) =& \frac{\frac{2\pi}{\sqrt{|\Delta|}}}{s-1} + \frac{2\pi}{\sqrt{|\Delta|}} \left(2\gamma + \log \frac{a}{|\Delta|}\right) \\ & - \frac{4\pi}{\sqrt{|\Delta|}} \left(\frac{1}{3} \log |\theta_1'(\tau)|^2 - \frac{2}{3} \log 2\right) + O(s-1). \end{aligned} \tag{2.5.39}$$

Cf. Landau (1903, p. 174, (155)).

Solution. As in (2.5.16) we have

$$\begin{aligned} -\sum_{n=1}^{\infty} \sigma_{-1}(n) q^{2n} &= \frac{1}{3} \log \theta_1'(\tau) - \frac{1}{12} \pi i \tau - \frac{1}{3} \log 2 \\ -\sum_{n=1}^{\infty} \sigma_{-1}(n) \bar{q}^{2n} &= \frac{1}{3} \log \theta_1'(-\bar{\tau}) - \frac{1}{12} \pi i (-\bar{\tau}) - \frac{1}{3} \log 2. \end{aligned} \tag{2.5.40}$$

Hence it follows that

$$\begin{aligned} & \sum_{n=1}^{\infty} \sigma_{-1}(n) q^{2n} + \sum_{n=1}^{\infty} \sigma_{-1}(n) \bar{q}^{2n} \\ & = -\frac{1}{3} \log |\theta_1'(\tau)|^2 + \frac{2}{3} \log 2 - \frac{1}{6} \pi y. \end{aligned} \tag{2.5.41}$$

The last term on the right of (2.5.41) after multiplying by $\frac{4\pi}{\sqrt{|\Delta|}}$ becomes $-\frac{\frac{2}{3}\pi^2}{\sqrt{|\Delta|}} y$ as in (2.5.36), canceling the term $\frac{\pi^2}{3a}$. □

Exercise 2.5.6 Prove that for a positive definite matrix $Y = \begin{pmatrix} a & b \\ b & c \end{pmatrix}$, with $|Y| = ac - b^2 > 0$ and $|d| = |Y| = \frac{1}{4} D$, let $Q = Q(m, n) = Y[\boldsymbol{a}] = am^2 + 2bmn + cn^2$, Theorem 2.5.1 reads

$$Z(s,Q) = 2c^{-s}\zeta(2s) + 2c^{-s}\left(\frac{c}{\sqrt{|d|}}\right)^{2s-1}\sqrt{\pi}\frac{\Gamma\left(s-\frac{1}{2}\right)}{\Gamma(s)}\zeta(2s-1)$$
$$+ 8\pi^s c^{-s}\left(\frac{c}{\sqrt{|d|}}\right)^{s-1/2}\frac{1}{\Gamma(s)}$$
$$\times \sum_{n=1}^{\infty} n^{s-1/2}\sigma_{1-2s}(n)\cos 2\pi n\frac{b}{c}K_{s-1/2}\left(2\pi n\frac{\sqrt{|d|}}{c}\right)$$

and that this is in conformity with the results of Theorem 2.5.1.

Solution. This occurs when we decompose $Q(m,n) = cn^2 + 2bmn + am^2$ as $|n+mz|^2$ and amounts to the result of Theorem 2.5.1, where in (1.1.2)–(1.1.6), a replaced by c resp. b by $2b$, so that $\Delta = 4d$. □

Equation (2.5.19) is often stated in slightly different form.

Exercise 2.5.7 Let

$$d = \frac{|\Delta|}{4} = ac - b^2,\ x = \frac{b}{a},\ y = \frac{\sqrt{d}}{a} = \frac{\sqrt{|\Delta|}}{2a} > 0.$$

Then show that

$$\frac{2^{2s}\sqrt{\pi}a^{s-1}}{|\Delta|^{s-\frac{1}{2}}} = 2\sqrt{\pi}\frac{\left(\frac{y}{\sqrt{d}}\right)^{1-s}}{d^{s-\frac{1}{2}}}$$

and that Theorem 2.5.1 may be written as (3.6.3) $\times a^{-s}$.

The limiting case $s \to 1$ gives rise to the Kronecker limit formula in the notation of Theorem 2.5.2.

$$\lim_{s\to 1}\left(Z(s) - \frac{\frac{2\pi}{\sqrt{|\Delta|}}}{s-1}\right) = \frac{2\pi}{\sqrt{|\Delta|}}\left(2\gamma + \log\frac{a}{|\Delta|}\right) - \frac{4\pi}{\sqrt{|\Delta|}}\log\eta(\omega_1)\eta(\omega_2). \tag{2.5.42}$$

The Chowla–Selberg formula or more precisely the Lerch–Chowla–Selberg formula in Landau's form reads

Theorem 2.5.2 (Lerch–Chowla–Selberg formula)

$$\frac{w}{2}\sum_{n=1}^{|\Delta|-1}\left(\frac{|\Delta|}{n}\right)\log\Gamma\left(\frac{n}{|\Delta|}\right) = \sum_{(a,b,c)}\log\left[\sqrt{\frac{2|\Delta|\pi}{c}}\eta(\omega_1)\eta(\omega_2)\right], \tag{2.5.43}$$

where $\Delta < 0$ is the discriminant of the imaginary quadratic field $\Omega = \mathbb{Q}(\sqrt{\Delta})$, w indicates the order of the group of roots of unity in Ω, $\omega_1 = \frac{-b+i\sqrt{|\Delta|}}{2c}$, $\omega_2 = \frac{-b-i\sqrt{|\Delta|}}{2c}$ (whence $\eta(\omega_1)\eta(\omega_2) = |\eta(\omega_1)|^2$) and on the right (a,b,c) runs through the complete set of representatives of classes of quadratic forms of discriminant Δ.

For more details, cf. Sect. 2.6.

Remark 2.5.2 (i) The Chowla–Selberg integral formula in matrix form is expounded in Kanemitsu and Tsukada (2014, pp. 136–138) and reviewed in Sect. 2.10. The case $\boldsymbol{g} = \boldsymbol{h} = z, n = m = 1$ is due to Chowla and Selberg (1949), Selberg and Chowla (1967) (cf. also Bateman and Grosswald, 1964), the case $m = 1$ is due to Berndt (1971) and the general case with $\boldsymbol{g} = \boldsymbol{h} = z$ is due to Terras (1985, p. 208, Example 4).

In Kanemitsu and Tsukada (2014), the Epstein zeta-function is treated as a quadratic form variation of the Fourier–Bessel expansion, $H_{1,1}^{1,1} \leftrightarrow H_{0,2}^{2,0}$ to the effect that the beta transform (or the perturbed Dirichlet series) corresponds to the K-Bessel series (or Fourier series). The K-Bessel function in (2.5.20) reduces to the exponential function for $s = 1$, whence the Bessel series is nothing but the Lambert series for the Dedekind eta-function and so the Kronecker limit formula arises most naturally in which (the even part of) the theta-function or the Dedekind eta-function shows up. Note that with $\phi_1(s, x)$ in RHB correspondence is $\vartheta(it, itx)$, (2.1.68).

In Zhang and Williams (1995), this natural deduction of eta-function expression is presented suggestive of the modular relation. Substituting $2ay$ for $\sqrt{|\Delta|}$, we see that the coefficient of the second term in (2.5.19) resp. in (2.5.20) is $2\sqrt{\pi}a^{-s}y^{1-2s}$ resp. $8\pi^s a^{-s} y^{\frac{1}{2}-s}$ in conformity with Zhang and Williams (1995, (2.11), (2.12)).
(ii) As is shown in the proof of Theorem 2.5.1, (2.5.38) reduces to a simpler form (2.5.23) if we use the eta-function instead of the theta-function. This is also motivated by the fact that Ramanujan's **discriminant function**

$$\Delta(q) = (2\pi)^{12} q^2 \prod_{n=1}^{\infty} (1 - q^{2n})^{24} = \left((2\pi)\eta(\tau)^2\right)^{12} = \left((2\pi)^{\frac{1}{2}} q^{\frac{1}{12}} \theta^e(\tau)\right)^{24}$$

the last equality being due to (2.5.15), is a cusp form of weight 12. This leads to the normalization

$$\mathbb{D}(\mathfrak{C}) = \mathbb{D}(\mathfrak{b}) = \sqrt{|\delta(\mathfrak{b})|}(2\pi)|\eta(\tau)|^2 = 2y(2\pi)|\eta(\tau)|^2 = \frac{2\sqrt{|d|}}{a}(2\pi)|\eta(\tau)|^2 \tag{2.5.44}$$

depending only on a, where $\delta(\mathfrak{b})$ is defined in (2.5.12). Under this, (2.5.23) leads to the class invariant form (2.5.46) of Meyer.

Now by (2.5.7), Theorem 2.5.1 may be paraphrased into the case of an imaginary quadratic field with discriminant d. *In what follows we often denote the discriminant by d instead of Δ.*

Theorem 2.5.3 (i) *Under the same conditions as in Theorem 2.5.1, we have the* **Chowla–Selberg integral formula**

$$\Gamma(s)\zeta(s,\mathfrak{C}) \tag{2.5.45}$$
$$= \frac{1}{w}\Big(2\Gamma(s)\zeta(2s)a^{-s} + \frac{2a^{s-1}\sqrt{\pi}}{|d|^{s-\frac{1}{2}}}\Gamma\left(s-\frac{1}{2}\right)\zeta(2s-1)$$
$$+ \frac{8\pi^s}{\sqrt{a}|d|^{\frac{s}{2}-\frac{1}{4}}}\sum_{n=1}^{\infty} n^{s-\frac{1}{2}}\sigma_{1-2s}(n)\cos 2\pi nx K_{s-\frac{1}{2}}(2\pi yn)\Big),$$

which is equivalent to the functional equation (2.5.22) *with* Δ *replaced by* d.
(ii) *The Laurent expansion at* $s = 1$ *in Meyer's form reads*

$$\zeta(s,\mathfrak{C}) = \frac{\frac{\pi}{w\sqrt{|d|}}}{s-1} + \frac{\pi}{w\sqrt{|d|}}\left(2\gamma + \log\frac{(2\pi)^2}{a}\right) - \frac{2\pi}{w\sqrt{|d|}}\log \mathbb{D}(\mathfrak{C}) + O(s-1), \tag{2.5.46}$$

where $\mathbb{D}(\mathfrak{C})$ *is the class invariant defined by* (2.5.44).

Proof Only (2.5.46) needs explanation. Substitution of (2.5.23) in (2.5.7) shows that the third term on the right of (2.5.45) is $-\frac{2\pi}{w\sqrt{|d|}}\log|\eta(\tau)|^2$.

Using (2.5.44), we arrive at (2.5.46). □

Corollary 2.5.1 *For an imaginary quadratic field with discriminant* d *and class number* h *we have the Laurent expansion*

$$\zeta_K(s) = \frac{\frac{\pi h}{w\sqrt{|d|}}}{s-1} + \frac{\pi h}{w\sqrt{|d|}}\left(2\gamma + \log\frac{(2\pi)^2}{a}\right) - \frac{2\pi h}{w\sqrt{|d|}}\sum_{\mathfrak{C}\in I/P}\log \mathbb{D}(\mathfrak{C}) \tag{2.5.47}$$
$$+ O(s-1).$$

In Theorem 2.5.3, (ii) we followed notation of Meyer whence so do we in Corollary 2.5.1. In other places, we use $\Delta = 4d$, which is more common. Siegel (1961) is also an essential reference but it is assumed that the class number is 1.

2.6 Lerch–Chowla–Selberg Formula

We are in a position to state the decomposition theorem which provides the other side of the class number formula (CNF) and (LCSF).

Lemma 2.6.1 (decomposition theorem) *For a quadratic field* $K = \mathbb{Q}(\sqrt{D})$ *of discriminant* d, *the multiplicative decomposition formula holds*

$$\zeta_K(s) = \zeta(s)L(s,\chi), \tag{2.6.1}$$

where $L(s,\chi)$ *is the Dirichlet* L*-function associated to the* **Kronecker (-Dirichlet) character** $\chi = \chi_d$ **attached to** K. *The Laurent expansion reads*

$$\zeta_K(s) = \frac{L(1,\chi)}{s-1} + \gamma L(1,\chi) + L'(1,\chi) + O(s-1).$$

For the decomposition of the Dedekind zeta-function of an Abelian field, cf. (5.3.4).

Hereafter we denote the Kronecker character χ_d by χ since the context will show we mean χ_d.

Comparing Corollary 2.5.1 and Lemma 2.6.1, we state and prove the Dirichlet class number formula and the Lerch–Chowla–Selberg formula for imaginary quadratic fields. For a proof of (2.6.3) cf. Sect. 2.9.

Theorem 2.6.1 *Let $h(d) = |I/P|$ denote the class number of the quadratic field $\mathbb{Q}(\sqrt{D})$ of discriminant d.*
(i) *The* **Dirichlet class number formula** *holds*

$$h(d) = \frac{w\sqrt{|d|}}{2\pi} L(1,\chi_d), \quad d < 0 \tag{2.6.2}$$

and

$$h(d) = \frac{\sqrt{d}}{2\log\varepsilon} L(1,\chi_d) \quad d > 0. \tag{2.6.3}$$

Cf. Sect. 5.3 for a cyclotomic field case.
(ii) *Suppose $d < 0$. The* **Lerch–Chowla–Selberg formula** *holds with class invariants,* (Meyer, 1981, p. 569, (**); Deninger, 1984, (4.5))*:*

$$(\prod_{\mathfrak{C}} \mathbb{D}(\mathfrak{C}))^2 = \left(\frac{2\pi}{\sqrt{|d|}}\right)^h \prod_{a=1}^{|d|-1} \Gamma\left(\frac{a}{|d|}\right)^{\frac{w}{2}\chi(a)}. \tag{2.6.4}$$

In terms of the Lerch zeta-function,

$$2\sum_{\mathfrak{C}\in I/P} \log \mathbb{D}(\mathfrak{C}) = h\left(2\gamma + \log\frac{(2\pi)^2}{\sqrt{|d|}}\right) - \frac{iw}{2\pi}\sum_{a=1}^{|d|-1} \chi(a)(\ell_1')^o\left(\frac{a}{M}\right). \tag{2.6.5}$$

Proof By Corollary 2.5.1, the residue of $\zeta_K(s)$ at $s = 1$ is $\frac{2\pi}{w\sqrt{|d|}}h$. Hence by the decomposition formula, Lemma 2.6.1, (2.6.2) follows.

By (2.6.2), (2.3.26) reads

$$\gamma L(1,\chi_d) + L'(1,\chi_d) = -\frac{\pi}{\sqrt{|d|}}\sum_{a=1}^{|d|-1} \bar{\chi}_d(a)\log\Gamma\left(\frac{a}{|d|}\right) + \frac{2\pi h}{w\sqrt{|d|}}(2\gamma + \log 2\pi). \tag{2.6.6}$$

Equating this with the Laurent constant in (2.5.47) and dividing both sides by $\frac{2\pi h}{w\sqrt{|d|}}$, we conclude that

$$2\sum_{\mathfrak{C}\in I/P} \log \mathbb{D}(\mathfrak{C}) = h \log \frac{2\pi}{\sqrt{|d|}} + \frac{w}{2}\sum_{a=1}^{|d|-1} \bar{\chi}(a) \log \Gamma\left(\frac{a}{|d|}\right). \tag{2.6.7}$$

Equation (2.6.4) is the exponentiated form of this.

In terms of the Lerch zeta-function, (2.6.6) amounts to (2.3.24).

Equating (2.5.47) and (2.3.24), we derive that

$$\frac{2\pi h}{w\sqrt{|d|}}\left(2\gamma + \log \frac{(2\pi)^2}{\sqrt{|d|}}\right) - \frac{4\pi}{w\sqrt{|d|}} \sum_{A\in I/P} \log \mathbb{D}(A)$$
$$= -\frac{i}{\sqrt{|d|}} \sum_{a=1}^{|d|-1} \bar{\chi}(a) (\ell_1')^o \left(\frac{a}{M}\right),$$

which leads to (2.6.5). □

2.7 Finite form of CNF and LCSF

Theorem 2.6.1, (i) gives the first stage of CNF in that the class number h is expressed in terms of the infinite series $L(1, \chi)$. Since h is finite, there was impetus to express the infinite component by a finite one. Theorem 2.3.3 gives this finite component. Thus, $A = B = C$:

Exercise 2.7.1 Prove the following finite form of the class number formula, where h means the class number of the quadratic field with discriminant Δ:

$$h = \frac{w\pi}{4\sqrt{|\Delta|}} \sum_{a=1}^{|\Delta|-1} \chi(a) \cot \frac{a}{|\Delta|}\pi = \frac{iw\tau(\chi)}{2\sqrt{|\Delta|}} \sum_{a=1}^{|\Delta|-1} \bar{\chi}(a) \bar{B}_1\left(\frac{a}{|\Delta|}\right) \tag{2.7.1}$$

for $\Delta < 0$ while

$$h = \frac{1}{2\sqrt{\Delta}\log\varepsilon} \sum_{a=1}^{\Delta-1} \hat{\chi}(a)\left(-\log 2 \sin\left(\frac{a}{\Delta}\pi\right)\right) \tag{2.7.2}$$

for $\Delta > 0$. Here χ means the Kronecker character.

Equation (2.7.1) will play a central in Chap. 5 with the first equality as Chowla's problem and the second as the Maillet–Demyanenko determinant (CPMD).

LCSF is in the similar spirit that the Laurent constant provided by the Kronecker limit formula is in finite form but is expressed in terms of infinite objects involving $L'(1, \chi)$, whence a finite form is wanted for the infinite objects. Kanemitsu and Tsukada (2014, pp. 12–15) streamlines the historical facts about the Lerch–Chowla–Selberg formula following Kurokawa (1994) in part and with reference to Deninger (1984) and Landau (1903). The main interest lies in the fact that there is an equality of two irrelevant-looking special functions—gamma and theta—at rational arguments

(Theorem 2.6.1). Here we shall give finite form for (2.7.4) as a preparation for the proof of LCSF for a real quadratic form in Sect. 2.9.

As is seen, it is a combination of two important formulas—Kronecker limit formula and the Lerch formula (by way of decomposition theorem). For the former there is enormous amount of literature (Meyer, 1981; Motohashi, 1968; Siegel, 1961; Zhang & Williams, 1995 (imaginary); Novikov, 1980; Zagier, 1975 (real)) and references therein. On the latter there is a detailed account in Weil (1975), esp. the references on pp. 67-68 are rather thorough. In Weil (1975) there is no mention of the first statement of the Lerch–Chowla–Selberg formula (Lerch, 1897, (26), p. 303), which is clarified in Kurokawa (1994). But earlier works by Berger *et al.* are not mentioned.

Both Kurokawa (1994) and Weil (1975) stress Lerch's formula (2.1.46) without resorting to its defining property as the principal solution to the difference equation

$$f(x+1) - f(x) = \log^{\alpha} x, \quad x > 0 \tag{2.7.3}$$

for each $\alpha \in \mathbb{N}$. Cf. Campbel (1966), Dufresnoy and Pisot (1963), Nörlund (1954), etc. This is essential for obtaining the LCSF in finite form for a real quadratic field. For the evaluation is needed of the cosine Fourier series (Landau, 1903, p. 180, (165.))

$$\Phi(x) = \sum_{n=1}^{\infty} \frac{\log n}{n} \cos 2\pi x n \tag{2.7.4}$$

as opposed to the Fourier sine series (or the Kummer Fourier series) (2.1.47). It had been studied by Lerch (1898) before Landau. It was taken up by Gut (1929) to derive the Kronecker limit formula for cyclotomic fields. It turns out that the function which Gut introduced is $-(R(x) + R(1-x))$, where $R(x) = R_2(s)$ is the Deninger R-function as the principal solution to (2.7.4) with $\alpha = 2$. Deninger's standpoint is that the R-function occupies its space as a proper special function in the same way as $\log \Gamma(x)$ is the principal solution to (2.7.4) with $\alpha = 1$ (Bohr–Mollerup theorem). Then from the uniqueness of the principal solution, it follows that

$$R(x) = -\zeta''(0, x), \tag{2.7.5}$$

where the prime means the derivative w.r.t. s.

As an analogy to Theorem 2.3.3 (Hashimoto et al., 2008, Theorem; Chakraborty et al., 2009c, Chap. 8), we have the following.

Theorem 2.7.1 *The finite expression*

$$L'(1,\chi) = (\gamma + \log 2\pi) L(1,\chi) + \begin{cases} \dfrac{1}{M} \displaystyle\sum_{\nu=1}^{M-1} R\left(\frac{\nu}{M}\right) \tau(\chi,\nu), & \chi \text{ even}, \\ \dfrac{\pi i}{M} \displaystyle\sum_{\nu=1}^{M-1} \log \Gamma\left(\frac{\nu}{M}\right) \tau(\chi,\nu), & \chi \text{ odd} \end{cases} \tag{2.7.6}$$

is equivalent to

$$R'\left(\frac{p}{q}\right) = -\gamma_1 - 2(\gamma + \log 2\pi q)\left(\gamma + \psi\left(\frac{p}{q}\right)\right) + 2\sum_{a=1}^{q-1} R\left(\frac{a}{q}\right)\cos\frac{2\pi ap}{q}$$
$$-2\pi\sum_{a=1}^{q-1}\log\Gamma\left(\frac{a}{q}\right)\sin\frac{2\pi ap}{q} - \log q(\log q + 2\log 2\pi),$$

where $\tau(\chi, \nu)$ *is the general Gauss sum* (2.3.4).

If χ is primitive, then by separability of the general Gauss sum, Lemma 2.3.1, (2.7.6) amounts to (2.3.25) in the case χ odd. For χ even, (2.7.6) reads

$$L'(1, \chi) = (\gamma + \log 2\pi)L(1, \chi) + \frac{\tau(\chi)}{M}\sum_{\nu=1}^{M-1}\bar{\chi}(n)R\left(\frac{\nu}{M}\right)$$

which leads to

$$L'(1, \chi_d) = (\gamma + \log 2\pi)L(1, \chi_d) + \frac{1}{\sqrt{d}}\sum_{\nu=1}^{d-1}\bar{\chi}(\nu)R\left(\frac{\nu}{d}\right), \tag{2.7.7}$$

(Deninger, 1984, (4.2)).

As the following table shows, the parity of imaginary and real quadratic field and the associated Fourier series are consistent (Table 2.5).

Here $\log\Gamma(x)$ for the Fourier sine series is up to some extra terms. Higher order principal solutions $R_\alpha(x)$ to (2.7.3) stimulated the study on higher derivatives of L-functions, cf. Chakraborty et al. (2009a) (Tables 2.6 and 2.7).

In the light of the above tables, Theorem 2.6.1 may be described roughly as follows. The class invariant on the left of (2.6.7) is the processed (even part of the) theta-function with one-time use of the functional equation while the right-hand side

Table 2.5 Fourier sine, cosine series and their parity

Quadr. field	Imaginary	Real
Fourier ser.	$\log\Gamma(x)$	$-\Phi(x)$
Hurwitz zeta	$\zeta'(0, x)$	$R(x) + R(1-x)$
Lerch zeta	$-\frac{i}{2}(\ell_1'(x) - \ell_1'(1-x))$	$-\frac{1}{2}(\ell_1'(x) + \ell_1'(1-x))$

Table 2.6 Lerch–Chowla–Selberg ingredients: Eisenstein ser

Name	RHB	β trans.	F.e.	Product f.
Eisenstein ser.	$\vartheta \leftrightarrow \zeta^e$	$\zeta_{\mathbb{Z}^2} \to \zeta^e$	$\zeta^e \to \ell^e$	ϑ^e

Table 2.7 Lerch–Chowla–Selberg ingredients: Fourier sine ser

Name	RHB	Diff .w.r.t. s	Termwise diff.
Der. of Hurwitz	$\vartheta, \tilde{\vartheta} \leftrightarrow \zeta(s,x)$	$\zeta'(0,x) \to (\ell_1^e)'$ (F.e.)	Fourier sine ser.
Der. of Lerch	$\tilde{\vartheta} \leftrightarrow \ell_s^o$	$(\ell_1^e)'$	Fourier sine ser.

of (2.6.5) is in terms of the Lerch zeta-function which reflects the property of the almost unprocessed conjugate theta-function.

In terms of the gamma function, the right side of (2.6.7) is processed once by the functional equation and in more orderly form than (2.6.5).

Thus in a sense, it is a relation between the processed theta-functions and its counterparts under RHB correspondence are linked by the decomposition theorem, Lemma 2.6.1, which in turn is a consequence of the Artin reciprocity law, (Takagi, 1971, pp. 196–199).

2.8 Chowla–Selberg Formula for Elliptic Functions

Let

$$\log \frac{2K}{\pi} = \sum_{p=1}^{\infty} \frac{4z^p}{p(1+z^p)}$$

be one of the elliptic modular functions, where p runs through odd integers (i.e. odd part) (Wang, 2011). In Selberg and Chowla (1967) we can find some results on $\frac{2K}{\pi}$. K is defined as an elliptic integral

$$K = K(k) = \int_0^{\frac{\pi}{2}} \frac{1}{\sqrt{1-k^2\sin^2\varphi}}\,\mathrm{d}\varphi \quad (0 < k < 1)$$

with conjugate integral

$$K' = K'(k') = \int_0^{\frac{\pi}{2}} \frac{1}{\sqrt{1-k'^2\sin^2\varphi}}\,\mathrm{d}\varphi \quad (k^2 + k'^2 = 1).$$

It is known that

$$\sqrt{\frac{2K}{\pi}} = 2^{\frac{1}{3}}(kk')^{-\frac{1}{6}} q^{\frac{1}{12}} \prod_{n=1}^{\infty}(1-q^{2n}) = 2^{\frac{1}{3}}(kk')^{-\frac{1}{6}}\eta(\tau),$$

$$\prod_{n=1}^{\infty}(1+q^{2n-1}) = 2^{\frac{1}{6}} q^{\frac{1}{24}} (kk')^{-\frac{1}{12}},$$

where $q = e^{\pi i \tau}$, (2.1.63) and $\eta(\tau)$ is the Dedekind eta-function, (2.5.15).

It is known that if $\frac{K'}{K} = \sqrt{n}$ for n a positive integer, then K may be evaluated. Selberg and Chowla consider the case $\frac{K'}{K} = \sqrt{p}$ for p an odd prime such that the class number of the field $\mathbb{Q}(\sqrt{-p})$ is 1 and $\tau = i\frac{K'}{K}$, so that $q = e^{-\pi\frac{K'}{K}}$.

Gauss wrote in his diary on Jan. 8, 1797 (when he was 20 years old)

$$\frac{\varpi}{2} = \int_0^1 \frac{1}{\sqrt{1-t^4}}\,\mathrm{d}t = \frac{1}{\sqrt{2}}\int_0^{\frac{\pi}{2}} \frac{1}{\sqrt{1-k^2\sin^2\varphi}}\,\mathrm{d}\varphi = 1.31102878\cdots,$$

where $k = \frac{1}{\sqrt{2}}$, the $\frac{1}{4}$th length of the lemniscate $((x^2+y^2)^2 = x^2 - y^2)$ as an analogue of the quadrant circle length

$$\frac{\pi}{2} = \arcsin 1 = \int_0^1 \frac{1}{\sqrt{1-t^2}}\,\mathrm{d}t.$$

Traditionally one writes

$$K\left(\frac{1}{\sqrt{2}}\right) = \int_0^{\frac{\pi}{2}} \frac{1}{\sqrt{1-k^2\sin^2\varphi}}\,\mathrm{d}\varphi$$

and calls this the elliptic integral of the first kind. There is an intriguing relation

$$K\left(\frac{1}{\sqrt{2}}\right) = \frac{1}{4\sqrt{\pi}}\Gamma^2\left(\frac{1}{4}\right).$$

The Lerch–Chowla–Selberg formula is stated as

$$K = \lambda\sqrt{\pi}\prod_{m=1}^{|d|}\Gamma\left(\frac{m}{|d|}\right)^{\left(\frac{d}{m}\right)\frac{h}{4w}},$$

where $\lambda \in \bar{\mathbb{Q}}$.

2.9 Real Quadratic Case

We give a brief account of CNF (2.6.3) for a real quadratic field. We appeal to the method of Hecke (1917) whose main idea is the following integral of the $\mathrm{SL}_2(\mathbb{Z})$ Eisenstein series over the **Heegner cycle** $\mathbb{R}_+/\mathfrak{O}_K^\times$, i.e. the fundamental domain of the unit group.

$$\zeta(s, \mathfrak{C}) = \frac{\Gamma(s)}{\Gamma^2\left(\frac{s}{2}\right)}\int_{\mathbb{R}_+/\mathfrak{O}_K^\times} \zeta_{Q(x)}(s)\,\frac{\mathrm{d}x}{x},$$

which is the same as Novikov (1980, (1.2)) and Bump and Goldfeld (1987, (3))

$$\zeta(s,\mathfrak{C}) = \frac{\Gamma(s)}{\Gamma^2\left(\frac{s}{2}\right)} \sum_{\lambda\in\mathfrak{b}/\mathfrak{O}_K^\times}{}' \int_0^\infty \frac{1}{\left(\frac{\lambda^2x^2+\lambda'^2x^{-2}}{N\mathfrak{b}}\right)^s}\frac{\mathrm{d}x}{x}, \tag{2.9.1}$$

which follows from (2.1.86).

Proof of the class number formula (2.6.3) *for a real quadratic field.*
We use Zagier's formula which depends on (2.1.86) and reads

$$\zeta(s,\mathfrak{C}) = \frac{1}{2}\frac{\Gamma(s)}{\Gamma^2\left(\frac{s}{2}\right)} \sum_{\lambda\in\mathfrak{b}/\mathfrak{O}_K^\times}{}' \int_{-\infty}^\infty \frac{1}{\left(\frac{\lambda^2e^v+\lambda'^2e^{-v}}{N\mathfrak{b}}\right)^s}\,\mathrm{d}v.$$

Recalling that $\mathfrak{O}_K^\times = \{\pm\varepsilon^n | n\in\mathbb{Z}\}$, we may divide the range of integration to obtain

$$\zeta(s,\mathfrak{C}) = \frac{1}{2}\frac{\Gamma(s)}{\Gamma^2\left(\frac{s}{2}\right)} \int_{-\log\varepsilon}^{\log\varepsilon} \zeta_{Q_v}(s)\,\mathrm{d}v, \tag{2.9.2}$$

where for $\lambda = mw + nw'$, where w' signifies the conjugate,

$$Q_v(m,n) = \frac{(mw+nw')^2e^v + (mw'+nw)^2e^{-v}}{N\mathfrak{b}}.$$

Since Q_v is a positive definite quadratic form of discriminant $-d<0$, we may appeal to (2.5.23) and conclude that the residue of $\zeta_{Q_v}(s)$ is $\frac{2\pi}{\sqrt{d}}$. Substituting this in (2.9.2), we see that the residue of $\zeta(s,\mathfrak{C})$ is

$$\frac{2\log\varepsilon}{\sqrt{d}}.$$

Hence by the same argument as above, we conclude (2.6.3), completing the proof. □

Corresponding to Corollary 2.5.1, we have the Kronecker limit formula for a real quadratic field. Here the class invariant $\mathbb{D}(\mathfrak{C})$ is defined by (2.5.44) with

$$\delta(\mathfrak{C}) = \begin{vmatrix} \alpha_1 & \alpha_1^\tau \\ \alpha_2 & \alpha_2^\tau \end{vmatrix} > 0,$$

where

$$\mathfrak{C}(x) = \mathbb{Z}\alpha_1(x)\oplus\mathbb{Z}\alpha_2(x), \quad \begin{pmatrix}\alpha_1(x)\\ \alpha_2(x)\end{pmatrix} = \begin{pmatrix}\alpha_1 & \alpha_1^\tau\\ \alpha_2 & \alpha_2^\tau\end{pmatrix} = \begin{pmatrix} x \\ (ix)^{-1}\end{pmatrix}$$

and τ is the non-trivial automorphism of K.

Lemma 2.9.1 *For a real quadratic field with discriminant d and class number h we have the Laurent expansion*

$$\zeta(s,\mathfrak{C}) = \frac{\frac{2\log\varepsilon}{\sqrt{d}}}{s-1} + \frac{2\log\varepsilon}{\sqrt{d}}\left(2\gamma + \log\frac{4(2\pi)^2}{\sqrt{d}}\right) - \frac{4}{\sqrt{d}}\int_1^{\varepsilon} \log \mathbb{D}(\mathfrak{C}(x))\, \mathrm{d}\log x + O(s-1) \tag{2.9.3}$$

or

$$\zeta_K(s) = \frac{\frac{2h\log\varepsilon}{\sqrt{d}}}{s-1} + \frac{2h\log\varepsilon}{\sqrt{d}}\left(2\gamma + \log\frac{4(2\pi)^2}{\sqrt{d}}\right) - \frac{4}{\sqrt{d}}\sum_{\mathfrak{C}\in I/P}\int_1^{\varepsilon} \log \mathbb{D}(\mathfrak{C}(x))\, \mathrm{d}\log x + O(s-1). \tag{2.9.4}$$

Remark 2.9.1 Novikov digresses from (2.9.1) into the use of a hypergeometric function, which method may be efficient in this direction (cf. Kanemitsu et al., 2008).

Now we compare (2.9.4) with (2.7.7) to deduce that

$$\frac{2h\log\varepsilon}{\sqrt{d}}\left(2\gamma + \log\frac{4(2\pi)^2}{\sqrt{d}}\right) - \frac{4}{\sqrt{d}}\sum_{\mathfrak{C}\in I/P}\int_1^{\varepsilon} \log \mathbb{D}(\mathfrak{C}(x))\, \mathrm{d}\log x$$
$$= (\gamma + \log 2\pi)L(1,\chi_d) + \frac{1}{\sqrt{d}}\sum_{a=1}^{d-1} \bar{\chi}(a) R\left(\frac{a}{d}\right)$$

or invoking the class number formula (2.6.3),

$$4\sum_{\mathfrak{C}\in I/P}\int_1^{\varepsilon} \log \mathbb{D}(\mathfrak{C}(x))\, \mathrm{d}\log x = 2h\log\varepsilon \log\frac{8\pi}{\sqrt{d}} - \sideset{}{^*}\sum_{a=1}^{d-1} \bar{\chi}(a) R\left(\frac{a}{d}\right), \tag{2.9.5}$$

(Deninger, 1984, (4.7)), where the * on the summation sign means the sum is over all reduced residue classes.

Now correspondingly to Theorem 2.6.1, (ii) we have

Theorem 2.9.1 (Deninger, 1984, (4.8)) *The Lerch–Chowla–Selberg formula for a real quadratic field K holds*

$$\left(\prod_{\mathfrak{C}} \mathbb{D}^{\varepsilon}(\mathfrak{C})\right)^4 = \left(\frac{8\pi}{\sqrt{d}}\right)^{2h\log\varepsilon} \sideset{}{^*}\prod_{a=1}^{d-1} \Gamma_2\left(\frac{a}{d}\right)^{-\chi(a)}, \tag{2.9.6}$$

Table 2.8 Ingredients for quadratic fields

Name	Imaginary	Real
Epstein	Pos. def.	Indef.
Class no. form.	Easier	More involved
Kronecker limit form.	Many authors	Hecke, Novikov, [KTT]

Table 2.9 Class number formula (CNF) for quadratic fields

Name	Imaginary	Real
Char.	Odd	Even
CNF	$\bar{B}_{1,\chi}$, cot	$\log \Gamma$, Clausen
Ref.	Hasse (1952), Meyer (1957)	Complications by units: Pell eqn

where

$$\mathbb{D}^{\varepsilon}(\mathfrak{C}) = \exp\left(\int_1^{\varepsilon} \log \mathbb{D}(\mathfrak{C}(x))\, \mathrm{d}\log x\right)$$

and

$$\Gamma_2 = e^R,$$

where $R = R(x)$ *is Deninger's R-function in* (2.7.5) *(Tables 2.8 and 2.9).*

Kanemitsu and Tsukada (2014) refers to Kanemitsu et al. (2008).

2.10 General Epstein Zeta-Functions

In this subsection, we reproduce part of the results in Kanemitsu and Tsukada (2007, Sect. 6.5, pp. 125–130) for general Epstein zeta-functions and deduce the Chowla–Selberg integral formula, Theorem 2.10.2 from the Fourier Bessel expansion, Theorem 2.10.1. Cf. also Chakraborty et al. (2009c, Sect. 6.7, pp. 126–130). This gives a smooth shift from the Fourier–Bessel expansion (2.12.42) to Theorem 3.6.1 in a more general way.

Notation. Let $\boldsymbol{g}, \boldsymbol{h} \in \mathbb{R}^n$ be n-dimensional real vectors which (in the first place) give rise to the perturbation and the (additive) characters, respectively.

Let $Y = (y_{ij})$ be a positive definite $n \times n$ real symmetric matrix. Define the Epstein zeta-function associated to the quadratic form

$$Y[\boldsymbol{a}] = \boldsymbol{a} \cdot Y\boldsymbol{a} = {}^t\boldsymbol{a}Y\boldsymbol{a} = \sum_{i,j=1}^{n} y_{ij}a_i a_j,$$

where $\boldsymbol{a} = (a_1, \ldots, a_n) \in \mathbb{R}^n$ and "$\cdot$" means the inner product, by

$$Z(Y, \boldsymbol{o}, \boldsymbol{o}, s) = \sum_{\substack{\boldsymbol{a}\in\mathbb{Z}^n \\ \boldsymbol{a}\neq\boldsymbol{o}}} \frac{1}{Y[\boldsymbol{a}]^s}, \quad \sigma > \frac{n}{2},$$

where $\sigma = \operatorname{Re} s$.

For $\boldsymbol{g},\ \boldsymbol{h} \in \mathbb{R}^n$ define the general *Epstein zeta-function* (with perturbation) by

$$Z(Y, \boldsymbol{g}, \boldsymbol{h}, s) = \sum_{\substack{\boldsymbol{a}\in\mathbb{Z}^n \\ \boldsymbol{a}+\boldsymbol{g}\neq\boldsymbol{o}}} \frac{e^{2\pi i \boldsymbol{h}\cdot\boldsymbol{a}}}{Y[\boldsymbol{a}+\boldsymbol{g}]^s}, \quad \sigma > \frac{n}{2},$$

and incorporate the *completion*

$$\Lambda(Y, \boldsymbol{g}, \boldsymbol{h}, s) = \pi^{-s}\Gamma(s)\, Z(Y, \boldsymbol{g}, \boldsymbol{h}, s),$$

which satisfies the functional equation of the form (2.12.2) with an additional factor and replacement of parameters (which follows from Theorem 2.10.2):

$$\Lambda(Y, \boldsymbol{g}, \boldsymbol{h}, s) = \frac{1}{\sqrt{|Y|}}\, e^{-2\pi i \boldsymbol{g}\cdot\boldsymbol{h}}\, \Lambda\left(Y^{-1}, \boldsymbol{h}, -\boldsymbol{g}, \frac{n}{2} - s\right).$$

Theorem 2.10.1 (Fourier–Bessel expansion) *Notation being as above, we have for* $b > 0$,

$$\pi^{-s}\,\Gamma(s) \sum_{\boldsymbol{a}\in\mathbb{Z}^n} \frac{e^{2\pi i \boldsymbol{h}\cdot\boldsymbol{a}}}{(Y[\boldsymbol{a}+\boldsymbol{g}]+b)^s} \tag{2.10.1}$$

$$= \frac{2}{\sqrt{|Y|}} \sum_{\substack{\boldsymbol{a}\in\mathbb{Z}^n \\ \boldsymbol{a}+\boldsymbol{h}\neq\boldsymbol{o}}} e^{-2\pi i \boldsymbol{g}\cdot(\boldsymbol{a}+\boldsymbol{h})} \sqrt{\frac{Y^{-1}[\boldsymbol{a}+\boldsymbol{h}]}{b}}^{\,s-\frac{n}{2}} K_{s-\frac{n}{2}}\left(2\sqrt{Y^{-1}[\boldsymbol{a}+\boldsymbol{h}]b}\,\pi\right)$$

$$+\,\delta(\boldsymbol{h})\,\frac{1}{\sqrt{|Y|}}\,\frac{\Gamma(s-\frac{n}{2})}{\pi^{s-\frac{n}{2}}}\,\frac{1}{b^{s-\frac{n}{2}}}.$$

To state Theorem 2.10.2, we introduce new notation.

Let $Y = \begin{pmatrix} A & B \\ {}^tB & C \end{pmatrix}$ be a block decomposition with A an $n \times n$ matrix and C an $m \times m$ matrix. Set

$$D = C - {}^tB^tA^{-1}B.$$

In accordance with this decomposition, we decompose the vectors $\boldsymbol{g} = \begin{pmatrix} \boldsymbol{g}_1 \\ \boldsymbol{g}_2 \end{pmatrix}, \boldsymbol{h} = \begin{pmatrix} \boldsymbol{h}_1 \\ \boldsymbol{h}_2 \end{pmatrix}$, $\boldsymbol{g}_1, \boldsymbol{h}_1 \in \mathbb{Z}^n, \boldsymbol{g}_2, \boldsymbol{h}_2 \in \mathbb{Z}^m$.

Theorem 2.10.2 (General Chowla–Selberg integral formula) *Under the above notation, we have*

$$\begin{aligned}
&\Lambda(Y, \boldsymbol{g}, \boldsymbol{h}, s) \qquad (2.10.2)\\
&= \delta(\boldsymbol{g}_2)\, e^{-2\pi i \boldsymbol{g}_2\cdot \boldsymbol{h}_2} \Lambda(A, \boldsymbol{g}_1, \boldsymbol{h}_1, s) + \delta(\boldsymbol{h}_1)\, \frac{1}{\sqrt{|A|}}\, \Lambda\Big(D, \boldsymbol{g}_2, \boldsymbol{h}_2, s-\frac{n}{2}\Big)\\
&+ \frac{2e^{-2\pi i \boldsymbol{g}_1\cdot \boldsymbol{h}_1}}{\sqrt{|A|}} \sum_{\substack{\boldsymbol{a}\in\mathbb{Z}^n\\ \boldsymbol{a}+\boldsymbol{h}_1\neq \boldsymbol{o}}} \sum_{\substack{\boldsymbol{b}\in\mathbb{Z}^m\\ \boldsymbol{b}+\boldsymbol{g}_2\neq \boldsymbol{o}}} e^{2\pi i(-\boldsymbol{g}_1\cdot\boldsymbol{a}+\boldsymbol{h}_2\cdot\boldsymbol{o})} e^{-2\pi i A^{-1}B(\boldsymbol{b}+\boldsymbol{g}_2)\cdot(\boldsymbol{a}+\boldsymbol{h}_1)}\\
&\times \sqrt{\frac{A^{-1}[\boldsymbol{a}+\boldsymbol{h}_1]}{D[\boldsymbol{b}+\boldsymbol{g}_2]}}^{\,s-\frac{n}{2}} K_{s-\frac{n}{2}}\Big(2\sqrt{A^{-1}[\boldsymbol{a}+\boldsymbol{h}_1]\, D[\boldsymbol{b}+\boldsymbol{g}_2]}\,\pi\Big).
\end{aligned}$$

Proof Noting that

$$\Lambda(Y, \boldsymbol{g}, \boldsymbol{h}, s) = \pi^{-s}\,\Gamma(s) \sum_{\substack{\boldsymbol{a}\in\mathbb{Z}^n, \boldsymbol{b}\in\mathbb{Z}^m\\ (\boldsymbol{a}+\boldsymbol{g}_1, \boldsymbol{b}+\boldsymbol{g}_2)\neq \boldsymbol{o}}} \frac{e^{2\pi i(\boldsymbol{h}_1\cdot\boldsymbol{a}+\boldsymbol{h}_2\cdot\boldsymbol{b})}}{Y[(\boldsymbol{a}+\boldsymbol{g}_1, \boldsymbol{b}+\boldsymbol{g}_2)]^s},$$

we divide the sum into three parts according as $\boldsymbol{b}+\boldsymbol{g}_2 = \boldsymbol{o}$ ($\boldsymbol{g}_2 = \boldsymbol{o}$ or not) and $\boldsymbol{b}+\boldsymbol{g}_2 \neq \boldsymbol{o}$:

$$\begin{aligned}
\Lambda(Y, \boldsymbol{g}, \boldsymbol{h}, s) &= \delta(\boldsymbol{g}_2)\,\pi^{-s}\,\Gamma(s) \sum_{\substack{\boldsymbol{a}\in\mathbb{Z}^n\\ \boldsymbol{a}+\boldsymbol{g}_1\neq \boldsymbol{o}}} \frac{e^{2\pi i(\boldsymbol{h}_1\cdot\boldsymbol{a}-\boldsymbol{h}_2\cdot\boldsymbol{g}_2)}}{Y[(\boldsymbol{a}+\boldsymbol{g}_1, \boldsymbol{o})]^s} \qquad (2.10.3)\\
&+ \pi^{-s}\,\Gamma(s) \sum_{\substack{\boldsymbol{b}\in\mathbb{Z}^m\\ \boldsymbol{b}+\boldsymbol{g}_2\neq \boldsymbol{o}}} \sum_{\boldsymbol{a}\in\mathbb{Z}^n} \frac{e^{2\pi i(\boldsymbol{h}_1\cdot\boldsymbol{a}+\boldsymbol{h}_2\cdot\boldsymbol{b})}}{Y[(\boldsymbol{a}+\boldsymbol{g}_1, \boldsymbol{b}+\boldsymbol{g}_2)]^s}.
\end{aligned}$$

Then apply the formula

$$Y[(\boldsymbol{a}, \boldsymbol{b})] = A[\boldsymbol{a}+A^{-1}B\boldsymbol{b}] + D[\boldsymbol{b}], \quad \boldsymbol{a}\in\mathbb{Z}^n, \boldsymbol{b}\in\mathbb{Z}^m$$

to transform (2.10.3) into

$$\begin{aligned}
&\Lambda(Y, \boldsymbol{g}, \boldsymbol{h}, s) \qquad (2.10.4)\\
&= \delta(\boldsymbol{g}_2)\,\pi^{-s}\,\Gamma(s)\, e^{-2\pi i \boldsymbol{h}_2\cdot\boldsymbol{g}_2} \sum_{\substack{\boldsymbol{a}\in\mathbb{Z}^n\\ \boldsymbol{a}+\boldsymbol{g}_1\neq \boldsymbol{o}}} \frac{e^{2\pi i \boldsymbol{h}_1\cdot\boldsymbol{a}}}{A[\boldsymbol{a}+\boldsymbol{g}_1]^s}\\
&+ \pi^{-s}\,\Gamma(s) \sum_{\substack{\boldsymbol{b}\in\mathbb{Z}^m\\ \boldsymbol{b}+\boldsymbol{g}_2\neq \boldsymbol{o}}} e^{2\pi i \boldsymbol{h}_2\cdot\boldsymbol{b}} \sum_{\boldsymbol{a}\in\mathbb{Z}^n} \frac{e^{2\pi i \boldsymbol{h}_1\cdot\boldsymbol{a}}}{(A[\boldsymbol{a}+\boldsymbol{g}_1+A^{-1}B(\boldsymbol{b}+\boldsymbol{g}_2)]+D[\boldsymbol{b}+\boldsymbol{g}_2])^s}.
\end{aligned}$$

The first sum on the right of (2.10.4) is $\Lambda(A, \boldsymbol{g}_1, \boldsymbol{h}_1, s)$ and to the inner sum in the second term, we apply Theorem 2.10.1. Then we may check that the second term on the right of (2.10.4) becomes the sum of the third and the second terms on the right of (2.10.2), whence the result follows. □

Proof of the case $m = 1$ based on Theorem 2.10.1 is due to Berndt (1971).

2.11 Product Expansion of Theta-Allied Functions

In previous sections, we worked either in $\sigma > 1$ or $\sigma < 0$ save for the Lerch zeta-function which can go into the critical strip $0 < \sigma < 1$. For the Riemann zeta-function to go into the critical strip, it needs the Euler product. Recall (1.1.1).

Euler products in general have been one of the most important objects of research from the time of Euler. As is known, the Euler product expresses the UFD property of the basic sequence $\{n\}$. Thus wherever there is a UFD, there is the Euler product.

Here we speak of the product expressions (2.5.14) for theta-functions, continuing the passages squeezing Lemma 2.1.1.

Corollary 2.11.1 (Infinite product for theta) *For* $|q| < 1$ *and* $z \neq 0$

$$\vartheta(\tau, x) = \prod_{n=1}^{\infty}(1 - q^{2n})(1 + zq^{2n-1})(1 + z^{-1}q^{2n-1}), \tag{2.11.1}$$

where $z = e^{2\pi i x}$.

Ogg's (1969a) proof of Corollary 2.11.1 amounts to proving that the inverse Mellin transform of the logarithm of the infinite product $\theta_0(\tau, x)$, say, satisfies the functional equation from which the same transformation formula as (2.1.64) is deduced. It therefore turns out that the quotient $\frac{\vartheta}{\vartheta_0}$, where ϑ is the additive expression in (2.1.62), is an elliptic function in x with periods $1, \tau$ with at most one singularity, so that it is a constant, cf. e.g. Chan (2020, pp. 53–54). Use is made of the second equality in (2.1.2) in the form

$$e^{-\pi \frac{\tau}{i} u} = \frac{1}{2\pi i}\int_{c}\left(\frac{\tau}{i}\right)^{-s}\Gamma(s)(\pi u)^{-s}\, ds, \quad c > 0, \quad u > 0. \tag{2.11.2}$$

Since the zeta-function is $\zeta(s)\zeta(s+1)$ (e.g. Weil, 1968) which is in RHB correspondence with $\log \eta$, it may be the case that if we employ an additive expression for the relevant theta-like function, we may argue in the same way as Ogg to derive the product expression. Hitherto only the product expression is used as the definition. This direction may be worth pursuing.

Thus we have seen two occurrences of the Fourier–Bessel expansion. It may be proper to ask if the Hurwitz formula may be another occurrence. But it seems that direct application leads to the K-Bessel function series (Kanemitsu & Tsukada, 2014, Example 4.3, pp. 121–122) which suggests a reason why Siegel's proof fails in the case of Hecke (Knopp, 2000, p. 207). As is elucidated in Li et al. (2020), it seems the Ewald expansion that leads most naturally to the Hurwitz formula.

Abundance principle was first formulated in Knopp (2000) but it was established by Hutchinson (1929) in the case of the Hurwitz zeta-function. Hence the functional equation alone is not enough to lead to a Hamburger-type theorem.

Glaeske (1966) studied the Integra-functional equation introduced by Maier (1936) for the completed Hurwitz zeta-function $\Gamma(s, z)$ and have a characterization. Maier and Kiesewetter (1971, pp. 109–127) contain an account of the Hurwitz-Lerch zeta-function in which Glaeske's result is mentioned.

2.12 Hecke's Functional Equation

For a more general case cf. Definition 3.2.1.

Definition 2.12.1 Let

$$0 < \lambda_1 < \lambda_2 < \cdots \to \infty,\ 0 < \mu_1 < \mu_2 < \cdots \to \infty$$

be increasing sequences of real numbers. For complex sequences $\{a_n\},\ \{b_n\}$ form the Dirichlet series

$$\varphi(s) = \sum_{n=1}^{\infty} \frac{a_n}{\lambda_n^s} \quad \text{and} \quad \psi(s) = \sum_{n=1}^{\infty} \frac{b_n}{\mu_n^s} \tag{2.12.1}$$

which we assume are absolutely convergent for $\sigma > \sigma_a^*$ and $\sigma > \sigma_b^*$, respectively. Then $\varphi(s)$ and $\psi(s)$ are said to satisfy **Hecke's functional equation (HFE)**

$$A^{-s}\Gamma(s)\varphi(s) = A^{-(r-s)}\Gamma(r-s)\psi(r-s), \tag{2.12.2}$$

where $A > 0$ is a constant, if there exists a regular function $\chi(s)$ outside of a compact set $\mathcal{S}$ such that

$$\chi(s) = A^{-s}\Gamma(s)\varphi(s), \quad \sigma > \alpha(\geq \sigma_a^*)$$

and

$$\chi(s) = A^{-(r-s)}\Gamma(r-s)\psi(r-s), \quad \sigma < \beta\,(\leq r - \sigma_b^*)$$

and such that $\chi(s)$ is convex in the sense that

$$e^{-\varepsilon|t|}\chi(\sigma + it) = O(1), \quad 0 < \varepsilon < \frac{\pi}{2}, \tag{2.12.3}$$

uniformly in σ, $\sigma_1 \leq \sigma \leq \sigma_2$, $|t| \to \infty$. We refer to the Dirichet series (2.12.1) as **Hecke L-functions (HLF)** which satisfy HFE, (2.12.2).

Following Bochner (1951), define the residual function

$$\mathrm{P}(x) = \frac{1}{2\pi i}\int_C \chi(s)x^{-s}ds, \tag{2.12.4}$$

where C encircles all the singularities of $\chi(s)$ in $\mathcal{S}$.

We introduce the modular type functions corresponding to Dirichlet series (2.12.1)

$$f(\tau) = \sum_{n=0}^{\infty} a_n e^{Ain\tau} \quad \text{and} \quad g(\tau) = \sum_{n=0}^{\infty} b_n e^{Ain\tau}, \quad \tau \in \mathcal{H}$$

which are absolutely convergent and satisfy the (modular) transformation formula

$$f(\tau) = C\left(\frac{\tau}{i}\right)^{-r} g\left(-\frac{1}{\tau}\right). \tag{2.12.5}$$

Lemma 2.12.1 (Hecke) *That the Dirichlet series* (2.12.1) *satisfy the condition that* $A^{-s}\Gamma(s)\varphi(s) + \frac{a_0}{s} + \frac{Cb_0}{r-s}$ *is BEV (bounded in every vertical strip) and satisfies the functional equation*

$$A^{-s}\Gamma(s)\varphi(s) = CA^{-(r-s)}\Gamma(r-s)\psi(r-s) \tag{2.12.6}$$

is equivalent to (2.12.5), *where* (2.12.6) *is* (2.12.2) *with* $\psi(r-s)$ *replaced by* $C\psi(r-s) = \sum_{n=1}^{\infty} \frac{Cb_n}{\mu_n^{r-s}}$. *We refer to* (2.12.6) *as the* **Hecke (type) functional equation (HTFE)**.

Lemma 2.12.1 is a slight generalization of Ogg (1969b, p. I-5, Theorem 1), which is a handy statement of Hecke's epoch-making discovery (Hecke, 1936, 1983). In most cases $C = 1$. For the level N case, cf. Miyake (1989, pp. 119–123). For a more general case, cf. (2.13.2) and Kanemitsu and Tsukada (2014).

Ogg states (Ogg, 1969b, p. I-7) that "Theorem 1 was a great step forward 75 years after the functional equation for the zeta-function, for it reduces a question about Dirichlet series to one about modular forms, which are easier to work with." This Ogg's view is in contrast to Weil's interpretation (Weil, 1971) that this is a revival of the theory of automorphic forms as extracted after Table 1.2.

As Hecke's Lemma 2.12.1 is the upper half-plane version, the Bochner relation

$$G_{0,1}^{1,0} \leftrightarrow G_{0,1}^{1,0}$$

is the right-half plane version, of the modular relation. One of the most famous examples of the Bochner modular relation is **Ramanujan's formula**

$$\sum_{n=1}^{\infty} \sigma_{-2\varkappa-1}(n)e^{-2\pi nx} + (-1)^{\varkappa+1}x^{2\varkappa}\sum_{n=1}^{\infty} \sigma_{-2\varkappa-1}(n)e^{-\frac{2\pi n}{x}} = \mathrm{P}(x), \tag{2.12.7}$$

where

$$
\begin{aligned}
&P_{2\varkappa+1}(x) \\
&= \sum_{j=1}^{\varkappa+1} \frac{(-1)^j}{(2j+1)!}\zeta(1-2j)\zeta(2\varkappa-2j+2)x^{2j-1} + \zeta(2\varkappa+2)x^{-1} - \frac{1}{2}\zeta(2\varkappa+1).
\end{aligned}
$$

Cf. Kanemitsu et al. (2002) and Dixit et al. (2020) for further developments.

Exercise 2.12.1 Prove that $\varphi(s) = \varphi(s, \alpha) = \zeta(s)\zeta(s+\alpha)$ is the generating function for the sum-of-divisors function $\sigma_{-\alpha}(n)$ in (2.5.21) and that for $\alpha 2 = \varkappa + 1$ an odd integer, it satisfies the Hecke functional equation

$$
\Gamma(1-s)\varphi(1-s) = (2\pi)^{-2s+\alpha+1}(-1)^{\frac{\alpha-1}{2}}\Gamma(\alpha+1-s)\varphi(s-\alpha).
$$

Deduce (2.12.7) from this.

In order to explain the cusp-form L-function in Example 2.12.1, we state basics of modular forms for the full modular group, cf. e.g. Apostol (1976b), Fomenko (1980), Lang (1976), Serre (1973), Shimura (1971), etc., and references therein. Cf. also Sect. 3.8 for the case of $SL_2(\mathbb{R})$.

Definition 2.12.2 Let

$$
SL_2(\mathbb{Z}) = \left\{ \gamma = \begin{pmatrix} a & b \\ c & d \end{pmatrix} \middle| a, b, c, d \in \mathbb{Z}, ad - bc = 1 \right\} \tag{2.12.8}
$$

be the **full modular group** (or the special linear group), which acts on the upper half-plane by the linear fractional transformation (or Möbius transformation):

$$
\gamma z = \frac{az+b}{cz+d}, \quad \gamma = \begin{pmatrix} a & b \\ c & d \end{pmatrix}, \quad z \in \mathcal{H}. \tag{2.12.9}
$$

Cf. the case of $SL_2(\mathbb{R})$, (3.8.1). By abuse of language we write $\Gamma = \Gamma(1) = PSL_2(\mathbb{Z})$ to mean the group of both matrices and linear fractional transformations $SL_2(\mathbb{Z})/\{\pm 1\}$.

Γ is generated by the translation and the Spiegelung

$$
Tz = z + 1, \quad Sz = -\frac{1}{z}
$$

and the translation generates the stabilizer of ∞.

$$
\Gamma_\infty = \langle T \rangle = \{T^n | n \in \mathbb{Z}\}.
$$

Then in the definition (3.2.6) of the Eisenstein–Maass series, the summation condition $\Gamma_\infty \backslash \Gamma$ means that the denominator is non-trivial and it makes sense.

Often congruence subgroups are studied. The **principal congruence subgroup** $\Gamma(N)$ of level N is defined to be the normal subgroup of Γ by

$$\Gamma(N) = \left\{\gamma = \begin{pmatrix} a & b \\ c & d \end{pmatrix} \equiv \begin{pmatrix} 1 & 0 \\ 0 & 1 \end{pmatrix} (\bmod N) \middle| a, b, c, d \in \mathbb{Z}, ad - bc = 1\right\}$$

for every positive integer N. Hence the notation $\Gamma = \Gamma(1)$ is consistent. Any subgroup of Γ containing $\Gamma(N)$ for some N is called a **congruence subgroup**. In this book we treat the congruence subgroup $\Gamma_0(N)$ defined by

$$\Gamma_0(N) = \left\{\gamma = \begin{pmatrix} a & b \\ c & d \end{pmatrix} \in \Gamma \middle| c \equiv 0 \pmod N\right\}. \tag{2.12.10}$$

Cf. Sect. 2.14.

A function f meromorphic resp. holomorphic (including ∞) on $\mathcal{H}$ is called a **modular function resp. modular form of weight** k if it satisfies the transformation formulas

$$f(Tz) = f(z+1) = f(z), \quad f(Sz) = f\left(-\frac{1}{z}\right) = z^k f(z).$$

The general transformation formula is given in (3.8.3) below. It follows that $k \geq 2$ is an even integer. Suppose f is holomorphic. Then since it is periodic of period 1, it has the power series expansion in $q = e^{\pi i \tau}$, (2.1.63):

$$f(\tau) = \sum_{n=0}^{\infty} a_n q^{2n} = \sum_{n=0}^{\infty} a_n e^{2\pi i n \tau}. \tag{2.12.11}$$

This is often called the q**-expansion**. If the constant term a_0 is 0, then f is called a **cups form** of weight k.

Exercise 2.12.2 Suppose $w = u + iv$ and $z = x + iy$ are related by $\gamma z = w, \gamma \in \Gamma$, where $\Gamma = \mathrm{SL}_2(\mathbb{R})$, cf. Sect. 3.8. Prove the following. (i) The imaginary part (or y-**part**) $y(\gamma z)$ of γz is given by

$$v(z) = y(\gamma z) = \operatorname{Im} \gamma(z) = \frac{y}{|j_\gamma(z)|^2} = \frac{y}{|cz+d|^2}, \tag{2.12.12}$$

where j is the **denominator**

$$j_\gamma(z) = cz + d \tag{2.12.13}$$

with a convention that $c \geq 0$ and if $c = 0$, then $d > 0$.

(ii) We have

$$\frac{dw}{dz} = \frac{d(\gamma z)}{dz} = \frac{1}{j_\gamma(z)^2} \tag{2.12.14}$$

and the Jacobian $J_\gamma(z)$ of $(u, v) = \gamma(x, y)$ is given by

$$J_\gamma(z) = |\gamma'(z)|^2 = \frac{v^2}{y^2}. \tag{2.12.15}$$

(iii) Let

$$\Delta^e = \frac{\partial^2}{\partial x^2} + \frac{\partial^2}{\partial y^2}$$

be the Euclidean Laplacian and let

$$\Delta = c\Delta^e. \tag{2.12.16}$$

Choose c so that Δ is Γ-invariant;

$$\Delta \cdot \gamma = \gamma \cdot \Delta, \quad \gamma \in \Gamma. \tag{2.12.17}$$

Equation (2.12.16) with $c = -1$ is the non-Euclidean Laplacian (3.4.14). Cf. Proposition 3.6.1 for another proof.

Solution. (ii) We prove (2.12.15).

$$J_\gamma(z) = \begin{vmatrix} u_x & v_x \\ u_y & v_y \end{vmatrix} = u_x v_y - u_y v_x = u_x^2 + v_x^2 = |\gamma'(z)|^2.$$

The second equality follows from (2.12.14). by the Cauchy–Riemann equation and (2.12.14).

By (2.12.12), we have

$$|\gamma(z)|^2 = |cz + d|^2 = \frac{y}{v},$$

whence the last equality in (2.12.14) and (2.12.15) follow. (iii) Since

$$\Delta f(x, y) = \Delta F(\gamma(x, y)) = \Delta F(u, v),$$

we need the following.

$$F_x = F_u u_x + F_v v_x,$$

whence

$$\begin{aligned} \frac{\partial}{\partial x} F_x &= \frac{\partial}{\partial x} F_u \cdot u_x + F_u u_{xx} + \frac{\partial}{\partial x} F_v \cdot v_x + F_v v_{xx} \\ &= F_{uu} u_x^2 + 2F_{uv} u_x v_x + F_{vv} v_x^2 + F_u u_{xx} + F_v v_{xx} \end{aligned} \tag{2.12.18}$$

and similarly for $\frac{\partial}{\partial y}F_y$. Adding (2.12.18) and its counterpart for $\frac{\partial^2}{\partial y^2}F$ and applying the Cauchy–Riemann equation, we conclude that

$$\Delta F(u, v) = cJ_\gamma(z)(F_{uu} + F_{vv}) = c\frac{v^2}{y^2}(F_{uu} + F_{vv})$$

by (2.12.15). Hence choosing $c = c'y^2$, we arrive at (2.12.17). □

Remark 2.12.1 In what follows we often define a function involving the complex power of $y(\gamma z)$, which are of course not holomorphic in z. Equations (2.12.14) and (2.12.14) imply the **change of variable** formulas

$$\mathrm{d}u\mathrm{d}v = \frac{v^2}{y^2}\mathrm{d}x\mathrm{d}y, \quad |\mathrm{d}z| = \frac{v}{y}|\mathrm{d}w|, \tag{2.12.19}$$

where $|\mathrm{d}z|$ means the arc length element. The first of (2.12.19), the change of variable formula $\mathrm{d}u\mathrm{d}v = J_\gamma^{-1}\mathrm{d}x\mathrm{d}y$ implies the invariance of the **hyperbolic measure**

$$\frac{\mathrm{d}x\mathrm{d}y}{y^2} = \frac{\mathrm{d}u\mathrm{d}v}{v^2}. \tag{2.12.20}$$

This will be applied in the proof of Proposition 3.6.1 where the invariance of a differential form is proved which in turn implies the invariance of the Laplacian and the outer normal derivative.

Exercise 2.12.3 Suppose f, g are modular forms of weight k w.r.t. Γ. Prove that the measure

$$f(z)\overline{g(z)}y^k\,\frac{\mathrm{d}x\mathrm{d}y}{y^2}$$

is invariant under Γ.

Solution. By definition,

$$f(\gamma z)\overline{g(\gamma z)}v^k\,\frac{\mathrm{d}u\mathrm{d}v}{v^2} = j_\gamma(z)^k\overline{j_\gamma(z)}^{-k}v^k f(z)\overline{g(z)}\,\frac{\mathrm{d}u\mathrm{d}v}{v^2} = f(z)\overline{g(z)}y^k\,\frac{\mathrm{d}x\mathrm{d}y}{y^2}$$

by (2.12.12) and (2.12.20). □

We introduce the **Hecke operator** by

$$(T_n f)(z) = n^\lambda \sum_{d|n}\left(\frac{n}{d}\right)^k \sum_{b \bmod d} f\left(\frac{\frac{n}{d}z + b}{d}\right), \tag{2.12.21}$$

where $\lambda = -1$ if f is a modular form of weight k. For more general treatment cf. Andrianov (1987, Chaps. 4 and 5). Here we unify the argument of Apostol (1976b, pp. 120–128) and Motohashi (1997, pp. 98–100) by (2.12.21). For non-holomorphic forms, cf. Sect. 3.8.

Exercise 2.12.4 Prove that the action of the Hecke operator on the Fourier expansion (2.12.11) is given by

$$(T_n f)(z) = \sum_{m=0}^{\infty} t_n(m) e^{2\pi i m z}, \quad f(z) = \sum_{m=0}^{\infty} a(m) e^{2\pi i m z},$$

where

$$t_n(m) = n^{\lambda+1} \sum_{d|(m,n)} d^{k-1} a\left(\frac{mn}{d^2}\right) \tag{2.12.22}$$

whose special case with $\lambda = -1$ reads

$$t_n(m) = \sum_{d|(m,n)} d^{k-1} a\left(\frac{mn}{d^2}\right). \tag{2.12.23}$$

This entails $t_n(m) = a(mn)$ for $(m, n) = 1$. Cf. (3.8.8) for non-holomorphic forms, where the formula is slightly different from (2.12.22) because of the presence of the factor $\sqrt{y}$.

Solution. Since

$$(T_n f)(z) = n^{\lambda+1} \sum_{d|n} \left(\frac{n}{d}\right)^{k-1} \sum_{m=0}^{\infty} a(m) e^{2\pi i \frac{mn}{d^2} z} \frac{1}{d} \sum_{b \bmod d} e^{2\pi i \frac{m}{d} b}$$

it follows by orthogonality that

$$(T_n f)(z) = n^{\lambda+1} \sum_{m=0}^{\infty} \sum_{d|m, d|n} \left(\frac{n}{d}\right)^{k-1} a(m) e^{2\pi i \frac{mn}{d^2} z}.$$

Writing $m = \ell d$, $\ell = 0, 1, \cdots$ and then $\delta = \frac{n}{d}$,

$$\begin{aligned}(T_n f)(z) &= \sum_{\ell=0}^{\infty} n^{\lambda+1} \sum_{d|n} \left(\frac{n}{d}\right)^{k-1} a(\ell d) e^{2\pi i \frac{\ell n}{d} z} \\ &= \sum_{\ell=0}^{\infty} \sum_{\delta|n} \delta^{k+\lambda} \left(\frac{n}{\delta}\right)^{\lambda+1} a\left(\frac{\ell n}{\delta}\right) e^{2\pi i \ell \delta z}.\end{aligned}$$

Collecting the terms with $\ell\delta = m$, we have

$$(T_n f)(z) = \sum_{m=0}^{\infty} \sum_{\delta|(m,n)} \delta^{k+\lambda} \left(\frac{n}{\delta}\right)^{\lambda+1} a\left(\frac{mn}{\delta^2}\right) e^{2\pi i m z},$$

whence the assertion follows. □

Lemma 2.12.2 *For $n \in \mathbb{N}$ let*

$$\Gamma_n = \left\{ A = \begin{pmatrix} a & b \\ c & d \end{pmatrix} \middle| a, b, c, d \in \mathbb{Z}, \det A = n \right\}.$$

Then $\Gamma_1 = \Gamma = \Gamma(1)$ is the full modular group, (2.12.8)*, which acts naturally on Γ_n and introduces the equivalence relation $A_1 \sim A_2$ if there exists a $V \in \Gamma$ such that $VA_1 = A_2$. A complete set of representatives of $\Gamma\backslash\Gamma_n$ is given by*

$$\Gamma\backslash\Gamma_n = \left\{ A = \begin{pmatrix} a & b \\ 0 & d \end{pmatrix} \middle| a, b, d \in \mathbb{Z}, ad = n, 0 \le b < d \right\}, \tag{2.12.24}$$

where b runs through any complete set of residues mod d and we denote equivalence classes and representatives by the same symbol. In particular $\sharp\Gamma\backslash\Gamma_n = \sum_{d|n} d = \sigma(n)$ and (2.12.21) *may be written as*

$$(T_n f)(z) = n^\lambda \sum_{ad=n} a^k \sum_{b \bmod d} f\left(\frac{az+b}{d}\right), \tag{2.12.25}$$

where A is as in (2.12.24).

Proof For every $A = \begin{pmatrix} a & b \\ c & d \end{pmatrix} \in \Gamma_n$ there exists a $V = \begin{pmatrix} e & f \\ g & h \end{pmatrix} \in \Gamma$ such that $VA \in \Gamma\backslash\Gamma_n$. Suppose $c \neq 0$. Since the $(2, 1)$-entry is $ga + ch$, we choose $g = \frac{c}{\delta}, h = -\frac{a}{\delta}$ and $\delta = (a, c)$ so that $ga + hc = 0$ and $(g, h) = 1$. Then choose e, f such that $eh - fg = 1$. Then this $V \in \Gamma$ satisfies the requirement.

To prove (2.12.24) it suffices to show that

$$A_1 = \begin{pmatrix} a_1 & b_1 \\ 0 & d_1 \end{pmatrix} \sim A_2 = \begin{pmatrix} a_2 & b_2 \\ 0 & d_2 \end{pmatrix}$$

if and only

$$a_1 = \pm a_2, d_1 = \pm d_2, b_1 \equiv \pm b_2 \bmod d_1. \tag{2.12.26}$$

Indeed, if $A_1 \sim A_2$, then there is a $V = \begin{pmatrix} e & f \\ g & h \end{pmatrix} \in \Gamma$ such that $VA_1 = A_2$. Hence $ga_1 = 0$, so that $g = 0$ since $a_1 d_1 = n$. Hence $1 = eh - fg = eh$. Hence $V = \pm\begin{pmatrix} 1 & f \\ 0 & 1 \end{pmatrix} \in \Gamma$ and the $(1, 2)$-enrty of VA_1 is $\pm d_1 + fd_1 = b_2$, i.e. (2.12.26) holds. Conversely, if (2.12.26) holds, then there is an f such that $b_2 = b_1 + fd_1$. Hence $V = \pm\begin{pmatrix} 1 & f \\ 0 & 1 \end{pmatrix} \in \Gamma$ satisfies $VA_1 = A_2$. □

Exercise 2.12.5 For $(m, n) = 1$ prove that

$$T_m T_n = T_{mn}. \tag{2.12.27}$$

Solution. We apply (2.12.25) with $A = \begin{pmatrix} a & b \\ 0 & d \end{pmatrix}$ resp. $B = \begin{pmatrix} \alpha & \beta \\ 0 & \delta \end{pmatrix}$ for T_n resp. T_m. Then

$$T_m T_n f(z) = (mn)^{\lambda} \sum_{A \in \Gamma\backslash\Gamma_n, B \in \Gamma\backslash\Gamma_m} (\alpha a)^k f((BA)z),$$

where

$$BA = \begin{pmatrix} \alpha a & \alpha b + \beta d \\ 0 & \delta d \end{pmatrix} = \begin{pmatrix} a' & \alpha b + \beta d \\ 0 & d' \end{pmatrix},$$

say. Since $(m, n) = 1$, we have $a' = \alpha a | mn$, $d' = \delta d | mn$ and $a'd' = mn$. Further $b' = \alpha b + \beta d$ runs through the complete set of representatives mod $d\delta$ since αb resp. βd run through the complete set of representatives mod d resp. mod δ and $(d, \delta) = 1$. Hence the map: $\Gamma\backslash\Gamma_m \times \Gamma\backslash\Gamma_n \to \Gamma\backslash\Gamma_{mn} : (B,\ A) \to BA$ is bijective and (2.12.27) follows. □

Theorem 2.12.1 *We have*

$$T_m T_n = \sum_{d|(m,n)} d^{k+2\lambda+1} T_{\frac{mn}{d^2}}. \tag{2.12.28}$$

In particular, for a prime p *and* $r \in \mathbb{N}$,

$$T_p T_{p^r} = T_{p^r} T_p = T_{p^{r+1}} + p^{k+2\lambda+1} T_{p^{r-1}}. \tag{2.12.29}$$

For holomorphic forms, $\lambda = -1$ *case applies*

$$T_m T_n = \sum_{d|(m,n)} d^{k-1} T_{\frac{mn}{d^2}}.$$

Proof First we prove (2.12.29). Note that (2.12.21) reads for a $n = p^r$

$$(T_{p^r} f)(z) = p^{r\lambda} \sum_{j=0}^{r} p^{(r-j)k} \sum_{b_j \bmod p^j} f\left(\frac{p^{r-j}z + b_j}{p^j}\right) \tag{2.12.30}$$

and in particular,

$$(T_p g)(z) = p^{k+\lambda} g(pz) + p^{\lambda} \sum_{b \bmod p} g\left(\frac{z+b}{p}\right).$$

Hence

$$(T_pT_{p^r}f)(z) = S_1 + S_2,$$
$$S_1 = p^{k+\lambda(r+1)}\sum_{j=0}^{r} p^{(r-j)k}\sum_{b_j \bmod p^j} f\left(\frac{p^{r-j+1}z + pb_j}{p^j}\right)$$
$$S_2 = p^{\lambda(r+1)}\sum_{b \bmod p}\sum_{j=0}^{r} p^{(r-j)k}\sum_{b_j \bmod p^j} f\left(\frac{p^{r-j}z + b_j + p^jb}{p^{j+1}}\right),$$

say. Writing

$$S_2 = p^{\lambda(r+1)}\sum_{j+1=1}^{r+1} p^{(r+1-(j+1))k}\sum_{b \bmod p}\sum_{b_j \bmod p^j} f\left(\frac{p^{r+1-(j+1)}z + b_j + p^jb}{p^{j+1}}\right),$$

we see that $b_j + p^jb$ runs through a complete set of residue classes mod p^{j+1} and is equal to $T_{p^{r+1}}f$ minus $j+1=0$ term, which is the $j=0$ term of S_1. Hence we obtain

$$(T_pT_{p^r}f)(z) = S_1' + T_{p^r}f,$$
$$S_1' = p^{k+\lambda(r+1)}\sum_{j=1}^{r} p^{(r-j)k}\sum_{b_j=0}^{p^j-1} f\left(\frac{p^{r-j}z + b_j}{p^{j-1}}\right).$$

We write

$$b_j = p^{j-1}q_j + r_j, \quad 0 \le q_j < p, \quad 1 \le j \le r,$$

where r_j runs through a complete set of residue classes mod p^{j-1}. Therefore we have $f\left(\frac{p^{r-j}z+b_j}{p^{j-1}}\right) = f\left(\frac{p^{r-j}z+r_j}{p^{j-1}}\right)$ for $0 \le q_j < p$ $(1 \le j \le r)$. Hence there are p such terms and

$$S_1' = p^{k+\lambda(r+1)}p^{-\lambda(r-1)}p^{\lambda(r-1)}\sum_{j-1=0}^{r-1} p^{(r-j)k}p\sum_{r_j \bmod p^{j-1}} f\left(\frac{p^{r-1-(j-1)}z + r_j}{p^{j-1}}\right)$$
$$= p^{k+2\lambda+1}T_{p^{r-1}}f.$$

Applying (2.12.30), we have

$$(T_{p^r}T_pf)(z)$$
$$= p^{k+\lambda}p^{r\lambda}\sum_{j=0}^{r} p^{(r-j)k}\sum_{b_j \bmod p^j} f\left(\frac{p^{r-j+1}z + b_j}{p^j}\right)$$
$$+ p^{\lambda}\sum_{b \bmod p} p^{r\lambda}\sum_{j=0}^{r} p^{(r-j)k}\sum_{b_j \bmod p^j} f\left(\frac{p^{r-j}z + p^{r-j}b + pb_j}{p^{j+1}}\right).$$

Dividing the second term into two, we deduce (2.12.29).

It remains to prove

$$T_{p^q}T_{p^r} = T_{p^r}T_{p^q} = \sum_{j=0}^{\min\{q,r\}} R^j T_{p^{q+r-2j}}, \quad R = p^{k+2\lambda+1}. \tag{2.12.31}$$

We may assume $q \geq r$ and prove (2.12.31) by induction on r. The $r = 1$ case is (2.12.29). We assume the case $r \geq 1$ and prove the case $r + 1$. First, we prove

$$\begin{aligned} T_{p^{r+1}}T_{p^q} &= T_pT_{p^r}T_{p^q} - RT_{p^{r-1}}T_{p^q} \\ &= T_{p^q}T_pT_{p^r} - RT_{p^q}T_{p^{r-1}} = T_{p^q}T_{p^{r+1}} \end{aligned}$$

Next, we consider $T_pT_{p^r}T_{p^q}$ in two ways using (2.12.29). On one hand,

$$T_pT_{p^r}T_{p^q} = T_{p^{r+1}}T_{p^q} + RT_{p^{r-1}}T_{p^q} = T_{p^{r+1}}T_{p^q} + R\sum_{j=0}^{r-1} R^j T_{p^{q+r-2j-1}}$$

and on the other

$$T_pT_{p^r}T_{p^q} = \sum_{j=0}^{r} R^j T_pT_{p^{q+r-2j}} = \sum_{j=0}^{r} R^j T_{p^{q+r-2j+1}} + R\sum_{j=0}^{r} R^j T_{p^{q+r-2j-1}},$$

both using induction hypothesis. Equating these and cancelling $\sum_{j=0}^{r-1} R^j \times T_{p^{q+r-2j-1}}$, we deduce that

$$T_{p^{r+1}}T_{p^q} = \sum_{j=0}^{r} R^j T_{p^{q+r-2j+1}} + RR^r T_{p^{q+r-2r-1}},$$

which is (2.12.31) with $r + 1$. Exercise 2.12.5 and (2.12.31) induce (2.12.28). □

Definition 2.12.3 If a modular form f satisfies

$$T_n f = \lambda(n) f$$

for a complex number $\lambda(n)$, then f is called an eigen-function or an eigen-form of the Hecke operator T_n and $\lambda(n)$ is called an eigen-value. If f is an eigen-function of all T_n, then it is called a simultaneous eigen-function of all Hecke operators. Every eigen-form is called normalized if the first Fourier coefficient $a(1) = 1$.

Lemma 2.12.3 *Let $k \geq 4$ be an even integer. If there is a simultaneous eigen-function f of weight k with its Fourier expansion $f(z) = \sum_{m=0}^{\infty} a(m)e^{2\pi imz}$, then $a(1) \neq 0$ and so we may always normalize f.*

Theorem 2.12.2 *In the setting of Exercise 2.12.4 suppose f is a cusp form of weight $k \geq 12$. Then f is a normalized simultaneous eigen-form of T_n if and only if*

$$a(m)a(n) = \sum_{d|(m,n)} d^{k-1} a\left(\frac{mn}{d^2}\right). \tag{2.12.32}$$

Proof By (2.12.22),

$$t_n(1) = n^{\lambda+1} a(n). \tag{2.12.33}$$

Suppose f is an eigen-function of T_n, Then $T_n f = \lambda(n) f$ is equivalent to

$$t_n(m) = \lambda(n) a(m), \tag{2.12.34}$$

which holds for all $m, n \in \mathbb{N}$. Equation (2.12.34) with $m = 1$ and (2.12.33) yield $\lambda(n)a(1) = t_n(1) = n^{\lambda+1}a(n)$, so that $\lambda(n) = n^{\lambda+1}a(n)$ if $a(1) = 1$. Hence $n^{\lambda+1}a(n)a(m) = t_n(m)$ and (2.12.32) follows. Conversely, if (2.12.32) holds, then we have $n^{\lambda+1}a(m)a(n) = t_n(m)$. As we saw, $\lambda(n) = n^{\lambda+1}a(n)$ if $a(1) = 1$. Hence (2.12.34) holds and the assertion follows. □

Example 2.12.1 Let f be a cusp form of weight k w.r.t. $\Gamma(1)$. Then we define the L-function corresponding to f by

$$L(s, f) = \sum_{n=1}^{\infty} \frac{a(n)}{n^s} \tag{2.12.35}$$

referring to it as a **cusp-form zeta-function** or cusp-form L-function.. It satisfies the Hecke functional equation

$$(2\pi)^{-s}\Gamma(s)L(s, f) = (-1)^{k/2}(2\pi)^{-(k-s)}\Gamma(k-s)L(k-s, f) \tag{2.12.36}$$

by Lemma 2.12.1. In case f is a modular form, one takes the effect into account of the constant term $a(0)$. To prove (2.12.36), this is subtracted and the Mellin transform of $f(iy) - f(0)$ is considered, cf., e.g. Serre (1973, p. 103). If the coefficient $a(n)$ satisfies the multiplication property (2.12.32), then it has the Euler product

$$L(s, f) = \prod_p \left(1 - a(p)p^{-s} + p^{k-1}p^{-2s}\right)^{-1}, \tag{2.12.37}$$

which is one of the goals of the theory of Hecke operators.

Exercise 2.12.6 Prove (2.12.37).

Solution. Instead of (2.12.32), we assume

$$a(m)a(n) = \sum_{d|(m,n)} d^{k-1} a\left(\frac{mn}{d^2}\right) \tag{2.12.38}$$

and in particular

$$a(p)a(p^n) = a(p^{n+1}) + p^{k-1}a(p^{n-1}). \tag{2.12.39}$$

By (2.12.38), $a(n)$ is multiplicative and it is well known that (2.12.35) has the Euler product

$$L(s, f) = \prod_p \left(1 + \sum_{n=1}^{\infty} a(p^n)\xi^n\right),$$

where $\xi = p^{-s}$, $|\xi| < 1$. We show that

$$(1 - a(p)\xi + p^{k-1}\xi^2)\left(1 + \sum_{n=1}^{\infty} a(p^n)\xi^n\right) = 1. \tag{2.12.40}$$

Indeed,

$$\begin{aligned}
&(1 + p^{k-1}\xi^2 - a(p)\xi)\sum_{n=1}^{\infty} a(p^n)\xi^n \\
&= \sum_{n=1}^{\infty} a(p^n)\xi^n + p^{k-1}\sum_{n=1}^{\infty} a(p^n)\xi^{n+2} - \sum_{n=1}^{\infty} a(p)a(p^n)\xi^{n+1} \\
&= \sum_{n=1}^{\infty} a(p^n)\xi^n - \sum_{n=1}^{\infty} a(p^{n+1})\xi^{n+1} + p^{k-1}\Big(\sum_{n=1}^{\infty} a(p^n)\xi^{n+2} - \sum_{n=1}^{\infty} a(p^{n-1})\xi^{n+1}\Big) \\
&= a(p)\xi - p^{k-1}\xi^2
\end{aligned}$$

by (2.12.39). Hence (2.12.40) follows. □

The following theorem constitutes the basis of results related to Hecke's functional equation (2.12.2), cf. Kanemitsu et al. (2002), Kanemitsu and Tsukada (2014, p. 10). The EBV property in Lemma 2.12.1 is incorporated in the following theorem in terms of convexity and the residual function. This will also work as a touchstone for modular relations in general and is referred to as the **Hecke modular relation (theorem)**.

Theorem 2.12.3 (Hecke modular relation (theorem)) *The functional equation* (2.12.2), *the modular transformation formula* (2.12.5), *the Bochner modular relation* (2.12.41), *the Fourier–Bessel expansion* (2.12.42), *the Riesz sum* (2.12.44), *and the Ewald expansion* (2.12.45) *are all equivalent:*

$$\sum_{n=1}^{\infty} a_n e^{-\lambda_n x} = \left(\frac{A}{x}\right)^r \sum_{n=1}^{\infty} b_n e^{-\mu_n \frac{A^2}{x}} + \mathrm{P}\left(\frac{x}{A}\right) \tag{2.12.41}$$

with $\mathrm{Re}\, x > 0$ *and*

$$A^{-s}\Gamma(s)\varphi(s,\alpha) = 2\alpha^{\frac{r-s}{2}}\sum_{n=1}^{\infty} b_n\mu_n^{-\frac{r-s}{2}} K_{s-r}(2A\sqrt{\alpha\mu_n}) + A^{-s}\int_0^{\infty} e^{-\alpha u}u^{s-1}\mathrm{P}\left(\frac{u}{A}\right)du \tag{2.12.42}$$

with $\alpha > 0, \sigma > \max\{r-\frac{1}{2}, -1\}, s \neq 0$*, where*

$$\varphi(s,\alpha) = \sum_{n=1}^{\infty}\frac{a_n}{(\lambda_n+\alpha)^s} \tag{2.12.43}$$

denotes the (Hurwitz type) **perturbed Dirichlet series** *associated to* φ.

The Riesz sum of order $\varkappa$

$$\frac{1}{\Gamma(\varkappa+1)}\sum_{\lambda_n\le x}{}' a_n(x-\lambda_n)^{\varkappa} = A^{-\varkappa}\sum_{n=1}^{\infty}\left(\frac{x}{\mu_n}\right)^{\frac{1}{2}(\varkappa+r)} b_n J_{\varkappa+r}(2A\sqrt{\mu_n x}) + \mathrm{P}_{\varkappa}(x), \tag{2.12.44}$$

$$\mathrm{P}_{\varkappa}(x) = \frac{1}{2\pi i}\int_{\mathcal{C}}\frac{\Gamma(s)\varphi(s)x^{s+\varkappa}}{\Gamma(s+\varkappa+1)}\,ds,$$

where $\mathcal{C}$ *is as in* (2.12.4). *Cf. Exercise 3.6.3 for the* J*-Bessel function.*

The Ewald expansion

$$A^{-s}\Gamma(s)\varphi(s) = A^{-s}\sum_{n=1}^{\infty}\frac{a_n}{\lambda_n^s}\Gamma(s, Aw\lambda_n) + A^{-(r-s)}\sum_{n=1}^{\infty}\frac{b_n}{\mu_n^{r-s}}\Gamma(r-s, Aw^{-1}\mu_n) + \int_0^{w}\mathrm{P}(x)x^{s-1}dx \tag{2.12.45}$$

with $\mathrm{Re}\, w > 0$.

Remark 2.12.2 The contents of Theorem 2.12.3 are expounded in far more general form in respective chapters of Kanemitsu and Tsukada (2014). The Bochner modular relation (2.12.41) is a right-half plane version of Hecke correspondence and is stated as a genesis of unprocessed modular relation, Chap. 3. The Fourier Bessel expansion (2.12.42) in Chap. 4. The Ewald expansion is in Chap. 5 and the Riesz sum, in Chap. 6. We shall give concrete examples of Ewald expansion in Sects. 2.14 and 2.15 and those of Riesz sum in Chap. 3 as well as in Sect. 2.16.

In Chandrasekharan and Narasimhan (1961, Lemma 6), equivalent assertions are the first three, and the Fourier–Bessel expansion (2.12.42) is replaced by Hardy's formula

$$\left(-\frac{1}{s}\frac{\mathrm{d}}{\mathrm{d}s}\right)^{\varkappa}\sum_{n=1}^{\infty} a_n \frac{1}{s} e^{-s\sqrt{\lambda_n}}$$
$$= 2^{3r+\varkappa}\Gamma\left(r+\varkappa+\frac{1}{2}\right)\pi^{r-\frac{1}{2}}\sum_{n=1}^{\infty}\frac{b_n}{(s^2+16\pi^2\mu_n)^{r+\varkappa+\frac{1}{2}}} \tag{2.12.46}$$
$$+\frac{1}{2\pi i}\int_C \frac{\Gamma(z)\varphi(z)\Gamma(2z+2\varkappa+1)2^{-\varkappa}}{\Gamma(z+\varkappa+1)} s^{-2z-2\varkappa-1}\mathrm{d}z.$$

It is first derived by the Hardy transform in Hardy (1908) without differentiation. It is remarked (Berndt, 1970, Theorem 8.1, p.342; Wang et al., 2022) that **the Hardy transform and the beta transform are reverse processes leading to the Fourier–Bessel expansion** as stated in Theorem 3.1.1.

The **Riesz sum of order** $\varkappa$ is defined by

$$A_{\lambda}^{\varkappa}(x) = \frac{1}{\Gamma(\varkappa+1)}\sum_{\lambda_n\le x}{}' a_n(x-\lambda_n)^{\varkappa} = \frac{1}{2\pi i}\int_{(c)}\frac{\Gamma(s)\varphi(s)x^{s+\varkappa}}{\Gamma(s+\varkappa+1)}\,\mathrm{d}s, \tag{2.12.47}$$

where $\varphi(s)=\sum_{n=1}^{\infty}\frac{a_n}{\lambda_n^s}$ and the prime on the summation means that the term corresponding to $x=[x]$ in case $\varkappa=0$ is to be halved (Kanemitsu & Tsukada, 2014, p. 171, (G-8-1)). $c>\sigma_a^*$, the abscissa of absolute convergence of φ. For some more details, cf. Exercise 2.13.1.

Although the method of Landau-Walfisz was faded because of the development of Bochner–Chandrasekharan–Narasimhan description, we note that Walfisz (1961a) establishes the identities of Hardy (for the circle problem, or the Dedekind zeta-function for the Gaussian field) and Voronoĭ and those for the imaginary and real quadratic fields and Walfisz (1961b) contains the case of the Epstein zeta-function with a positive definite binary quadratic form. Cf. Kanemitsu and Tsukada (2014, pp. 33–35). The method remains the same in all the cases, i.e. work for the higher order Riesz sums and then reduce to the order 0 (or at least lower order) case by **differencing** due to Landau.

The following lemma reveals the situation surrounding equating the Riemann resp. Hecke type as the Hecke resp. Riemann-type functional equation.

Lemma 2.12.4 (Kanemitsu & Tsukada, 2014, p. 120, Corollary 4.1) *Under similar assumptions as in Definition 2.12.1, the functional equation*

$$\Gamma(A_1 s)\varphi(s) = \Gamma(A_1(r-s))\psi(r-s)$$

is equivalent to

$$\frac{1}{A_1}\Gamma(A_1 s)\sum_{k=1}^{\infty}\frac{\alpha_k}{\left(\lambda_k^{\frac{1}{A_1}}+z^{\frac{1}{A_1}}\right)^{A_1 s}} \tag{2.12.48}$$

$$= 2{A_1}^{-1}z^{\frac{1}{2}A_1(r-s)}\sum_{k=1}^{\infty}\frac{\beta_k}{{\mu_k}^{\frac{1}{2}A_1(r-s)}}K_{A_1(r-s)}\left(2(\mu_k z)^{\frac{1}{2A_1}}\right)+\mathrm{P}(s),$$

where $\mathrm{P}(s)=\sum_{k=1}^{L}\mathrm{Res}\Big(\Gamma(A_1(s-w))\Gamma(A_1 s)\varphi(s)\,z^{w-s},\, w=s_k\Big)$ *and where* s_k *are all the poles of* $\Gamma(A_1(s-w))\Gamma(A_1 s)\varphi(s)\,z^{w-s}$ *in* $\mathcal{S}$.

Equation (2.12.50) in the next exercise is a consequence of Lemma 2.12.4 with $A_1=\frac{1}{2}$ and Theorem 2.12.3. It entails Hamburger's partial fraction expansion (4.5.12). Cf. Theorem 4.5.2.

Exercise 2.12.7 The Riemann functional equation (FER) reads **when $\zeta(s)$ is viewed as the Epstein zeta-function of a unary quadratic form**

$$\pi^{-s}\Gamma(s)\zeta(2s)=\pi^{-(\frac{1}{2}-s)}\Gamma\left(\frac{1}{2}-s\right)\zeta(1-2s). \tag{2.12.49}$$

Prove that the following are equivalent to (2.12.49).
The theta-transformation formula (cf. (2.1.64))

$$\sum_{n=1}^{\infty}e^{-\pi n^2 x}=x^{-\frac{1}{2}}\sum_{n=1}^{\infty}e^{-\frac{\pi n^2}{x}}+\frac{1}{2}\left(x^{-\frac{1}{2}}-1\right),\quad \mathrm{Re}(x)>0.$$

Watson's formula (Watson, 1931)

$$\pi^{-s}\Gamma(s)\sum_{n=-\infty}^{\infty}\frac{1}{(n^2+\alpha^2)^s}=2\alpha^{\frac{1}{2}-s}\sum_{n=-\infty}^{\infty}|n|^{s-\frac{1}{2}}K_{s-\frac{1}{2}}(2\pi\alpha|n|), \tag{2.12.50}$$

where the term corresponding to $n=0$ on the right-hand side is to be understood to mean (3.6.14). Cf. the passage before Exercise 3.6.8.
Ewald expansion

$$\pi^{-s}\Gamma(s)\zeta(2s)=\pi^{-s}\sum_{n=1}^{\infty}\frac{1}{n^{2s}}\Gamma(s,\pi n^2 w) \tag{2.12.51}$$

$$+\pi^{s-\frac{1}{2}}\sum_{n=1}^{\infty}\frac{1}{n^{1-2s}}\Gamma\left(\frac{1}{2}-s,\frac{\pi n^2}{w}\right)+\frac{w^{s-\frac{1}{2}}}{2s-1}-\frac{w^s}{2s},\quad \mathrm{Re}(w)>0.$$

For Hamburger's modular relations for the original (FER), cf. Corollary 4.5.1.

We shall *elucidate the genesis of Siegel's proof of Hamburger's theorem* (Siegel, 1922). Knopp (2000, pp. 201–211) contains a rather transparent description of the

Siegel–Hecke proof of the Hamburger theorem (esp. the translation of introduction of Hecke, 1944). Cf. Hamburger (1921), Siegel (1922), Hecke (1936, 1944, 1983), etc. However, in neither Knopp (2000) nor Culp-Ressler et al. (2000) there is a source given of Siegel's proof.

We appeal to the following which is (an extract of) Kanemitsu and Tsukada (2014, pp. 120–121, Example 4.2).

Lemma 2.12.5 *Suppose the Dirichlet series φ and ψ satisfy the Riemann-type functional equation*

$$\Gamma\left(\frac{1}{2}s\right)\varphi(s) = \Gamma\left(\frac{1}{2}(r-s)\right)\psi(r-s)$$

with a simple pole at $s = r > 0$ with residue ρ. Then the Fourier–Bessel expansion $H^{1,1}_{1,1} \leftrightarrow H^{2,0}_{0,2}$ entails

$$\begin{aligned}
&2\Gamma\left(\frac{1}{2}s\right)\sum_{n=1}^{\infty}\frac{\alpha_n}{\left(\lambda_n^2+z^2\right)^{\frac{1}{2}s}}\\
&= 4z^{\frac{1}{2}(r-s)}\sum_{n=1}^{\infty}\frac{\beta_n}{\mu_n^{\frac{1}{2}(r-s)}}K_{\frac{1}{2}(r-s)}(2\mu_n z) + 2\Gamma\left(\frac{1}{2}s\right)\varphi(0)\,z^{-s}\\
&+\rho\Gamma\left(\frac{1}{2}(s-r)\right)z^{r-s}
\end{aligned}$$

which is a generalization of Watson's formula.

We specialize Lemma 2.12.5 to the case

$$\lambda_n = \mu_n = \sqrt{\pi}n, \quad r = 1, \quad z = \sqrt{\pi}t, \quad s = 1$$

and simplify the resulting expression to derive

$$\sum_{n=1}^{\infty}\alpha_n\left(\frac{1}{t+in}+\frac{1}{t-in}\right)\frac{1}{2} - \pi t H(t) = 2\pi\sum_{n=1}^{\infty}\beta_n e^{-2\pi nt}$$

which is Siegel (1922, (12)). After this, Siegel's argument applies. Indeed, both Siegel (1922, (5)) and Culp-Ressler et al. (2000, formula before (6)) are the reduction of the K-Bessel function to the exponential function, (3.1.6).

We state the Shimura–Taniyama–Weil conjecture established through tremendous efforts by many great minds as an example of the HFE. Cf. Breuil et al. (2001), Liu (2022), etc.

Theorem 2.12.4 (Wiles, Breuil–Conrad–Diamond–Taylor) *If $E/\mathbb{Q}$ is an elliptic curve of conductor N_E, then E is modular, i.e. there is an eigen-form f for $\Gamma_0(N_E)$ (of weight 2) such that*

$$L(s, E) = L(s, f).$$

Corollary 2.12.1 *Suppose $E/\mathbb{Q}$ is an elliptic curve of conductor N_E and we define the completed L-function (cf.* (2.15.2)*)*

$$\Lambda_E(s) = \left(\frac{2\pi}{\sqrt{N_E}}\right)^{-s} \Gamma(s)L(s, E), \tag{2.12.52}$$

then $L(s, E)$ can be entirely continued over the whole plane and satisfies the (Hecke type) functional equation

$$\Lambda_E(1-s) = \Lambda_E(s).$$

2.13 G- and H-Functions and Hierarchy

We make a frequent use of the G-(and H-)function according to the notation of Kanemitsu and Tsukada (2014), Prudnikov et al. (1986), Wolfram (1998–2021) which have slight difference from that of Erdélyi et al. (1953) in that z is to read z^{-1}.

The Fox H-function is defined for $(0 \le n \le p, 0 \le m \le q, A_j, B_j > 0)$ by

$$H^{m,n}_{p,q}\left(z \,\middle|\, \begin{matrix} (1-a_1, A_1), \dots, (1-a_n, A_n), (a_{n+1}, A_{n+1}), \dots, (a_p, A_p) \\ (b_1, B_1), \dots, (b_m, B_m), (1-b_{m+1}, B_{m+1}), \dots, (1-b_q, B_q) \end{matrix}\right)$$

$$= \frac{1}{2\pi i}\int_L \frac{\prod_{j=1}^{m} \Gamma\big(b_j + B_j s\big) \prod_{j=1}^{n} \Gamma\big(a_j - A_j s\big)}{\prod_{j=n+1}^{p} \Gamma\big(a_j + A_j s\big) \prod_{m+1}^{q} \Gamma\big(b_j - B_j s\big)} z^{-s} \mathrm{d}s,$$

under a certain convergence condition, where L is a (deformed) Bromwich contour $\sigma = c$. We use known results on the H-, G-functions freely in what follows, cf. Kanemitsu and Tsukada (2014). We only state the inversion formula which often appears in Chap. 3.

$$H^{m,n}_{p,q}\left(\frac{1}{z} \,\middle|\, \begin{matrix} \{(1-a_j, A_j)\}_{j=1}^{n}, \{(a_j, A_j)\}_{j=n+1}^{p} \\ \{(b_j, B_j)\}_{j=1}^{m}, \{(1-b_j, B_j)\}_{j=m+1}^{q} \end{matrix}\right)$$
$$= H^{n,m}_{q,p}\left(z \,\middle|\, \begin{matrix} \{(1-b_j, B_j)\}_{j=1}^{m}, \{(b_j, B_j)\}_{j=m+1}^{q} \\ \{(a_j, A_j)\}_{j=1}^{n}, \{(1-a_j, A_j)\}_{j=n+1}^{p} \end{matrix}\right). \tag{2.13.1}$$

The Meijer G-functions is a special case with $A_j = 1, B_j = 1$:

$$G^{m,n}_{p,q}\left(z\,\middle|\,\begin{matrix}a_1,\dots,a_n,a_{n+1},\dots,a_p\\ b_1,\dots,b_m,b_{m+1},\dots,b_q\end{matrix}\right)$$
$$=H^{m,n}_{p,q}\left(z\,\middle|\,\begin{matrix}(a_1,1),\dots,(a_n,1),(a_{n+1},1),\dots,(a_p,1)\\ (b_1,1),\dots,(b_m,1),(b_{m+1},1),\dots,(b_q,1)\end{matrix}\right)$$
$$=\frac{1}{2\pi i}\int_L\frac{\prod\limits_{j=1}^{m}\Gamma\big(b_j+s\big)\prod\limits_{j=1}^{n}\Gamma\big(1-a_j-s\big)}{\prod\limits_{j=n+1}^{p}\Gamma\big(a_j+s\big)\prod\limits_{m+1}^{q}\Gamma\big(1-b_j-s\big)}z^{-s}\mathrm{d}s.$$

The integrals are absolutely convergent if $m+a>\frac{1}{2}(p+q)$, which condition is satisfied in almost all the cases appearing below.

It helps to have a better grasp of the degree of specifications if one notices the G-function hierarchy involved in the expressions, including the Fourier–Bessel expansions, (2.5.24). We restate (Kanemitsu & Tsukada, 2014, p. 227, (8.7)) in the setting of general modular relation for Hecke type zeta-functions.

$$\sum_{k=1}^{\infty}\frac{\alpha_k}{\lambda_k^s}G^{m+1,n}_{p,q+1}\left(z\lambda_k\,\middle|\,\begin{matrix}1-a_1,\dots,1-a_n,a_{n+1},\dots,a_p\\ s,b_1,\dots,b_m,1-b_{m+1},\dots,1-b_q\end{matrix}\right)\tag{2.13.2}$$
$$=\sum_{k=1}^{\infty}\frac{\beta_k}{\mu_k^{r-s}}G^{n+1,m}_{q,p+1}\left(\frac{\mu_k}{z}\,\middle|\,\begin{matrix}1-b_1,\dots,1-b_m,b_{m+1},\dots,b_q\\ r-s,a_1,\dots,a_n,1-a_{n+1},\dots,1-a_p\end{matrix}\right)$$
$$+\sum_{k=1}^{L}\mathrm{Res}\Big(\Gamma(w-s\mid\Delta)\,\chi(w)\,z^{s-w},\,w=s_k\Big).$$

Exercise 2.13.1 Interpret the contents of the Hecke modular relation theorem, Theorem 2.12.3 as follows.

The Bochner modular relation (2.12.41): $G^{1,0}_{0,1}\leftrightarrow G^{1,0}_{0,1}$. This is the source of all subsequent studies on the modular relations.

The Fourier–Bessel expansion (2.12.42): $G^{1,1}_{1,1}\leftrightarrow G^{2,0}_{0,2}$.

The Ewald expansion (2.12.45): $G^{2,0}_{1,2}\leftrightarrow G^{1,1}_{1,2}$.

This is expounded in more general form in Kanemitsu and Tsukada (2014, Chap. 5, Theorems 5.2, 5.3). The latter implies the incomplete gamma expansion (Ewald expansion) from the confluent hypergeometric function expansion.

The Riesz sums or the first J-Bessel expansion (2.12.44): $G^{1,0}_{1,1}\leftrightarrow G^{1,0}_{0,2}$. The Bessel function of the first kind is defined in Sect. 3.6.3.

Solution. We only deal with the Riesz sum case.

$$\sum_{k=1}^{\infty} \frac{\alpha_k}{\lambda_k^s} G_{1,1}^{1,0}\left(z\lambda_k \,\middle|\, \begin{matrix} a \\ s \end{matrix}\right) = \sum_{k=1}^{\infty} \frac{\beta_k}{\mu_k^{r-s}} G_{0,2}^{1,0}\left(\frac{\mu_k}{z} \,\middle|\, \begin{matrix} - \\ r-s, 1-a \end{matrix}\right)$$
$$+ \sum_{k=1}^{L} \mathrm{Res}\left(\frac{\chi(w)\, z^{s-w}}{\Gamma(a-s+w)}, w = s_k\right) \quad (z > 0),$$

which by

$$G_{1,1}^{1,0}\left(z \,\middle|\, \begin{matrix} a \\ b \end{matrix}\right) = 0 \qquad |z| > 1.$$

and

$$G_{0,2}^{1,0}\left(z \,\middle|\, \begin{matrix} - \\ a, b \end{matrix}\right) = z^{\frac{1}{2}(a+b)} J_{a-b}(2\sqrt{z})$$

amounts to

$$\frac{z^s}{\Gamma(a-s)} \sum_{\lambda_k < \frac{1}{z}} \alpha_k (1 - z\lambda_k)^{a-s-1}$$
$$= z^{\frac{1}{2}(s-r-1+a)} \sum_{k=1}^{\infty} \frac{\beta_k}{\mu_k^{\frac{1}{2}(r-s-1+a)}} J_{r-s-1+a}\left(2\sqrt{\frac{\mu_k}{z}}\right)$$
$$+ \sum_{k=1}^{L} \mathrm{Res}\left(\frac{\chi(w)\, z^{s-w}}{\Gamma(a-s+w)}, w = s_k\right) \quad (z > 0).$$

□

2.14 Riemann–Siegel Formula for HLF à la Kuz'min

In this and the next section we present the Riemann–Siegel formula for HLF as examples of the Ewald expansion and in Sect. 2.16 we state examples of the Riesz sums.

Kuz'min's method is very closely related to the Ewald expansion for the L-function in question.

Recall the functional equation (2.12.36) for $L(s, f)$ (f being a cusp form of weight k) defined by (2.12.35) and the Ewald expansion, which now takes the form

$$(2\pi)^{-s}\Gamma(s)L(s, f) = (2\pi)^{-s} \sum_{n=1}^{\infty} \frac{a_n}{n^s} \Gamma(s, 2\pi n w) \tag{2.14.1}$$
$$+ (-1)^{\frac{k}{2}} (2\pi)^{-(k-s)} \sum_{n=1}^{\infty} \frac{a_n}{n^{k-s}} \Gamma(k-s, 2\pi n/w)$$

valid for $\mathrm{Re}\, w > 0$.

Kuz'min's argument lies in the observation that the series on the right of (2.12.51) corresponding to (2.14.1) are absolutely convergent not only for $\mathrm{Re}\, w > 0$ but also for $\arg w = \frac{\pi}{2}$ (i.e. to (RHP) (2.1.80)), so that we may take w to be a purely imaginary number. It should be noted that although Kuz'min (1934) treated the Dirichlet L-function, he restricted to the case where the gamma factor is $\Gamma(s/2)$, i.e. the case of Riemann's functional equation (cf. Lavrik, 1990).

Our subsequent discussion is rather relevant to Akiyama and Tanigawa (1999) (cf. also Ferminger, 1992) in the sense that both follow by the choice $\arg w = \frac{\pi}{4}$ in the absolutely convergent region $\mathrm{Re}\, w > 0$, who gave the expression as an Ewald expansion for the L-function $L(s, f_0)$ associated to a cusp form f_0 of weight k with respect to the congruence subgroup $\Gamma_0(N)$, (2.12.10). They truncated the series at large M and made an estimate of the error term, proceeding to conduct numerical calculation of various zeta- and L-functions for some range of $\mathrm{Im}\, s$. To make their calculation practical, Akiyama and Tanigawa made a delicate choice of the parameter w (r in their notation, they take $r = e^{i(\frac{\pi}{2}-\delta(t))}$, $0 < \delta(t) \le \frac{\pi}{2}$ and choose $\delta(t)$ in a delicate manner). The first paragraph of p. 204 Akiyama and Tanigawa (1999) is quite instructive when compared with the remark made by Lavrik (1968).

Since we may also prove the Riemann–Siegel integral formula for $L(s, f_0)$ in the same vein as we prove it for $L(s, f)$, there is a hope that we may get an asymptotic formula, which then may enable us to avoid the use of incomplete gamma function values which they remark are difficult to calculate and are time-consuming (Akiyama and Tanigawa, 1999, the first line of Sect. 4, p. 1208).

The **congruence-subgroup L-function (CSL)** $L(s, f_0)$ satisfies HFE

$$\left(\frac{2\pi}{\sqrt{N}}\right)^{-s} \Gamma(s)L(s, f_0) = \mu \left(\frac{2\pi}{\sqrt{N}}\right)^{-(k-s)} \Gamma(k-s)L(k-s, f_0) \qquad (2.14.2)$$

$\mu = (-1)^{\frac{k}{2}} = \pm 1$. (2.12.36) is the case with $N = 1$, $\mu = 1$, and we state the result for this $L(s, f_0)$.

Our aim is to prove the following Theorem 2.14.1 in two ways (another proof is given in Sect. 2.15). To state it we introduce some notation.

For a cusp form $f_0(z) = \sum_{n=1}^{\infty} a_n e^{2\pi i n z}$ of weight k with respect to $\Gamma_0(N)$, we put

$$\varphi(z, \xi) = \sum_{n=1}^{\infty} \frac{a_n}{z-n} e^{-\frac{2\pi n \xi}{\sqrt{N}}} \qquad (2.14.3)$$

for $\mathrm{Re}\, \xi > 0$.

Theorem 2.14.1 *Let $\varepsilon = e^{\frac{\pi i}{4}}$ and $0 < \mathrm{Re}\, s < k$. Then we have the Riemann–Siegel integral formula*

$$\left(\frac{2\pi}{\sqrt{N}}\right)^{-s}\Gamma(s)L(s,f_0)=\left(\frac{2\pi}{\sqrt{N}}\right)^{-s}\frac{\Gamma(s)}{2\pi i}\int_{0\swarrow 1}\varphi(z,\varepsilon)e^{\frac{2\pi i}{\sqrt{N}}\bar{\varepsilon}z}z^{-s}dz \qquad (2.14.4)$$
$$-\mu(-1)^{\frac{k}{2}}\left(\frac{2\pi}{\sqrt{N}}\right)^{-(k-s)}\frac{\Gamma(k-s)}{2\pi i}\int_{0\nwarrow 1}\varphi(z,\bar{\varepsilon})e^{-\frac{2\pi i}{\sqrt{N}}\varepsilon z}z^{-(k-s)}dz,$$

where the path $0\swarrow 1$ *and* $(0\nwarrow 1)$ *are well known ones due to Siegel, i.e. the directed lines making the angle* $\frac{5}{4}\pi$ *and* $\frac{3}{4}\pi$ *respectively, with the positive direction of the real axis and intersecting with it between the interval* $[0,1]$.

Proof On account of the form (2.14.2) of the functional equation, the expression (2.14.1) as the Ewald expansion takes the form

$$\left(\frac{2\pi}{\sqrt{N}}\right)^{-s}\Gamma(s)L(s,f_0)=\sum_{n=1}^{\infty}\frac{a_n}{\left(\frac{2\pi}{\sqrt{N}}n\right)^{s}}\Gamma\left(s,\frac{2\pi nw}{\sqrt{N}}\right) \qquad (2.14.5)$$
$$+\mu(-1)^{\frac{k}{2}}\sum_{n=1}^{\infty}\frac{a_n}{\left(\frac{2\pi}{\sqrt{N}}n\right)^{k-s}}\Gamma\left(k-s,\frac{2\pi nw^{-1}}{\sqrt{N}}\right).$$

Since w is at our disposal as long as it remains in the sector $|\arg w|<\frac{\pi}{2}$, we may take $w=\varepsilon$ to deduce that

$$\left(\frac{2\pi}{\sqrt{N}}\right)^{-s}\Gamma(s)L(s,f_0)=\left(\frac{2\pi}{\sqrt{N}}\right)^{-s}\sum_{n=1}^{\infty}\frac{a_n}{n^s}\Gamma\left(s,\frac{2\pi n\varepsilon}{\sqrt{N}}\right) \qquad (2.14.6)$$
$$+\mu(-1)^{\frac{k}{2}}\left(\frac{2\pi}{\sqrt{N}}\right)^{-(k-s)}\sum_{n=1}^{\infty}\frac{a_n}{n^{k-s}}\Gamma\left(k-s,\frac{2\pi n\bar{\varepsilon}}{\sqrt{N}}\right).$$

It remains to replace the incomplete gamma functions with the integral in (2.14.4). This is made in two steps.

First putting $w=\frac{2\pi n\varepsilon}{\sqrt{N}}$ and $\frac{2\pi n\bar{\varepsilon}}{\sqrt{N}}$, and changing the variables $u\leftrightarrow\frac{2\pi x}{\sqrt{N}}$ in (2.14.7), we obtain

$$\Gamma\left(s,\frac{2\pi n\varepsilon}{\sqrt{N}}\right)=(\varepsilon n)^s e^{-\frac{2\pi n\varepsilon}{\sqrt{N}}}\frac{\Gamma(s)\sin\pi s}{\pi}\int_0^{\infty}\frac{e^{-\frac{2\pi x}{\sqrt{N}}}}{x+\varepsilon n}x^{-s}dx$$

and

$$\Gamma\left(s,\frac{2\pi n\bar{\varepsilon}}{\sqrt{N}}\right)=(\bar{\varepsilon}n)^s e^{-\frac{2\pi n\bar{\varepsilon}}{\sqrt{N}}}\frac{\Gamma(s)\sin\pi s}{\pi}\int_0^{\infty}\frac{e^{-\frac{2\pi x}{\sqrt{N}}}}{x+\bar{\varepsilon}n}x^{-s}dx.$$

Now we are to replace the last integral with a contour integral along the line from right to left intersecting with the imaginary axis between 0 and $-i$. Kuz'min (1930) did not give details, and we give some details.

Choosing $0 < c < \frac{1}{\sqrt{2}}$ and putting $v = \bar{\varepsilon}n$ in (2.14.9) and $v = \varepsilon n$ in (2.14.10), we obtain

$$\sin \pi s \int_0^\infty \frac{e^{-\frac{2\pi}{\sqrt{N}}x}}{x+\varepsilon n} x^{-s}\,dx = \frac{\varepsilon^{-s}}{2i} \int_{0\swarrow 1} \frac{e^{\frac{2\pi i}{\sqrt{N}}\bar{\varepsilon}z}}{z-n} z^{-s}\,dz$$

and

$$\sin \pi s \int_0^\infty \frac{e^{-\frac{2\pi}{\sqrt{N}}x}}{x+\bar{\varepsilon}n} x^{-s}\,dx = -\frac{\varepsilon^{s}}{2i} \int_{0\nwarrow 1} \frac{e^{-\frac{2\pi i}{\sqrt{N}}\varepsilon z}}{z-n} z^{-s}\,dz.$$

Note that the integrals on the right-hand sides of the above two equations are absolutely convergent for $\mathrm{Re}\, s > 0$.

Substituting these results into (2.14.6), we complete the proof of Theorem 2.14.1. □

Exercise 2.14.1 Prove

$$\Gamma(s, w) = w^s e^{-w} \frac{\Gamma(s) \sin \pi s}{\pi} \int_0^\infty \frac{e^{-u} u^{-s}}{u+w} du, \tag{2.14.7}$$

which holds for $\mathrm{Re}\, w > 0$ and $0 < \mathrm{Re}\, s < 1$.

Solution. As in Akiyama and Tanigawa (1999, Formula (5)), we have

$$\Gamma(s, w) = w^s \int_1^\infty e^{-wt} t^{s-1} dt$$

for $\mathrm{Re}\, w > 0$. Hence, as in Kuz'min (1934), writing $t+1$ for t in the integral above, we have

$$\Gamma(s, w) = w^s e^{-w} \int_0^\infty e^{-wt} (1+t)^{s-1} dt.$$

Suppose $0 < \mathrm{Re}\, s < 1$ and replace the factor $(1+t)^{s-1}$ by

$$\frac{1}{\Gamma(1-s)} \int_0^\infty e^{-(1+t)u} u^{-s} \mathrm{d}u$$

(the HGT (2.1.3)) to obtain

$$\Gamma(s, w) = w^s e^{-w} \frac{1}{\Gamma(1-s)} \int_0^\infty e^{-wt} dt \int_0^\infty e^{-(1+t)u} u^{-s} du. \tag{2.14.8}$$

By absolute convergence, we may change the order of integration in (2.14.8), and therefore after integration, we deduce (2.14.7). □

Exercise 2.14.2 Let $0 < \mathrm{Re}\, s < 1$ and $b, c > 0$. By Cauchy's theorem, prove that

$$\int_{-\infty-ic}^{\infty-ic} \frac{e^{2\pi i b z}}{z-v} z^{-s} dz = -2ie^{\frac{\pi i s}{2}} \sin \pi s \int_0^\infty \frac{e^{-2\pi b x}}{x+iv} x^{-s} dx \tag{2.14.9}$$

for $\operatorname{Im} v < -c$, and

$$\int_{-\infty+ic}^{\infty+ic} \frac{e^{-2\pi ibz}}{z-v} z^{-s}dz = 2ie^{\frac{-\pi is}{2}} \sin \pi s \int_0^{\infty} \frac{e^{-2\pi bx}}{x-iv} x^{-s}dx \tag{2.14.10}$$

for $\operatorname{Im} v > c$.

2.15 Riemann–Siegel Formula for HLF: Koshlyakov's Method

Koshlyakov (1954) developed his theory of Dedekind zeta-function $\zeta_\Omega(s)$ of a rational or a quadratic field Ω, real or imaginary. For the Dedekind zeta-function, cf. Sect. 2.5, Sect. 4.1. In the sense that he was successful in deriving the Riemann–Siegel integral formula for $\zeta_\Omega(s)$ for a real quadratic field Ω, Koshlyakov's achievement remains the record till now along with the case of $\zeta^2(s)$ by Motohashi (1987) (cf. Guthman, 1999).

The case of $\zeta_\Omega(s)$ with Ω an imaginary quadratic field is almost verbatim to the case of a congruence-subgroup L-function $L(s, f_0)$ (cf. Sect. 2.12), and we treat both cases together at a stretch to give another proof of Theorem 2.14.1.

For a cusp form f_0 of weight k w.r.t. $\Gamma_0(N)$ we put

$$A = \frac{2\pi}{\sqrt{N}}.$$

Then the constant A for an imaginary quadratic field formally arises by replacing N by $|\Delta|$: $A = \frac{2\pi}{\sqrt{|\Delta|}}$.

Then the Dedekind zeta-function $\zeta_\Omega(s)$ of the imaginary quadratic field Ω and the congruence-subgroup L-function $L(s, f_0)$ satisfy the functional equation of the form

$$\xi(s, f_0) = C\xi(r-s, f_0), \tag{2.15.1}$$

where $\xi(s, f_0)$ is the **completed zeta-function** (cf. (2.12.52), (2.1.9))

$$\xi(s, f_0) = \begin{cases} A^{-s}\Gamma(s)\zeta_\Omega(s) \\ A^{-s}\Gamma(s)L(s, f_0) \end{cases} \tag{2.15.2}$$

according as the zeta-function is the Dedekind zeta-function or the congruence-subgroup L-function. Hence by (2.14.2) $C = \mu = (-1)^{\frac{k}{2}}, r = k$ for $L(s, f_0)$ and $C = 1, r = 1$ for the Dedekind zeta-function.

A remarkable feature of Koshlyakov's method is his ingenious use of the (solutions of) differential equations, foreseeing the twenty-first century interaction of number theory and differential equations. Only through his method, it is clearly understood why we should have cylinder functions.

Koshlyakov uses five functions K, L, X, Y and Z which are closely related to solutions of the Bessel differential equation, but we only need X, Y and Z in this section. The K-function $K_{r_1,r_2}(x)$ plays an essential role in Sect. 4.3. The X-function is the inverse Mellin transform of the gamma factor appearing in the functional equation and the general will be defined by (4.5.5), whose special case reads

$$X(x) = X_{0,1}(x) = \frac{1}{2\pi i}\int_{(c)} \Gamma(z)x^{-z}\,\mathrm{d}z. \tag{2.15.3}$$

The Y-**function** is defined by

$$Y(x) = Y_{r_1,r_2}(x) = \frac{1}{2\pi i}\int_{(c)} \frac{\pi}{2\sin\frac{\pi z}{2}\Gamma^{r_1}\left(\frac{2-z}{2}\right)\Gamma^{r_2}(2-z)}x^{-z}\,\mathrm{d}z \tag{2.15.4}$$

with $\operatorname{Re} x > 0, 0 < c < 2$.

Exercise 2.15.1 Check that in our case, $(r_1, r_2) = (0, 1)$ we have

$$X(x) = e^{-x} \tag{2.15.5}$$

and

$$Y(x) = \frac{\sin x}{x}. \tag{2.15.6}$$

Equation (2.15.5) is the fundamental Mellin transform pair. Equation (2.15.6) follows by formulas for the gamma function. Cf. Koshlyakov (1954, Table on p. 122).

Exercise 2.15.2 Prove the formula (Koshlyakov, 1954, I, (9.10), (9.11))

$$\int_1^\infty X(ax)Y(bx)x\,\mathrm{d}x = \frac{Z(a,b)}{a^2+b^2},\quad a>0, b>0, \tag{2.15.7}$$

where

$$\begin{aligned} Z(a,b) =& \frac{X(a)}{2^{2-2r_2}}\frac{\mathrm{d}}{\mathrm{d}x}\left\{x^{r_2}Y(bx)\right\}^{(r_1+2r_2-1)}\Big|_{x=1} \\ &+ \frac{Y(b)}{2^{2-2r_2}}\frac{\mathrm{d}}{\mathrm{d}x}\left\{(-1)^{(r_1+2r_2-1)}X(ax)\right\}^{(r_1+2r_2-1)}\Big|_{x=1}. \end{aligned} \tag{2.15.8}$$

In our case, (2.15.7) and (2.15.8) together read

$$\begin{aligned}&\int_1^\infty X(ax)Y(bx)x\,\mathrm{d}x \\ &= \frac{1}{a^2+b^2}\left\{X(a)\frac{\mathrm{d}}{\mathrm{d}x}(xY(bx))_{x=1} - Y(b)\frac{\mathrm{d}}{\mathrm{d}x}(X(ax))_{x=1}\right\} \\ &= \frac{e^{-a}}{a^2+b^2}\left(\cos b + a\frac{\sin b}{b}\right) = \frac{e^{-a}}{a^2+b^2}Z_1(a,b),\end{aligned} \tag{2.15.9}$$

say.

We are in a position to give another proof of Akiyama and Tanigawa (1999, Theorem 4) by Koshlyakov's method, featuring hereafter the completed congruence-subgroup L-function (2.15.2).

Another proof of Theorem 2.14.1.

Let $c > k$ in (2.15.4). With the above-mentioned notation (recall $\varepsilon = e^{\frac{\pi i}{4}}$, $A = \frac{2\pi}{\sqrt{N}}$ etc.), we let

$$\Phi_{f_0}(\varepsilon, s) := \varepsilon^s \sum_{n=1}^{\infty} a_n I_n(s) \tag{2.15.10}$$

for $0 < \sigma < k$, where

$$I_n = I_n(s) = \int_1^\infty X(\varepsilon A n x)\, x^{s-1}\,\mathrm{d}x \tag{2.15.11}$$

Then first we have (2.15.18) by Exercise 2.15.3.

Secondly we have an expression for $I_n(s)$ in terms of the function $Z_1(a,b)$ introduced in (2.15.9):

$$I_n = \frac{2}{\pi}A^{-s}\sin\frac{\pi s}{2}\Gamma(s)e^{-\varepsilon A n}J_n, \tag{2.15.12}$$

where

$$J_n = J_n(s) = \int_0^\infty \frac{G_n(u)}{u^2+in^2}u^{1-s}\,\mathrm{d}u.$$

For a proof, cf. Exercise 2.15.4.

It remains to express J_n as a contour integral, which can be done by the same procedure as we adopted in Sect. 2.14. Then, we have, for $0 < \xi < 1$,

$$\begin{aligned}&\int_{\infty-i\xi}^{-\infty-i\xi} \frac{G_n(u)}{u^2+in^2}u^{1-s}\,\mathrm{d}u \\ &= -\int_0^\infty \frac{G_n(u)}{u^2+in^2}u^{1-s}\,\mathrm{d}u + e^{\pi i s}\int_0^\infty \frac{G_n(u)}{u^2+in^2}u^{1-s}\,\mathrm{d}u \\ &= e^{\frac{\pi i s}{2}} 2i\sin\frac{\pi s}{2}\int_0^\infty \frac{G_n(u)}{u^2+in^2}u^{1-s}\,\mathrm{d}u,\end{aligned} \tag{2.15.13}$$

which leads, by the change of variables $u = \bar{\varepsilon}x$ on the left, to

$$\int_{0\swarrow 1} \frac{G_n(\bar{\varepsilon}x)}{-i(x^2-n^2)}(\bar{\varepsilon}x)^{1-s}\bar{\varepsilon}\,\mathrm{d}x = \varepsilon^{2s}2i\sin\frac{\pi s}{2}\int_0^\infty \frac{G_n(u)}{u^2+in^2}u^{1-s}\,\mathrm{d}u. \quad (2.15.14)$$

Using (2.15.14) in (2.15.12), we conclude that

$$\begin{aligned}\varepsilon^s I_n &= \frac{1}{\pi i}A^{-s}\Gamma(s)e^{-n\varepsilon A}\int_{0\swarrow 1}\frac{G_n(\bar{\varepsilon}x)}{x^2-n^2}x^{1-s}\,\mathrm{d}x \quad (2.15.15)\\ &= \frac{1}{2\pi i}A^{-s}\Gamma(s)e^{-\varepsilon An}\int_{0\swarrow 1}x^{-s}\left(\frac{1}{x-n}e^{i\bar{\varepsilon}Ax}+\frac{1}{x+n}e^{-i\bar{\varepsilon}Ax}\right)\mathrm{d}x\\ &= \frac{1}{2\pi i}A^{-s}\Gamma(s)e^{-\varepsilon An}\int_{0\swarrow 1}x^{-s}\frac{1}{x-n}e^{i\bar{\varepsilon}Ax}\,\mathrm{d}x\end{aligned}$$

whose left-hand side member is a holomorphic function of s in $\sigma > 0$ and whose right-hand side integral is convergent for $\sigma > 0$, and so (2.15.15) is valid in $\sigma > 0$.

Substituting (2.15.15) in (2.15.10), we finally obtain

$$\Phi_{f_0}(\varepsilon, s) = \frac{1}{2\pi i}A^{-s}\Gamma(s)\int_{0\swarrow 1}x^{-s}\left(\sum_{n=1}^{\infty}\frac{a_n e^{-\varepsilon An}}{x-n}\right)e^{i\bar{\varepsilon}Ax}\,\mathrm{d}x, \quad (2.15.16)$$

which coincides with the first integral on the right of (2.14.4).

We may adopt a similar reasoning to the counterpart

$$\Phi_{f_0}(\bar{\varepsilon}, k-s) = \bar{\varepsilon}^{k-s}\sum_{n=1}^{\infty}a_n I_n',$$

with

$$I_n' = \int_1^\infty X\left(\bar{\varepsilon}Ax\right)x^{k-1-s}\,\mathrm{d}s.$$

Indeed I_n' can be expressed as

$$\frac{2}{\pi}A^{-(k-s)}\sin\frac{\pi(k-s)}{2}\Gamma(k-s)e^{-\bar{\varepsilon}An}J_n',$$

where

$$J_n' = \int_0^\infty \frac{\bar{G}_n(u)}{u^2-in^2}u^{1-k+s}\,\mathrm{d}u$$

and where

$$\bar{G}_n(u) = \overline{G_n(u)} = \frac{n\bar{\varepsilon}}{u}\sin Au + \cos Au.$$

J_n' may be transformed into a contour integral by a similar way as above.

In this way we obtain the counterpart of (2.15.15) valid in $0 < \sigma < k$:

$$\bar{\varepsilon}^{k-s} \int_1^\infty X\left(\bar{\varepsilon} A x n\right) x^{k-1-s}\, \mathrm{d}x \tag{2.15.17}$$
$$= -\frac{1}{2\pi i} A^{-(k-s)} \Gamma(k-s) X\left(\bar{\varepsilon} A n\right) \int_{0 \nwarrow 1} x^{-(k-s)} \frac{1}{x-n} e^{-i\varepsilon A x}\, \mathrm{d}x$$

whence we finally deduce, corresponding to (2.15.16),

$$\Phi_{f_0}(\bar{\varepsilon}, k-s) = -\frac{1}{2\pi i} A^{-(k-s)} \Gamma(k-s) \int_{0 \nwarrow 1} x^{-(k-s)} \left(\sum_{n=1}^{\infty} \frac{a_n e^{-\bar{\varepsilon} A n}}{x-n} e^{-i\varepsilon A x} \right) \mathrm{d}x,$$

which is seen to coincide with the second integral on the right of (2.14.4), and the proof is complete. □

Exercise 2.15.3 Prove that

$$\xi(s, f_0) = \Phi_{f_0}(\varepsilon, s) + C\Phi_{f_0}(\bar{\varepsilon}, k-s) \tag{2.15.18}$$

for $0 < \sigma < k$.

Solution. Substituting (2.15.3) in (2.15.10) and changing the order of integration, which we may, we see that the inner integral is

$$\int_1^\infty x^{s-z-1}\, \mathrm{d}x,$$

which is $1/(z-s)$.

Hence

$$\Phi_{f_0}(\varepsilon, s) = \frac{1}{2\pi i} \int_{(c)} \xi(z, f_0) \frac{\varepsilon^{s-z}}{z-s}\, \mathrm{d}z. \tag{2.15.19}$$

We shift the line of integration to the left to obtain

$$\Phi_{f_0}(\varepsilon, s) = \xi(s, f_0) + \frac{1}{2\pi i} \int_{(k-c)} \xi(z, f_0) \frac{\varepsilon^{s-z}}{z-s}\, \mathrm{d}z. \tag{2.15.20}$$

Changing the variables $z = k - w$ and using the functional equation (2.15.1), we derive that

$$\frac{1}{2\pi i} \int_{(k-c)} \xi(z, f_0) \frac{\varepsilon^{s-z}}{z-s}\, \mathrm{d}z = -C\bar{\varepsilon}^{k-s} \sum_{n=1}^{\infty} a_n \int_1^\infty X\left(\bar{\varepsilon} A n x\right) x^{k-1-s}\, \mathrm{d}x \tag{2.15.21}$$
$$= -C\Phi_{f_0}(\bar{\varepsilon}, k-s)$$

by reversing the argument that led us from (2.15.10) to (2.15.19), whence (2.15.18) follows. □

Exercise 2.15.4 Prove (2.15.12).

Solution. First from (2.15.4) we have for $0 < \sigma < 2$

$$x^{s-1} = \frac{2}{\pi} \sin \frac{\pi s}{2} \Gamma(s) \int_0^\infty Y(xt) x t^{1-s} \, dt. \tag{2.15.22}$$

We substitute (2.15.22) for x^{s-1} in the definition of I_n in (2.15.11) and change the order of integration to infer that

$$\begin{aligned} I_n &= \frac{2}{\pi} \sin \frac{\pi s}{2} \Gamma(s) \int_0^\infty t^{1-s} \int_1^\infty X(\varepsilon A n x) \, Y(xt) x \, dx \, dt \qquad (2.15.23) \\ &= \frac{2}{\pi} \sin \frac{\pi s}{2} \Gamma(s) \int_0^\infty (Au)^{1-s} \int_1^\infty X(\varepsilon A n x) \, Y(Aux) \, x \, dx \, A \, du \end{aligned}$$

by the change of variables $t = Au$.

The inner integral on the right-hand side of the last equality in (2.15.23) is

$$\frac{e^{-\varepsilon A n}}{(\varepsilon A n)^2 + (Au)^2} G_n(u), \tag{2.15.24}$$

in view of (2.15.9), where we put for simplicity

$$G_n(u) = Z_1(\varepsilon A n, Au) = \frac{n\varepsilon}{u} \sin Au + \cos Au. \tag{2.15.25}$$

Hence (2.15.12) follows. □

2.16 Some Applications of the Riesz Sum

In this section we give some interesting applications of the Riesz sum which are not widely known. The first example is the weighted character sum and the second is the Riesz sum of the von Mangoldt function, both being of order 1. Some more examples will be given in Sect. 4.5.

Weighted character sums have been considered extensively and Yamamoto (1977) is the decisive work which has closed further research, cf. Wang et al. (2020) for its elucidation and historical account. Here we concentrate on the weighted sum (5.1.5) for χ odd and primitive and prove the following

Proposition 2.16.1 *Suppose χ is odd and primitive* mod M. *Then with S_1 denoting the sum in* (2.16.3)

$$S_1 = -\frac{\tau(\chi)}{\pi i}L(1, \bar{\chi}), \tag{2.16.1}$$

which is in conformity with Washington (1983, Theorem 4.9, p. 37) *(being the first generalized Bernoulli number* (2.3.15)*) and*

$$R^1(x) = x\sum_{a\le x}\chi(a) - \sum_{a\le x} a\chi(a) = L(0, \chi)x + \frac{i\tau(\chi)}{2\pi^2\sqrt{M}}\sum_{n=1}^{\infty}\frac{\bar{\chi}(n)}{n^2}\sin(2\pi xn). \tag{2.16.2}$$

Proof Yamamoto (1977, Example 5.2) with $\alpha = \frac{t}{u} = \frac{1}{1}$ reads

$$S_1 = \frac{1}{M}\sum_{a\le\alpha M}{}' a\chi(a) = \frac{\tau(\chi)}{2\pi i}\sum_{n=1}^{\infty}\frac{b_1(n)\chi(n)}{n} + \frac{\tau(\chi)}{(2\pi i)^2}\sum_{n=1}^{\infty}\frac{b_2(n)\chi(n)}{n}, \tag{2.16.3}$$

where

$$b_1(n) = \chi(-1)\eta^{nt} - \eta^{-nt} = -2$$
$$b_2(n) = \chi(-1)(1 - \eta^{nt}) + 1 - \eta^{-nt} = 0$$

since $\eta = e^{\frac{2\pi i}{t}} = 1$. Hence (2.16.1) follows.

Another natural way is to consider the Riesz sum of order 1 (Kanemitsu and Tsukada, 2014, Chap. 6). We have $(x > 0)$

$$R^1(x) = \sum_{a\le x}(x-a)\chi(a) = \frac{1}{2\pi i}\int_{(c)}\frac{\Gamma(s)}{\Gamma(s+2)}L(s, \chi)x^{s+1}\,ds,$$

$c > 1$. As a common procedure, we shift the integration path to $\sigma = -d, 0 < d < 1$, encountering the simple pole at $s = 0$, and apply the functional equation (2.3.11). Thus we deduce that

$$\begin{aligned} &R^1(x) - L(0, \chi)x \qquad\qquad (2.16.4)\\ &= \frac{\tau(\chi)x}{\pi i}\frac{1}{2\pi i}\int_{(-d)}\frac{\Gamma(s)}{\Gamma(s+2)}\Gamma(1-s)\cos\left(\frac{\pi s}{2}\right)L(1-s, \bar{\chi})\left(\frac{2\pi x}{M}\right)^s ds\\ &= \frac{\tau(\chi)x}{\pi i}\sum_{n=1}^{\infty}\frac{\bar{\chi}(n)}{n}\frac{1}{2\pi i}\int_{(-d)}\frac{\Gamma(s)}{\Gamma(s+2)}\Gamma(1-s)\cos\left(\frac{\pi s}{2}\right)\left(\frac{2\pi nx}{M}\right)^s ds\\ &= \frac{\tau(\chi)x}{\pi i}\sum_{n=1}^{\infty}\frac{\bar{\chi}(n)}{n}I(z), \end{aligned}$$

on substituting the Dirichlet series for $L(1-s, \bar{\chi})$, where

$$I(z) = \frac{1}{2\pi i}\int_{(-d)} \frac{1}{s(s+1)}\Gamma(1-s)\cos\left(\frac{\pi s}{2}\right) z^s \, ds.$$

Using $\Gamma(1-s) = (-s)(-s-1)\Gamma(-s-1)$, we see that

$$I(z) = \frac{1}{2\pi i}\int_{(-d)} \Gamma(-s-1)\cos\left(\frac{\pi s}{2}\right) z^s \, ds. \tag{2.16.5}$$

Putting $w = -s-1$, (2.16.5) becomes

$$\begin{aligned} I(z) &= -z^{-1}\frac{1}{2\pi i}\int_{(d-1)} \Gamma(w)\sin\left(\frac{\pi w}{2}\right) z^{-w} \, dw \\ &= -z^{-1}\frac{1}{2\pi i}\int_{(d)} \Gamma(w)\sin\left(\frac{\pi w}{2}\right) z^{-w} \, dw \end{aligned}$$

after shifting the path to $\sigma = d$. Note that this is the sinus cardinals function which plays an essential role in signal restoration (Chakraborty et al., 2016, pp. 163–168). Substituting $\sin\left(\frac{\pi w}{2}\right) = \frac{e^{\frac{\pi}{2}iw} - e^{-\frac{\pi}{2}iw}}{2i}$, we deduce that

$$\begin{aligned} I(z) &= -\frac{z^{-1}}{2i}\left(\frac{1}{2\pi i}\int_{(d)} \Gamma(w)(-iz)^{-w} \, dw - \frac{1}{2\pi i}\int_{(d)} \Gamma(w)(iz)^{-w} \, dw\right) \\ &= z^{-1}\frac{e^{iz} - e^{-iz}}{2i} = z^{-1}\sin z \end{aligned} \tag{2.16.6}$$

by the Mellin transform. Substituting (2.16.6) in (2.16.4), we obtain

$$\begin{aligned} R^1(x) &= L(0,\chi)x + \frac{\tau(\chi)x}{\pi i}\sum_{n=1}^{\infty} \frac{\bar{\chi}(n)}{n}\frac{M}{2\pi n x}\sin\left(\frac{2\pi n x}{M}\right) \\ &= L(0,\chi)x + \frac{\tau(\chi)M}{2i\pi^2}\sum_{n=1}^{\infty} \frac{\bar{\chi}(n)}{n^2}\sin\left(\frac{2\pi n x}{M}\right) \end{aligned}$$

whence we conclude (2.16.2). □

Choosing $x = M$, we infer that

$$S_1(x) = L(0,\chi)$$

which is (2.16.1) in view of (2.3.13).

Now we turn to the Riesz sum of the **von Mangoldt function** $\Lambda(n)$, which is defined by

$$-\frac{\zeta'(s)}{\zeta(s)} = \sum_{n=1}^{\infty} \frac{\Lambda(n)}{n^s}, \quad \sigma > 1. \tag{2.16.7}$$

The results below are developed in Yang et al. (2023). For basic knowledge on zeta-functions, we refer to Davenport (1980), Ingham (1932), Ivić (1985), Prachar (1957), Titchmarsh (1951) etc. In particular, Ingham (1932) may be considered as one of the earliest occurrences of the Riesz sum method devoted to the case where the explicit formulas play an essential role for zeta-functions which may not satisfy the functional equation, as in (2.16.7). Explicit formula refers to an expression in terms of non-trivial zeros of the associated zeta-function. We restrict to non-trivial zeros of the Riemann zeta-function, which are infinitely many and all lie in the critical strip $0 < \sigma < 1$.

Although we work with $L(s) = -\frac{\zeta'(s)}{\zeta(s)}$, the argument applies to $L(s)$ of the form

$$L(s) = \sum_{n=1}^{\infty} \frac{f(n)}{n^s} = \frac{Z(s)}{\zeta(s)}, \quad Z(s) = \sum_{n=1}^{\infty} \frac{g(n)}{n^s}$$

absolutely convergent for $\sigma > 1$ and $Z(s)$ is a meromorphic function which has no zeros at the zeros of $\zeta(s)$ (this corresponds to simplicity of zeros of the Riemann zeta-function, which we assume) and such that

$$L(s) = O(\log |s|) \tag{2.16.8}$$

for $|s|$ large (save for neighborhoods of trivial zeros). Equation (2.16.8) is satisfied by $L(s) = -\frac{\zeta'(s)}{\zeta(s)}$, (Prachar, 1957, p. 227, Satz 4.3; Davenport, 1980, p. 108) etc.

We call the (formal) expansion

$$\sum_{n=1}^{\infty} a_n \bar{B}_{\varkappa}(nx) = -\frac{\varkappa!}{(2\pi i)^{\varkappa}} \sum_{\ell=1}^{\infty} (e^{2\pi i \ell x} + (-1)^{\varkappa} e^{-2\pi i \ell x}) c_\ell, \quad c_\ell = \ell^{-\varkappa} \sum_{d|\ell} d^k a_d$$

the **Davenport expansion** and we refer to the left-hand side series as the **Davenport series**, where $\{a_n\}$ are arithmetical functions which appear in PNT. Since $\bar{B}'_{\varkappa}(x) = \varkappa \bar{B}_{\varkappa-1}(x)$, we may work with the integrated Davenport series and shift to the lower order by termwise differentiation as long as the differentiated series is uniformly convergent. Cf. Davenport (1937a, b), Walum (1991), etc.

Lemma 2.16.1 (explicit formula) *For $x \geq 1$ we have the explicit formula*

$$\sum_{n \leq x} \frac{\Lambda(n)}{n}(x - n)$$
$$= (-1 + \log x - \gamma)x + \log 2\pi - \sum_{\rho} \frac{x^\rho}{\rho(\rho - 1)} - \sum_{k=1}^{\infty} \frac{1}{2k(2k+1)} x^{-2k}.$$

Proof To treat (2.16.13), we shall find an explicit formula for

$$\sum_{n\le x}\frac{\Lambda(n)}{n}(x-n)=\frac{1}{2\pi i}\int_{(c)}\frac{1}{s(s+1)}\left(-\frac{\zeta'(s+1)}{\zeta(s+1)}\right)x^{s+1}\,\mathrm{d}s \tag{2.16.9}$$

in analogy with Ingham (1932, p. 21, (14)). We apply the Cauchy residue theorem to the rectangle with vertices at $(c\pm iT)$, $(-N\pm iT)$. The horizontal integrals tend to 0 as $T\to\infty$ and the vertical one also tends to 0 as $N\to\infty$ by (2.16.8). Hence the integral in (2.16.9) is the sum of residues at poles in the strip $-N<\sigma<c$. The residue at the double pole at $s=0$ is

$$(-1+\log x-\gamma)x$$

and the residue at $s=-1$ is $\log 2\pi$, where we use $\zeta'(0)=-\frac{1}{2}\log 2\pi$ and

$$-\frac{\zeta'(s)}{\zeta(s)}=\frac{1}{s-1}-\gamma+O(s-1),$$

cf. e.g. Davenport (1980, p.81).

If ρ is a zero of $\zeta(s)$, the denominator $\zeta(s+1)$ has a simple pole at $s=\rho-1$ and is of the form $\zeta'(\rho)(s+1-\rho)+\cdots$ and $\zeta'(\rho)$ cancels the numerator. If it is a non-trivial zero ρ, then the residue is $-\frac{x^\rho}{\rho(\rho-1)}$. If it is a trivial zero $s=-2k$, $k=1,2,\cdots$, the residue is $-\frac{1}{2k(2k+1)}x^{-2k}$. □

Lemma 2.16.2 *As a special case of Mikolás formula* (Mikolás, 1956), *we have Li et al.* (2010b, Proposition 1)

$$\bar{B}_\varkappa(x)=-(-1)^\varkappa\varkappa!\frac{1}{2\pi i}\int_{(c)}\frac{\Gamma(s)}{\Gamma(s+\varkappa)}\zeta(1-s-\varkappa)x^{-s}\,\mathrm{d}s, \tag{2.16.10}$$

where $0<c<1$, *which entails*

$$\frac{1}{2}\left(\bar{B}_2(x)-\frac{1}{6}\right)=-\frac{1}{2\pi i}\int_{(b)}\frac{1}{s(s+1)}\zeta(-s-1)x^{-s}\,\mathrm{d}s,\quad -\frac{1}{2}<b<0,$$

where $\bar{B}_2(x)-\frac{1}{6}=\{x\}^2-\{x\}$.

As the formula (Chakraborty et al., 2009c, (4.2))

$$\frac{1}{\varkappa}\left(B_\varkappa(x)-\bar{B}_\varkappa(x)\right)=\sum_{n\le x}{}'(x-n)^{\varkappa-1}, \tag{2.16.11}$$

where the prime on the summation sign means that the term corresponding to $n=x$ is to be halved in the case $\varkappa=0$, suggests, we divide the sum on the left of (2.16.15) into two: $n\le x$ and $n>x$, i.e.

$$\frac{1}{2}\sum_{n=1}^{\infty}\frac{\Lambda(n)}{n^2}\left(\bar{B}_2\left(\frac{n}{x}\right)-\frac{1}{6}\right)=\frac{1}{2}\sum_{n\le x}+\frac{1}{2}\sum_{n>x}. \tag{2.16.12}$$

The first sum on the right is

$$\frac{1}{2}\sum_{n\le x}\frac{\Lambda(n)}{n^2}\left(\bar{B}_2\left(\frac{n}{x}\right)-\frac{1}{6}\right)=-\frac{1}{2x^2}\sum_{n\le x}\frac{\Lambda(n)}{n}(x-n), \tag{2.16.13}$$

the Riesz sum in Lemma 2.16.1.

Thus the following principle comes out.

The whole sum over $n=1,2,\cdots$ *in* (2.16.14) *with periodic Bernoulli polynomial weight consists of* $n\le x$ *part (Riesz sum of order* 1*) and* $n>x$ *part (Popov's formula). The whole sum has two expressions—the (integrated) Davenport expansion which is a consequence of the functional equation (Theorem* 2.16.4*) and the explicit formula involving trivial zeros (Theorem* 2.16.1*). Comparison of them yields the Kummer Fourier series in the case of the von Mangoldt function.*

We state the explicit formula for the Davenport sum of the von Mangoldt function, cf. Theorem 2.16.1. We are to use

$$\begin{aligned}&\frac{1}{2}\sum_{n=1}^{\infty}\frac{f(n)}{n^{r+1}}\left(\bar{B}_2\left(\frac{n}{x}\right)-\frac{1}{6}\right) \qquad (2.16.14)\\ &=-\frac{1}{2\pi i}\int_{(b)}\frac{1}{s(s+1)}\zeta(-s-1)L(s+r+1)x^s\,\mathrm{d}s\\ &=-\frac{1}{2\pi i}\int_{(c)}\frac{1}{(1-s)(2-s)}\zeta(1-s)L(s+r-1)x^{s-2}\,\mathrm{d}s,\end{aligned}$$

where $\frac{3}{2}<c:=b+2<2$. As long as the series $J:=\frac{1}{2}\sum_{n=1}^{\infty}\frac{f(n)}{n^{r+1}}$ is convergent, we may think of (2.16.14) as the Davenport expansion (with $\bar{B}_2$ weight) with the series $\frac{1}{2}B_2J$ subtracted. Cf. e.g. (2.16.20) in Theorem 2.16.1.

Instead of going on general lines, we specify to the case of the von Mangoldt function and state the explicit formula for the Davenport sum, cf. Theorem 2.16.1.

Lemma 2.16.3 *For* $x>1$ *we have*

$$\begin{aligned}&\frac{1}{2}\sum_{n=1}^{\infty}\frac{\Lambda(n)}{n^2}\left(\bar{B}_2\left(\frac{n}{x}\right)-\frac{1}{6}\right) \qquad (2.16.15)\\ &=-\frac{1}{2x}\left(\left(\frac{1}{x}-1\right)\log 2\pi+1+\log x-\gamma\right)+\sum_{k=1}^{\infty}\frac{\zeta(2k+1)}{(2k+1)(2k+2)}x^{-2k-2}.\end{aligned}$$

Exercise 2.16.1 Prove Lemma 2.16.3.

Solution. We specify $r=1$ and $L(s)=-\frac{\zeta'(s)}{\zeta(s)}$ and apply the Cauchy residue theorem to the rectangle as in the proof of Lemma 2.16.1. $s=1$ is a double pole and the residue is

$$\frac{1}{2x}(-\log 2\pi + 1 + \log x + \gamma).$$

The residue at $s = 0$ is $\frac{1}{2}(\log 2\pi)x^{-2}$. At a zero ρ of $\zeta(s)$, the denominator $\zeta(s)$ has a simple pole and is of the form $\zeta'(\rho)(s-\rho)+\cdots$. If it is a non-trivial zero, then it is canceled by the zero of $\zeta(1-s)$. If it is a trivial zero $s = -2k$, $k = 1, 2, \cdots$, the residue is $\frac{\zeta(2k+1)}{(2k+1)(2k+2)}x^{-2k-2}$. Hence the sum of residues is $\sum_{k=1}^{\infty}\frac{\zeta(2k+1)}{(2k+1)(2k+2)}x^{-2k-2}$. Hence the assertion follows. □

Lemma 2.16.4 *The Davenport expansion for the von Mangoldt function, Möbius function, and Liouville function are consequences of the functional equation (FER):*

$$\frac{1}{2}\sum_{n=1}^{\infty}\frac{f(n)}{n^{r+1}}\bar{B}_2(nx) = \frac{1}{2\pi^3}\sum_{n=1}^{\infty}\frac{\tilde{f}(n)}{n^2}\cos 2\pi nx,$$

where

$$\tilde{f}(n) = \tilde{f}_r(n) = \sum_{d|n} d^{1-r} f(d).$$

The differentiated form is the traditional Davenport expansion

$$\sum_{n=1}^{\infty}\frac{f(n)}{n^r}\bar{B}_1(nx) = -\frac{2}{\pi}\sum_{n=1}^{\infty}\frac{\tilde{f}(n)}{n}\sin 2\pi nx. \tag{2.16.16}$$

Proof For simplicity, we treat the case $r = 1$. We use (2.16.14) in the form

$$\frac{1}{2}\sum_{n=1}^{\infty}\frac{f(n)}{n^2}\left(\bar{B}_2(nx) - \frac{1}{6}\right) \tag{2.16.17}$$
$$= -\frac{1}{2\pi i}\int_{(c)}\frac{1}{(1-s)(2-s)}\zeta(1-s)L(s)x^{2-s}\,\mathrm{d}s,$$

where $1 < c < 2$. We substitute

$$\zeta(1-s)L(s) = \pi^{\frac{1}{2}-s}\frac{\Gamma\left(\frac{s}{2}\right)}{\Gamma\left(\frac{1-s}{2}\right)}\zeta(s)L(s) = \pi^{\frac{1}{2}-s}\frac{\Gamma\left(\frac{s}{2}\right)}{\Gamma\left(\frac{1-s}{2}\right)}\sum_{n=1}^{\infty}\frac{\tilde{f}(n)}{n^s}$$

in (2.16.17) and change the order of summation and integration to deduce that

$$\frac{1}{2}\sum_{n=1}^{\infty}\frac{f(n)}{n^2}\left(\bar{B}_2(nx) - \frac{1}{6}\right) \tag{2.16.18}$$
$$= -\frac{1}{\pi^{\frac{3}{2}}}\sum_{n=1}^{\infty}\frac{\tilde{f}(n)}{n^2}\frac{1}{2\pi i}\int_{(c)}\frac{\Gamma\left(\frac{s}{2}\right)}{(2-s)\Gamma\left(\frac{3-s}{2}\right)}(\pi nx)^{2-s}\,\mathrm{d}s.$$

In order to evaluate the resulting integral, we use the differentiated form

$$\sum_{n=1}^{\infty}\frac{f(n)}{n}\bar{B}_1(nx)=-\pi^{\frac{1}{2}}x\sum_{n=1}^{\infty}\tilde{f}(n)\frac{1}{2\pi i}\int_{(c)}\frac{\Gamma\left(\frac{s}{2}\right)}{\Gamma\left(\frac{3-s}{2}\right)}(\pi nx)^{-s}\,ds. \tag{2.16.19}$$

The integral is the G-function

$$G_{0,2}^{1,0}\left(z\,\middle|\,\begin{matrix}-\\ b,b-\frac{1}{2}\end{matrix}\right)=\frac{z^{b-\frac{1}{2}}}{\sqrt{\pi}}\sin 2\sqrt{z}$$

with $z=(\pi nx)^2$ and $b=0$. Hence substituting this, (2.16.19) leads to (2.16.16). □

Lemma 2.16.3 in conjunction with Lemma 2.16.4 leads to the main result:

Theorem 2.16.1 *For* $0<x<1$ *we have*

$$\begin{aligned}&\frac{1}{2\pi^2}\sum_{n=1}^{\infty}\frac{\log n}{n^2}\cos 2\pi nx+\frac{1}{12}\frac{\zeta'}{\zeta}(2) \qquad (2.16.20)\\ &=\frac{1}{2}\sum_{n=1}^{\infty}\frac{\Lambda(n)}{n^2}\left(\bar{B}_2(nx)-\frac{1}{6}\right)\\ &=-\frac{x}{2}((x-1)\log 2\pi+1-\log x-\gamma)+\sum_{k=1}^{\infty}\frac{\zeta(2k+1)}{(2k+1)(2k+2)}x^{2k+2}.\end{aligned}$$

By differentiation, we have

$$\begin{aligned}-\frac{1}{\pi}\sum_{n=1}^{\infty}\frac{\log n}{n}\sin 2\pi nx&=\sum_{n=1}^{\infty}\frac{\Lambda(n)}{n}\bar{B}_1(nx) \qquad (2.16.21)\\ &=-x\log 2\pi-\frac{1}{2}(-\log 2\pi-\log x-\gamma)+\sum_{k=1}^{\infty}\frac{\zeta(2k+1)}{2k+1}x^{2k+1}.\end{aligned}$$

The first equality in (2.16.21) is by the Davenport expansion.
Incorporating Lemmas 2.16.3 and 2.16.1 in (2.16.12) we obtain Popov's result.

Corollary 2.16.1 (Popov) *For* $|x|>1$ *we have*

$$\begin{aligned}\sum_{n>x}\frac{\Lambda(n)}{n^2}\left(\frac{1}{6}-\bar{B}_2(nx)\right)&=\frac{1}{x}(2-\log 2\pi)+\sum_{\rho}\frac{x^{\rho-2}}{\rho(\rho-1)}\\ &+\sum_{k=1}^{\infty}\frac{1}{2k(2k+1)}x^{-2k-2}-\sum_{k=1}^{\infty}\frac{2\zeta(2k+1)}{(2k+1)(k+1)}x^{-2k-2}.\end{aligned}$$

To prove Corollary 2.16.2 we need

Lemma 2.16.5 (Ramanujan) *For* $0 \le \lambda \in \mathbb{Z}$ *and* $|z| < |\alpha|$ *we have*

$$\sum_{m=2}^{\infty} \frac{\zeta(m,\alpha)}{m+\lambda} z^{m+\lambda} = \sum_{k=0}^{\lambda} \binom{\lambda}{k} \zeta'(-k, \alpha - z)\, z^{\lambda-k} - \zeta'(-\lambda, \alpha) - \sum_{k=1}^{\lambda} \frac{1}{k} \zeta(k-\lambda, \alpha)\, z^k + \frac{1}{\lambda+1} (\psi(\alpha) - H_\lambda)\, z^{\lambda+1}. \tag{2.16.22}$$

Here $\zeta(s,\alpha)$, $\psi(\alpha)$, H_λ *denote the Hurwitz zeta-function, the Euler digamma function, harmonic number, respectively.*

Lemma 2.16.5 with $\lambda = 0$ and Lerch's formula (2.1.46) yield a special case of Wilton's formula (1923)

$$\sum_{n=1}^{\infty} \zeta(n,a) \frac{x^n}{n} = \log\Gamma(a-x) - \log\Gamma(a) + \psi(a)x, \quad |x| < |a|.$$

The *odd part* is

$$\sum_{k=1}^{\infty} \zeta(2k+1, a) \frac{x^{2k+1}}{2k+1} = \frac{1}{2} (\log\Gamma(a-x) - \log\Gamma(a+x)) + \psi(a)x$$

cf. Srivastava and Choi (2001, p. 159). Hence the last sum in (2.16.21) is

$$\sum_{k=1}^{\infty} \zeta(2k+1) \frac{x^{2k+1}}{2k+1} = \frac{1}{2} (\log\Gamma(1-x) - \log\Gamma(1+x)) - \gamma x. \quad |x| < 1. \tag{2.16.23}$$

Corollary 2.16.2 *Theorem 2.16.1 implies Kummer's Fourier series (cf. Corollary 2.1.1)*

$$\log\Gamma(x) = \log\sqrt{\pi} - (\gamma + \log 2\pi) B_1(x) - \frac{1}{2} \log\sin\pi x + \sum_{n=1}^{\infty} \frac{\log n}{\pi n} \sin 2\pi n x, \quad 0 < x < 1. \tag{2.16.24}$$

The Davenport expansion for the von Mangoldt function is equivalent to the functional equation for the Riemann zeta-function (up to Ramanujan's formula (2.16.22)*).*

Proof Equation (2.16.23) becomes on using the reciprocal relation $\Gamma(x)\Gamma(1-x) = \frac{\pi}{\sin \pi x}$,

$$\sum_{k=1}^{\infty}\zeta(2k+1)\frac{x^{2k+1}}{2k+1}=\frac{1}{2}(-\log\sin\pi x-\log x+\log\pi)-\log\Gamma(x)-\gamma x, \tag{2.16.25}$$

$$|x|<1.$$

Substituting this in (2.16.21) proves (2.16.24). The second assertion follows from the fact that Kummer's Fourier series is equivalent to the functional equation and Lemma 2.16.4. □

We consider the case of the Möbius function based on Titchmarsh (1943) and state the results corresponding to Lemma 2.16.4 as the following

Exercise 2.16.2 (Patkowski, 2019, Theorem 2.1) Prove the following.

$$\frac{1}{2\pi^2}\cos 2\pi x-\frac{1}{12}\frac{1}{\zeta(2)}=\frac{1}{2}\sum_{n=1}^{\infty}\frac{\mu(n)}{n^2}\left(\bar{B}_2(nx)-\frac{1}{6}\right) \tag{2.16.26}$$

$$=\frac{1}{2\pi^2}\sum_{k=1}^{\infty}\frac{(-1)^k}{(2k)!}(2\pi x)^{2k}.$$

The differentiated form amounts to the classical Davenport expansion

$$-\frac{\sin 2\pi x}{\pi}=\sum_{n=1}^{\infty}\frac{\mu(n)}{n}\bar{B}_1(nx)=-\frac{1}{\pi}\sum_{k=0}^{\infty}\frac{(-1)^k}{(2k+1)!}(2\pi x)^{2k+1}.$$

Apply (2.16.17) as

$$\frac{1}{2}\sum_{n=1}^{\infty}\frac{\mu(n)}{n^2}\left(\bar{B}_2(nx)-\frac{1}{6}\right)$$

$$=-\frac{1}{2\pi i}\int_{(c)}\frac{1}{(1-s)(2-s)}\zeta(1-s)\zeta^{-1}(s)x^{2-s}\,ds.$$

In computing residues (at $s=-2k$) in this case, we apply the formula

$$\zeta'(-2k)=\frac{(-1)^k\zeta(2k+1)(2k)!}{\pi^{2k}2^{2k+1}},\quad k\in\mathbb{N}.$$

Note that the residue at $s=0$ is $-2x^2$.

Unlike Corollary 2.16.2, this gives only the power series expansions for the sine and cosine functions.

Exercise 2.16.3 Via the Riesz sum (2.12.47), deduce (2.16.11) from (2.16.10).

Cf. Exercise 4.5.5 for another application of the Riesz sum.

References

Akiyama, S., & Tanigawa, Y. (1999). Calculation of values of L-functions associated to elliptic curves. *Mathematical Components, 68*, 1201–1231.

Andrianov, A. N. (1987). *Quadratic forms and Hecke operators*. Berlin etc: Springer.

Apostol, T. M. (1976a). *Introduction to analytic number theory*. Berlin etc: Springer.

Apostol, T. M. (1976b). *Modular functions and Dirichlet series in number theory*. Berlin etc: Springer.

Bateman, P. T., & Grosswald, E. (1964). On Epstein's zeta function. *Acta Arithmetica, 9*, 365–373.

Berndt, B. C. (1969). Identities involving the coefficients of a class of Dirichlet series III. *Transactions of the American Mathematical Society, 146*, 323–348.

Berndt, B. C. (1970). Identities involving the coefficients of a class of Dirichlet series IV. *Transactions of the American Mathematical Society, 149*, 179–185.

Berndt, B. C. (1971). Identities involving the coefficients of a class of Dirichlet series VI. *Transactions of the American Mathematical Society, 160*, 157–167.

Bochner, S. (1951). Some properties of modular relations. *Annals of Mathematics* (2) *53*, 332–363= Collected Papers of Salomon Bochner. Part II. American Mathematical Society Providence, RI, 1991 (pp. 665–696).

Breuil, Chr, Conrad, B., Diamond, F., & Taylor, R. (2001). On the modularity of elliptic curves over $\mathbb{Q}$: Wild 3-aduic exercises. *Journal of the American Mathematical Society, 14*, 843–939.

Briggs, W. E. (1961). The irrationality of γ or of sets of similar constants. *Norske Videnskabers Selskabs Forh, 34*, 25–28.

Bump, D., & Goldfeld, D. (1987). A Kronecker limit formula for cubic fields. In Rankin, R. (Ed.) Modular Forms (pp. 43–49). Ellis Harwood.

Campbel, R. (1966). *Les intégrales Euléirennes et leurs applications-Étude aprofondie de la fonction gamma*. Paris: Dunod.

Chakraborty, K., Kanemitsu, S., & Kuzumaki, T. (2009a). Finite expressions for higher derivatives of the Dirichlet L-function and the Deninger R-function. *Hardy-Ramanujan Journal, 32*, 38–53.

Chakraborty, K., Kanemitsu, S., & Kuzumaki, T. (2016). *A quick introduction to complex analysis*. Singapore etc: World Scientific Publishing.

Chakraborty, K., Kanemitsu, S., Li, J.-H., & Wang, X.-H. (2009b). Manifestations of the Parseval identity. *Proceedings of the Japan Academy Series A: Mathematical Sciences, 85*, 149–154.

Chakraborty, K., Kanemitsu, S., & Tsukada, H. (2009c). *Vistas of special functions II*. Singapore etc: World Scientific Publishing.

Chan, H. H. (2020). *Theta functions, elliptic functions and π*. Berlin-Boston: de Gruyter.

Chandrasekharan, K., & Narasimhan, R. (1961). Hecke's functional equation and arithmetical identities. *Annals of Mathematics*, 2(74), 1–23.

Chen, N.-X. (2010). *Möbius inversion in physics*. New Jersey-London-Singapore etc: World Scientific Publishing.

Chowla, S., & Selberg, A. (1949). On Epstein's zeta-function (I). *Proceedings of the National Academy of Sciences of the United States of America, 35*, 371–374; Collected Papers of Atle Selberg I, Springer, 1989, 367–370. *The Collected Papers of Sarvadaman Chowla II, CRM, 1999*, 719–722.

Culp-Ressler, W., Flood, K., Ann Heath, S. R., & de Azevedo Pribitikin, W. (2000). On solutions to Riemann's functional equation. *Ramanujan Journal, 4*, 5–9.

Davenport, H. (1937b). On some infinite series involving arithmetic functions II, Quarterly Journal of Mathematics 8, 313–320=Collected papers of H. Davenport, IV, Academic Press London etc., 1977, 1787–1794.

Davenport, H. (1937a). On some infinite series involving arithmetic functions, Quarterly Journal of Mathematics 8, 8–13=Collected papers of H. Davenport IV, Academic Press London etc., *1977*, 1781–1786.

Davenport, H. (1980). *Multiplicative number theory*, 1st edn. Markham, Chicago; 1967, 2nd ed. Springer, New York etc.

de la Vallée Poussin, Ch. -J. (1896). Recherches analytiques de la théorie des nombres premiers. *Annales de la Société scientifique de Bruxelles***20 B**, 1896, 183–256, 281–362 363–397, **21 B**, 251–342.

Deninger, C. (1984). On the analogue of Chowla-Selberg formula for real quadratic fields. *Journal für die Reine und Angewandte Mathematik, 351*, 171–191.

Dixit, A., Gupta, R., Kumar, R., & Maji, B. (2020). Generalized Lambert series, Raabe's cosine transform and a two-parameter generalization of Ramanujan's formula for $\zeta(2m+1)$. *Nagoya Mathematical Journal, 239*, 232–293.

Dufresnoy, J., & Pisot, Ch. (1963). Sur la relation fonctionnelle $f(x+1)^{f}(x) = \varphi(x)$. *Bulletin of the Society Mathematical Belgaum, 15*, 259–270.

Ehrenpreis, L. (1987). Fourier analysis, partial differential equations and automorphic forms. In *Proceedings of Symposia in Pure Mathematics* (Vol. 49, pp. 45–100), Part 2, American Mathematical Society, Providence, RI.

Eisenstein, G. (1844). Aufgaben und Lehrsätze. *The Journal für die Reine und Angewandte Mathematik,***27**, 281–283=*Mathematische Werke*, Vol. 1, 1975, Chelsea, pp. 108–110.

Erdélyi, A., Magnus, W., Oberhettinger, F., & Tricomi, F. G. (ed.), (1953). *Higher transcendental functions*, Vol. I–III. New York: McGraw-Hill.

Feng, J.-W., Kanemitsu, S., & Kuzumkai, T. (2021). On Fourier transforms and Hilbert space. *Hardy-Ramanujan Journal, 44*, 143–151.

Ferminger, S. (1992). Zéros des fonction L de courbres elliptiques. *Experimental Mathematics, 1*, 167–173.

Fomenko, O. M. (1980). Applications of the theory of modular forms to number theory, Itogi Nauki, Tekhniki, Algebraic and Geometric Topology 15 (1977), 5–91=*Journal of Soviet Mathematics, 14*, 1307–1362.

Frazier, M. W. (1999). *An introduction to wavelettes through linear algebra*. New York etc: Springer.

Funakura, T. (1990). On Kronecker's limit formula for Dirichlet series with periodic coefficients. *Acta Arithmetica, 55*, 59–73.

Glaeske, H.-J. (1966). Über eine Charakterisierung der Hurwitzschen Zetafunktion. *Publicationes Mathematicae Debrecen, 13*, 139–144.

Goldstein, L. J. (1971). *Analytic number theory*. New Jersey: Prentice-Hall.

Gut, M. (1929). Die Zetafunktion, die Klassenzahl und die Kronecker'shce Grenzformel eines beliebiges Kreiskörpers. *Commentarii Mathematici Helvetici, 1*, 160–226.

Guthman, A. (1999). Asymptotic expansions for Dirichlet series associated to cusp forms. *Publications de l'Institut Mathématique* (Beograd) (N.S.) **65** (79), 69–96.

Hamburger, H. (1921). Über die Riemannschen Funktionalgleichung der ζ-Funktion. *Mathematische Zeitschrift, 10*, 240–254. (Erste Mitteilung)

Hardy, G. H. (1908). Some multiple integrals. *Quarterly Journal of Mathematics (Oxford)* (2) *5*, 357–375; *Collected Papers. Vol. V* (1972), 434–452, Comments 453.

Hashimoto, M., Kanemitsu, S., & Toda, M. (2008). On Gauss' formula for ψ and finite expressions for the L-series at 1. *Journal of the Mathematical Society of Japan, 60*, 219–236.

Hasse, H. (1952). *Über die Klassenzahl abelscher Zahlkörper*. Akademie Verl., 1952, 2nd ed. 1985 (with Appendix by J. Martinet).

Hecke, E. (1923). *Vorlesungen über die Theorie der algebraischen Zahlen*. Leipzig: Akad. Verl.; English transl. *Lectures on algebraic number theory*. Berlin etc: Springer. 1981.

Hecke, E. (1944). *Herleitung des Euler-Produktes der Zetafunktion und einiger L-Reihen aus ihrer Funktionalgleichung, Mathematische Annalen 119, 266–287=Mathematische Werke, 919–940* (p. 1959). Ruprecht, Göttingen: Vandenhoeck u.

Hecke, E. (1983). *Lectures on Dirichlet Series, Modular Functions and Quadratic Forms*. In B. Schoenberg in coll. with W. Maak, Vandenhoeck u. Ruprecht, Göttingen (first edition: *Dirichlet Series*, Planographed Lecture Notes, Princeton IAS, Edwards Brothers, Ann Arbor, 1938).

Hecke, E. (1917). Über die Kroneckersche Grenzformel für reelle quadratische Körper und die Klassenzahl relativ-abelscher Körper, Verh. Naturforsch. Ges. i. Basel 28, 363–372= Math. Werke, Vandenhoeck u. *Ruprecht, Göttingen, 1970*, 198–207.

Hecke, E. (1936). *Über die Bestimmung Dirichletscher Reihen durch ihre Funktionalgleichung, Mathematische Annalen 112, 664–669=Mathematische Werke, 591–626* (p. 1959). Ruprecht, Göttingen: Vandenhoeck u.

Hutchinson, J. I. (1929). Properties of functions represented by Dirichlet series $\sum (a\nu + b)^{-s}$ or by linear combinations of such series. *Transactions of the American Mathematical Society, 31*, 322–344.

Ingham, A. E. (1932). *The distribution of prime numbers, Cambridge UP, Cambridge* (2nd ed. New York: Hafner, 1971).

Ishibashi, M., & Kanemitsu, S. (1999). Dirichlet series with periodic coefficients. *Results in Mathematics, 35*, 70–88.

Ivić, A. (1985). *The Riemann Zeta-Function*. New York: Wiley.

Iwasawa, K. (1975). *On a theorem of S. Chowla*, unpublished.

Kanemitsu, S. (1987). Evaluation of certain limits in closed form. In *Proceedings of the International Conference Number Theory, University Laval* (pp. 459–474), de Gruyter, Berlin.

Kanemitsu, S., Tanigawa, Y., & Yoshimoto, M. (2002). Ramanujan's formula and modular forms. In S. Kanemitsu, & C. Jia (Eds.) *Number-theoretic Methods – Future Trends, Proceedings of a conference held in Iizuka* (pp. 159–212). Dordrecht: Kluwer.

Kanemitsu, S., & Kuzumaki, T. (2001). On a generalization of the Maillet determinant II. *Acta Arithmetica, 99*, 343–361.

Kanemitsu, S., Tanigawa, Y., & Tsukada, H. (2008). On Kronecker's limit formula and the hypergeometric function. *Hardy-Ramanujan Journal, 31*, 28–40.

Kanemitsu, S., & Tsukada, H. (2007). *Vistas of special functions*. Singapore etc: World Scientific.

Kanemitsu, S., & Tsukada, H. (2014). *Contributions to the theory of zeta-functions: The modular relation supremacy*. Singapore etc: World Scientific Publishing.

Kimura, T. (1998). *Prehomogeneous vector spaces*. Tokyo: Iwanami-shoten.

Kitaoka, Y. (2025). *An introduction to algebraic number theory*. Singapore etc: World Scientific Publishing.

Knopp, M., et al. (2000). Hamburger's theorem on $\zeta(s)$ and the abundance principle for Dirichlet series with functional equations. In R. P. Bambah (Ed.), *Number Theory* (pp. 201–216). Hindustan Book Agency: New Delhi.

Koshlyakov, N. S. (1954). Investigation of some questions of analytic theory of the rational and quadratic fields, I-III (Russian). *Izvestiya Akademii Nauk SSSR. Seriya Matematicheskaya, 18*, 113–144, 213–260, 307–326, Errata: ibid. *19* (1955), 271.

Kumagai, H. (2002). On unified Kronecker limit formulae. *Kyushu Journal of Mathematics, 56*, 41–51.

Kurokawa, N. (1994). One hundred years of the zeta regularized products. In *The 39th Algebra Symposium* (pp. 153–166).

Kuz'min, R. (1930). Contribution to the theory of a class of Dirichlet series (Russian). *Izvestiya Akademii Nauk SSSR. Seriya Matematicheskaya Natural Science, 7*, 115–124.

Kuz'min, R. (1934). On the roots of Dirichlet series. *Izvestiya Akademii Nauk SSSR. Seriya Matematicheskaya Natural Science, 7*, 1471–1491.

Kuzumaki, T. (2011). Asymptotic expansions for a class of zeta-functions. *Ramanujan Journal, 24*, 331–343.

Landau, E. (1903). Über die zu einem algebraischen Zahlkörper gehörige Zetafunktion und die Ausdehnung der Tschebyscheffschen Primzahltheorie auf das Problem der Verteilung der Primideale. *Journal für die Reine und Angewandte Mathematik,125*, 176–188=Collected Works, Vol. 1. *Thales Verl. Essen,1985*, 201–325.

Landau, E. (1921). *Vorlesungen über Zahlentheorie*, Teubner, Leipzig 1921 and reprint Chelsea, New York 1967.

Lang, S. (1976). *Introduction to modular forms*. Berlin etc: Springer.

Lavrik, A. F. (1968). An approximate functional equation for the Dirichlet L-function. *Trudy Moskov. Mat. Obšč,18*, 91–104 =Trans. *Moscow Mathematical Society,18*, 101–115.

Lavrik, A. F. (1990). Functional equations with a parameter for zeta-functions (Russian). *Izvestiya Akademii Nauk SSSR. Seriya Matematicheskaya,54*, 501–521; English translation in *Mathematics of the USSR-Izvestiya, 36*, 519–540 (1991).

Lehmer, D. H. (1975). Euler constants for arithmetic progressions, Acta Arithmetica **27**, 125–142; Selected Papers of D. H. Lehmer, Vol. II, 591-6oe, Charles Babbage Res. Center, Manitoba 1981.

Lerch, M. (1897). Sur quelques formules du nombre des classes. *Bulletin des Sciences Mathematiques, 21*, 290–304.

Lerch, M. (1898). Bemerkungen über trigonometrischen Reihen mit positive Coeffizienten. *Sitzungsber. König. Böhm. Ges. Wiss. Math. Naturwiss. Kl. No., 23*, 1–19.

Li, H. -Y., Kuzumaki, T., & Kanemitsu, S. (2022). On zeta functions and allied theta-functions, to appear *Advances in Applied Analysis and Number Theory*, Chapter 4, (pp. 51–97). World Scientific Publishing. https://doi.org/10.1142/9789811272608_0004

Li, W. -B., Li, H. -Y., & Mehta, J. (2020). Around the Lipschitz summation formula. *Mathematical Problems in Engineering,2020*, Article ID 5762823, 16. https://doi.org/10.1155/2020/5762823

Li, H.-L. (2005). On generalized Euler constants and an integral related to the Piltz divisor problem. *Siaulai Mathematical Physics Seminar, 8*, 81–93.

Li, H.-L., Hashimoto, M., & Kanemitsu, S. (2010a). Structural elucidation of Eisenstein's formula. *Science China Mathematics, 53*, 2341–2350.

Li, H.-L., Li, F.-H., Wang, N.-L., & Kanemitsu, S. (2017). *Number theory and its applications II*. London-Singapore-New Jersey: World Scientific Publishing.

Li, H.-L., Ma, J., & Zhang, W.-P. (2010b). On some Diophantine Fourier series. *Acta Mathematica Sinica (English Series), 26*, 1125–1132.

Liu, T. -X. (2022). L-functions of elliptic curves and modular forms. https://www.sas.rochester.edu.

Li, F.-H., Wang, N.-L., & Kanemitsu, S. (2013). *Number theory and its applications*. London-Singapore-New Jersey: World Scientific Publishing.

Maier, W., & Kiesewetter, H. (1971). *Funktionalgleichungen mit analytischen Lösungen* Vandenhoeck & Ruprecht, Gïttingen - Zürich.

Maier, W. (1936). Gitterfunktionen in Zahlebene. *Mathematische Annalen, 113*, 363–379.

Meyer, C. (1957). *Die Berechnung der Klassenzahl Abelscher Körper über quadratischen Zahlkörpern*, 1957 Akademie-Verl., Berlin.

Meyer, C. (1981). *L-Funktionen quadratischer Zahlkörper*. Köln: Vorlesung.

Mikolás, M. (1956). Mellinsche Transformation und Orthogonalität bei $\zeta(s, u)$; Verallgemeinerung der Riemannschen Funktionalgleichung von $\zeta(s)$. *Acta Scientific Math. (Szeged), 17*, 143–164.

Milnor, J. (1983). On polylogarithms, Hurwitz zeta functions, and the Kubert identities. *Enseignement des Mathématiques, 2*(29), 281–322.

Miyake, T. (1989). *Modular forms*. New York etc: Springer.

Motohashi, Y. (1968). A new proof of the limit formula of Kronecker. *Proceedings of the Japan Academy Series A. Mathematical Sciences, 66*, 614–616.

Motohashi, Y. (1987). *Lectures on the Riemann-Siegel formula*. Colorado Press, Boulder: University.

Motohashi, Y. (1997). *Spectral theory of the Riemann zeta-function*. Cambridge etc: Cambridge UP.

Nörlund, N. E. (1954). *Vorlesungen über Differenzenrechnung*. New York: Chelsea.

Novikov, A. P. (1980). Kronecker's limit formula in a real quadratic field. *Izvestiya Akademii Nauk SSSR. Seriya Matematicheskaya, 44*, 147–176.

Oberhettinger, F. (1974). *Tables of Mellin transforms*. Berlin-Heidelberg-New York: Springer.

Ogg, A. (1969a). On product expansions of theta-functions. *Proceedings of the American Mathematical Society, 21*, 365–368.

Ogg, A. (1969b). *Modular Forms and Dirichlet Series*. New York: Benjamin.

Paris, R. B., & Kaminski, D. (2001). *Asymptotics and Mellin-Barnes integrals*. Cambridge: Cambridge UP.

Patkowski, A. (2019). On Popov's formula involving the von Mangoldt function. *The Pi Mu Epsilon Journal, 15*(1), 45–47.

Prachar, K. (1957). Primzahlverteilung. Berlin-Göttingen-Heidelberg: Springer (Second printing, Springer, 1978).

Prudnikov, A. P., Bychkov, Yu. A., & Marichev, D. I. (1986). *Integrals and series*. Supplementary chapters. Moscow: Nauka.

Rademacher, H. (1973). *Topics in analytic number theory*. Berlin-Heidelberg: Springer.

Salerno, S., Vitolo, A., & Zannier, U. (1991). Functional equations for Dirichlet series with periodic coefficients. *Ricerche di Matematica, 40*, 222–229.

Sato, F. (1992). *Searching for the origin of prehomogeneous vector spaces*. Japan (in Japanese). Keynote address at annual meeting of the Mathematics Society

Schnee, W. (1930). Die Funktionalgelichung der Zetafunktion und der Dirichletschen Reihen mit periodischen Koeffizienten. *Mathematische Zeitschrift, 31*, 378–390.

Selberg, A., & Chowla, S. (1967). On Epstein's zeta-function. *Journal für die Reine und Angewandte Mathematik,227*, 86–110; Collected Papers of Atle Selberg I, Springer, 1989 (pp. 521–545); The Collected Papers of Sarvadaman Chowla II. *CRM,1999* (pp. 1101–1125).

Serre, J.-P. (1973). *A course in arithmetic*. New York-Heidelberg-Berlin: Springer.

Shimura, G. (1971). *Introduction to the theory of automorphic functions*. Princeton UP: Princeton.

Shirasaka, S. (2002). On the Laurent coefficients of a class of Dirichlet series. *Results in Mathematics, 42*, 128–138.

Siegel, C. L. (1922). Bemerkungen zu einem Satz von Hamburger über die Funktionalgleichung der Riemannschen Zetafunktion, Mathematische Annalen 85, 276–279= Ges. Abh., Bd. I. Springer. Berlin-Heidelberg, 1966 (pp. 154–156).

Siegel, C. L. (1961). *Lectures on advanced analytic number theory*. Bombay: Tata Institute.

Sneddon, I. N. (1972). *The use of integral transforms*. New York etc: MacGraw-Hill.

Spira, R. (1970). Residue class characters. *Duke Mathematical Journal, 37*, 633–637.

Srivastava, H. M., & Choi, J.-S. (2001). *Series associated with the zeta and related functions*. Dordrecht etc: Kluwer Academic Publisher.

Stankus, E. P. (1983). A remark on the Laurent coefficients of the Riemann zeta-function. *Zapiski Naučnykh Seminarov LOMI,121*, 103–107= English transl., *Journal of Soviet Mathematics,29*, 1302–1305 (1985).

Takagi, T. (1971). *Algebraic number theory—number theory and class field theory*, 2nd ed. Tokyo: Iwanami-shoten; 2st ed. 1947, ditto.

Terras, A. (1985). *Harmonic analysis on symmetric spaces and applications I*. Berlin etc: Springer.

Titchmarsh, E. C. (1951). *The Theory of the Riemann Zeta-Function*, (second edition revised by D. R. Heath-Brown), OUP 1986.

Titchmarsh, E. C. (1943). Some properties of the Riemann zeta-function. *Quarterly Journal of Mathematics (Oxford Ser.), 2*(14), 16–26.

Walfisz, A. A. (1961a). On sums of coefficients of some Dirichlet series. *Soobščenija Akademii Nauk Gruzinskoj SSR,26*, 9–16.

Walfisz, A. A. (1961b). On the theory of a class of Dirichlet series. *Soobščenija Akademii Nauk Gruzinskoj SSR,27*, 9–16.

Walum, H. (1991). Multiplication formulae for periodic functions. *Pacific Journal of Mathematics, 149*, 383–396.

Wang, X.-H., Wang, N.-L., & Kanemitsu, S. (2022). *An extension of Maass theory to general Dirichlet series*. Integral Transforms and Special Functions. https://doi.org/10.1080/10652469.2022.2057483

Wang, N.-L. (2011). On Riemann's posthumous Fragment II on the limit values of elliptic modular functions. *Ramanujan Journal, 24*, 129–145.

Wang, N.-L., Agarwal, K., & Kanemitsu, S. (2020). Limiting values and functional and difference equations. *Mathematics, Difference and Differential Equations, 8*, 407. https://doi.org/10.3390/math8030407

Washington, L. (1983). *Introduction to cyclotomic fields*. New York etc: Springer.

Watson, G. N. (1931). Some self-reciprocal functions. *Quarterly Journal of Mathematics (Oxford Ser.), 2*, 298–309.

Weil, A. (1968). Sur une formule classique, *Journal of the Mathematical Society of Japan, 20*, 400–402 = *Coll. Papers, III*, Springer, 1980, Berlin etc., 198–200.

Weil, A. (1971). Dirichlet series and automorphic forms. LNM **189**. Berlin-Heidelberg-New York: Springer.

Weil, A. (1975). *Elliptic functions according to Eisenstein and Kronecker*. Berlin-Heidelberg-New York: Springer.

Whittaker, E. T., & Watson, G. N. (1927). *A course in modern analysis* (4th ed.). Cambridge: Cambridge UP.

Yamamoto, Y. (1977). Dirichlet series with periodic coefficients. In *Proceedings of the International Symposium on Algebraic Number Theory*, Kyoto 1976 (pp. 275–289). JSPS, Tokyo.

Yang, Q., Mehta, K., & Kanemitsu, S. (2023). On Popov's explicit formula and the Davenport expansion, to appear.

Yoshida, H. (1999). Special functions appearing in the limit formulas and reciprocity laws. *RIMS Kokyuroku, 1091*, 45–55.

Zagier, D. (1975). A Kronecker limit formula for real quadratic fields. *Mathematische Annalen, 213*, 153–184.

Zhang, N.-X., & Williams, K. (1995). On the Epstein zeta function. *Tamkang Journal of Mathematics, 26*, 165–176.

Zhu, P.-Y., & Mehta, J. (2021). A new proof of the functional equation for the Riemann zeta-function. *Hardy-Ramanujan Journal, 44*, 143–151.

Chapter 3
Unified Theory of Epstein Zeta-Functions and Zeta-Functions Associated with Real-Analytic Automorphic Forms

In this chapter, the main figures are Maass forms (EMS), ramified functional equation (RFEM), Fourier–Whittaker expansion (FWE), and Whittaker function.

3.1 The Hardy–Hecke Transform: Streamlining of History

Notation and formulas. We assemble data on special functions which are frequently used throughout. The **modified Bessel function of the third kind of index** ν, referred to as the K**-Bessel function** $K_\nu(z)$, is defined by (Watson 1944, p. 183, (15)):

$$K_\nu(z) = \frac{1}{2}\left(\frac{z}{2}\right)^\nu \int_0^\infty e^{-t-\frac{z^2}{4t}} t^{-\nu}\,\frac{\mathrm{d}t}{t}, \quad \operatorname{Re}\nu > -\frac{1}{2}, \quad |\arg z| < \frac{\pi}{4}. \tag{3.1.1}$$

By the change of variable, we arrive at a handy formula which leads to Hardy's formula (3.1.21)

$$\int_0^\infty t^s e^{-\alpha t^p - \frac{\beta}{t^p}}\,\frac{\mathrm{d}t}{t} = \frac{2}{p}\left(\frac{\beta}{\alpha}\right)^{\frac{s}{2p}} K_{\frac{s}{p}}(2\sqrt{\alpha\beta}), \quad \operatorname{Re}\alpha > 0, \operatorname{Re}\beta > 0, 0 \neq p \in \mathbb{R}. \tag{3.1.2}$$

As the Mellin inverse we have the **inverse Heaviside integral** which most often appears in the context of Fourier–Bessel expansion (2.12.3), cf. the hierarchy (3.2.4).

$$\begin{aligned} 2K_\nu(2z) &= \frac{1}{2\pi i}\int_L \Gamma\left(s+\frac{\nu}{2}\right)\Gamma\left(s-\frac{\nu}{2}\right) z^{-2s}\,\mathrm{d}s \\ &= G^{2,0}_{0,2}\left(z^2 \,\middle|\, {- \atop \frac{\nu}{2}, -\frac{\nu}{2}}\right), \end{aligned} \tag{3.1.3}$$

K. Chakraborty et al., *Modular Relations and Parity in Number Theory*,
Infosys Science Foundation Series, https://doi.org/10.1007/978-981-96-6471-9_3

where L is a suitable Bromwich path.

We state Hecke's generalization of the Hardy transform unifying (3.1.1) and (3.1.3) as is expounded in Rademacher (1973, pp. 55–58). For $\alpha, \beta \in \mathbb{C}$ and $a, b > 0$, we have

$$\int_0^\infty t^{\beta-\alpha} e^{-\frac{a}{t}-bt} \frac{\mathrm{d}t}{t} = \frac{1}{2\pi i} \int_{(c)} \Gamma(s+\alpha)\Gamma(s+\beta) a^{-s-\alpha} b^{-s-\beta} \, \mathrm{d}s, \tag{3.1.4}$$

where (c) separates the poles of the integrand (in Hecke' paper (Hecke 1921, (12),(13)), $c > -\operatorname{Re}\alpha, -\operatorname{Re}\beta$). A prototype of (3.1.4) with $b = 1$ can be found in Mellin (1910, p. 325, (65)). We refer to a procedure involving (3.1.4) the **Hecke transform** or the **Hardy–Hecke transform**.

Hecke has made important applications of (3.1.4) in the theory of real quadratic fields, in particular, the Kronecker limit formula in Hecke (1921) and Hecke (1923). It is further generalized by Rademacher (1927, p. 352, Hilfssatz 13) to algebraic number fields of higher degree and played an important role.

For a penetrating remark on the meaning of the beta transform and the Hardy–Hecke transform, cf. Remark 3.1.1.

The following properties of $K_s(z)$ are often used.

The symmetry property

$$K_{-\nu}(z) = K_\nu(z) \tag{3.1.5}$$

holds true. This is a consequence of the counterpart of Kummer's transformation, cf. (3.3.4) and it will turn out to be part of the functional equation (in the Fourier–Bessel expansion), cf. Theorem 2.5.1.

For a half-integer index, the K-Bessel function reduces to the exponential function times a rational function, i.e. for $0 \le \nu \in \mathbb{Z}$ we have

$$K_{\nu+\frac{1}{2}}(z) = \left(\frac{\pi}{2z}\right)^{1/2} e^{-z} \sum_{m=0}^{\nu} (2z)^{-m} \frac{\Gamma(\nu+m+1)}{m!\,\Gamma(\nu+1-m)} \tag{3.1.6}$$

and, in particular,

$$K_{\pm\frac{1}{2}}(z) = \sqrt{\frac{\pi}{2z}} e^{-z}. \tag{3.1.7}$$

The **Whittaker function** $W_{\kappa,\mu}$ is one of two independent solutions to the **Whittaker differential equation** (cf. (Erdélyi et al. 1953, p. 248, (4))) (of polynomial growth).

$$\frac{\mathrm{d}^2 w}{\mathrm{d}x^2} + \left(-\frac{1}{4} + \frac{\kappa}{x} + \frac{\frac{1}{4}-\mu^2}{x^2}\right) w = 0. \tag{3.1.8}$$

This is the last entry in G-function hierarchy (3.2.4) and will play the fundamental role in the theory of Fourier–Whittaker expansion.

As the closest to the Whittaker function, we have the **confluent hypergeometric function** $\Psi(a, c; x)$ (also denoted $U(a, c; x)$), which is a solution of the confluent hypergeometric differential equation

$$x\frac{\mathrm{d}^2 w}{\mathrm{d}x^2} + (c - x)\frac{\mathrm{d}w}{\mathrm{d}x} - aw = 0. \tag{3.1.9}$$

It has the integral representation (Erdélyi et al. 1953, I, p. 255 (2))

$$\Gamma(a)\Psi(a, c; x) = \Gamma(a)U(a, c; x) = \int_0^\infty e^{-xt} t^{a-1}(1+t)^{c-a-1}\,\mathrm{d}t, \tag{3.1.10}$$

where the last integral may be viewed as the Mellin transform of $e^{-x\cdot}(1+\cdot)^{c-a-1}$ at a. Cf. (Abramowitz and Stegun (1965, p. 505, 13.2.5), Erdélyi et al. (1953, I, p. 255, (2), p. 269, (4))). The U-notation will be mainly used in Chap. 4. These are connected by the identity (Erdélyi et al. 1953, I, p. 264, (4)), (Prudnikov et al. 1986, p. 797)

$$\Psi(a, c; x) = x^{-\frac{1}{2}-\mu} e^{\frac{1}{2}x} W_{\kappa,\mu}(x), \tag{3.1.11}$$

where $\kappa = \frac{1}{2}c - a$ and $\mu = \frac{1}{2}c - \frac{1}{2}$. In view of this, many results on the Whittaker function are stated in terms of the confluent hypergeometric function $\Psi(a, c; x)$ and vice versa.

$W_{\kappa,\mu}(x)$, resp., $\Psi(a, c; x)$ is the solution which is of polynomial growth as $\operatorname{Im} x \to \infty$. The other solution which goes to infinity is $M_{\kappa,\mu}(x)$ resp. $\Phi(a, c; x)$, cf. toward the end of Sect. 4.1. To find results on the Whittaker function, Whittaker and Watson (1927, pp. 337–354) may be the most convenient since the results are stated in terms of the Whittaker function rather than the confluent hypergeometric function. For confluent hypergeometric function, one may consult (Erdélyi et al. 1953, I, pp. 248–295) and other modern references. Hereafter we mainly use the Whittaker function $W_{\kappa,\mu}(x)$ and the K-Bessel function as its special case with the confluent hypergeometric function $\Psi(a, b; x)$ supplementarily, e.g. in Exercise 3.1.1. In this chapter, we use the notation $U(a, c; z)$ for the confluent hypergeometric function.

Exercise 3.1.1 *Prove that by the transformation* (3.1.11), (3.1.8), *and* (3.1.9) *shift to each other.*

From Erdélyi et al. (1953, p. 265, (13) and (14)), we also have

$$\begin{aligned} W_{0,\nu}(x) &= \frac{1}{\sqrt{\pi}} x^{\frac{1}{2}} K_\nu\left(\frac{1}{2}x\right) \\ &= e^{-\frac{1}{2}x} x^{\nu+\frac{1}{2}}\, \Psi\left(\nu + \frac{1}{2}, 2\nu + 1;\ x\right). \end{aligned} \tag{3.1.12}$$

We make a frequent use of the following reduction formula for the G-function (Erdélyi et al. 1953, I, p. 216, (8))

$$G^{2,1}_{1,2}\left(x \,\middle|\, \begin{matrix} a \\ b,\ c \end{matrix}\right) = \Gamma(b-a+1)\Gamma(c-a+1)x^{\frac{1}{2}(b+c-1)}\, e^{\frac{1}{2}x}\, W_{\kappa,\mu}(x), \quad (3.1.13)$$

where $\kappa = a - \frac{1}{2}(b+c+1),\ \mu = \frac{1}{2}(b-c)$.

Exercise 3.1.2 *Prove the Mellin transform formula* (Oberhettinger 1974, p. 115, 11.1) *(*$a > 0$*)*

$$\int_0^\infty x^s K_\nu(ax)\,\frac{\mathrm{d}x}{x} = 2^{s-2}a^{-s}\Gamma\left(\frac{s+\nu}{2}\right)\Gamma\left(\frac{s-\nu}{2}\right). \quad (3.1.14)$$

Solution. It suffices to prove (3.1.14) with $a = 1$. Although the proof is simple, we try to show a few different forms of (3.1.14) and its Mellin inversion. In (3.1.3) make the change of variable $s \leftrightarrow \frac{s}{2}$ and apply the Mellin inversion to deduce that

$$\int_0^\infty x^s 2K_\nu(2x)\,\frac{\mathrm{d}x}{x} = \frac{1}{2}\Gamma\left(\frac{s+\nu}{2}\right)\Gamma\left(\frac{s-\nu}{2}\right), \quad (3.1.15)$$

which leads to (3.1.14) by the change of variable.

The second way is to use (Erdélyi et al. 1954, I, p. 331) ($0 \le \operatorname{Re}\nu < c$)

$$2K_\nu(x) = \frac{1}{2\pi i}\int_{(c)} \Gamma(s)\Gamma(s-\nu)\left(\frac{x}{2}\right)^{\nu-2s}\mathrm{d}s = x^\nu G^{2,0}_{0,2}\left(x^2 \,\middle|\, \begin{matrix} - \\ 0,\ -\nu \end{matrix}\right). \quad (3.1.16)$$

By the change of variable $\nu - 2s = -z$, (3.1.16) with $\frac{x}{2}$ replaced by x leads to (3.1.3).

We note that (3.1.3) with $z = \sqrt{x}$ reads

$$2K_\nu(2\sqrt{x}) = \frac{1}{2\pi i}\int_L \Gamma\left(s+\frac{\nu}{2}\right)\Gamma\left(s-\frac{\nu}{2}\right)x^{-s}\,\mathrm{d}s$$

or

$$\int_0^\infty x^s 2K_\nu(2\sqrt{x})\,\frac{\mathrm{d}x}{x} = \Gamma\left(s+\frac{\nu}{2}\right)\Gamma\left(s-\frac{\nu}{2}\right),$$

which is Prudnikov et al. (1986, p. 665, 8.4.23.1). □

Bellman introduces the Voronoǐ function (or the generalized Bessel function) (Bellman 1956, (1.1)) by

$$V_a(x,y) = \int_0^\infty e^{-\pi x^2 t - \frac{\pi y^2}{t}}\, t^{-a}\,\mathrm{d}t = 2\left(\frac{y}{x}\right)^{1-a} K_{1-a}(2\pi xy), \quad (3.1.17)$$

which reduces for $\operatorname{Re} x > 0$ to Bellman (1950, (3))

$$W_a(x) = V_a\left(\sqrt{\frac{x}{\pi}}, \frac{1}{\sqrt{\pi}}\right) = \int_0^\infty e^{-xt-\frac{1}{t}} t^{-a}\,\mathrm{d}t = 2x^{\frac{a-1}{2}} K_{a-1}(2\sqrt{x}). \tag{3.1.18}$$

He states that (3.1.17) *was first introduced by Voronoĭ* in Voronoĭ (1904) and has been treated by Hardy (1908) and Steen (1930).

But Voronoĭ states (Voronoĭ 1904, p. 211) that the function $\xi(x)$

$$\xi(x) = 2\int_1^\infty \frac{e^{-2\sqrt{x}t}}{\sqrt{t^2-1}}\,\mathrm{d}t, \quad \sqrt{x} > 0$$

was studied by several renowned authors and some properties indicated by Riemann (1855) were proved by Stieltjes (1886), among other things,

$$\Gamma^2(s) = \int_0^\infty t^s \xi(t)\,\frac{\mathrm{d}t}{t}, \quad \operatorname{Re} s > 0.$$

It turns out that

$$\xi(x) = 2K_0(2\sqrt{x}) = W_1(x), \tag{3.1.19}$$

whose first equality is a special case $s = 0$ of (Erdélyi et al. 1953, II, p. 18, (15))

$$\Gamma\left(\frac{1}{2} - s\right) K_s(z) = \sqrt{\pi}\left(\frac{1}{2}z\right)^{-s} \int_1^\infty \frac{e^{-zt}}{(t^2-1)^{s+\frac{1}{2}}}\,\mathrm{d}t, \quad z > 0. \tag{3.1.20}$$

He mentions two special cases $a = \frac{3}{2}$ and $a = 1$ of (3.1.18) and remarks that *the latter case introduces a cylinder function, used by Maass in the construction of his functions*, which is partly confirmed by (3.1.19). The former case reads

$$\frac{1}{2}\int_0^\infty e^{-xt-\frac{1}{4t}} t^{-\frac{3}{2}}\,\mathrm{d}t = W_{\frac{3}{2}}\left(\frac{x}{4}\right) = \sqrt{x}e^{-\sqrt{x}}.$$

Both Bellman (1950) and Bellman (1956) are very enlightening but at the same time misleading because Maass does not use $W_1(x) = 2K_0(2\sqrt{x})$ but $K_{ir}(z)$ in the form of (3.1.20).

What Bellman suggested was the use of the K-Bessel function in general, which he expressed by attributing to the Voronoĭ function $W_1(x)$.

Hardy's formula Hardy (1908) reads

$$\psi_a(z) = z^{\frac{a+1}{2}} K_{\frac{a+1}{2}}(z) = \int_0^\infty e^{-t^2-\frac{z^2}{t^2}} t^{a+1}\,\frac{\mathrm{d}t}{t}, \quad \operatorname{Re} z > 0 \tag{3.1.21}$$

and we call an operation involving this the **Hardy transform**. As shown in comments on the paper in Hardy's collected papers, this paper was overlooked by Watson in his treatise Watson (1944) although it is essentially (3.1.1).

Recall the beta transform in Lemma 2.5.1. The terminology beta transform arises in contrast to the Hecke gamma transform (2.1.3) above.

Koshlyakov used the beta transform extensively in Koshlyakov (1954) which may be described as *the theory of special functions related to the K-Bessel function.*

Independently of Koshlyakov, Berndt 1967, 1970, resp., Berndt (1969), made an extensive study of perturbed Dirichlet series (2.12.43) and the beta transform, resp., the Hardy transform, deriving the Fourier–Bessel expansion (2.12.42). (Berndt, 1970, Theorem 3.1) establishes the equivalence of the Hecke functional equation and the Fourier–Bessel expansion for the case of the general gamma factor.

Berndt (1971) develops the theory of Epstein zeta-functions by the beta transform method. For the case of Eisenstein series, cf. Kuzumaki (2011). Cf. Chowla and Selberg (1949), Selberg and Chowla (1967), etc.

Then the Hardy transform is used in Kanemitsu et al. (2004) to deal with the Epstein zeta-function. The remark that Hardy (1908) seems to be the first who used the beta transform (in treating lattice sums) is to be interpreted to mean the Hardy transform as well as the beta transform leads to the Fourier–Bessel expansion (Theorem 3.1.1).

The method was also adopted by Katsurada and Matsumoto under the name of the Mellin-Barnes integrals in the context of applications of Atkinson's dissection, Katsurada and Matsumoto (1997), Katsurada and Matsumoto (2002). But this terminology usually refers to more general integrals as synonymously as the Fox H-function (Erdélyi et al. 1953, I, p. 39) and Paris and Kaminski (2001).

Results on the beta transform are culminated in Chakraborty et al. (2015) and is a basis of the Fourier–Bessel expansion (Kanemitsu and Tsukada 2014).

Following Liu et al. (2025), we shall reveal the secret power of the Hecke gamma transform as the method of separation of variables.

Proposition 3.1.1 *The Hecke transform, or the Hecke integral, is the Mellin transform pair of the K-Bessel function ($\nu = \alpha - \beta$)*

$$\int_0^\infty t^{\beta-\alpha} e^{-\frac{a}{t}-bt} \frac{\mathrm{d}t}{t} \tag{3.1.22}$$
$$= \left(\frac{b}{a}\right)^{\frac{\nu}{2}} 2K_\nu(2\sqrt{ab})$$
$$= \frac{1}{2\pi i}\int_{(c)} \Gamma(s+\alpha)\,\Gamma(s+\beta)\,a^{-s-\alpha}b^{-s-\beta}\,\mathrm{d}s.$$

Hardy's formula is a special case

$$\int_0^\infty t^\mu e^{-\frac{z^2}{t}-t} \frac{\mathrm{d}t}{t} = z^\mu 2K_\mu(2z) = 2\psi_{2\mu-1}(z) \tag{3.1.23}$$
$$= \frac{1}{2\pi i}\int_{(c)} \Gamma(s+\alpha)\,\Gamma(s+\alpha+\mu)\,z^{-2s-2\alpha}\,\mathrm{d}s.$$

Remark 3.1.1 We shall elucidate the reason why the beta transform, Lemma 2.5.1, has been so much used in treating shifted sequences, which we denote $1 + x$ symbolically.

The key lies in the merit of the gamma transform (2.1.3). It puts x resp. $1 + x$ in the denominator to the numerator in the exponential function. In the case of $1 + x$, the merit is more visible; it being additive and the exponential function decomposes. If we deal with a sum, then we often need to adopt dissection of the sum so as to separate the 1-part and $x \neq 0$-part. This is the origin of the dissections. In the beta transform, it can separate 1 and x in the inverse Mellin transform as can be seen:

$$\begin{aligned} &\Gamma(s) \int_0^\infty \xi^s e^{-\xi} e^{-x\xi} \frac{d\xi}{\xi} \\ &= \Gamma(s)(1+x)^{-s} \\ &= \frac{1}{2\pi i} \int_{(c)} \Gamma(z+s)\,\Gamma(-z)\, x^z dz = G_{1,1}^{1,1}\left(x^{-1} \,\middle|\, \begin{matrix} 1 \\ s \end{matrix}\right). \end{aligned} \tag{3.1.24}$$

In the light of the governing principle $G_{1,1}^{1,1} \leftrightarrow G_{0,2}^{2,0}$, we may say that in the beta transform, one needs to apply the functional equation to transform the integral into $G_{0,2}^{2,0}$. Cf. the proof of Theorem 2.5.1.

On the other hand, in the Hardy–Hecke transform, one starts from the gamma transform, the Mellin transform, and incorporates a modular relation to arrive at the Hecke integral—the final form. Cf. the proof of Theorem 3.7.1.

We summarize the contents of the above remark in the following theorem and describe it symbolically as "The Hardy and beta transforms are reverse processes."

Theorem 3.1.1 *The Hardy transform resp. the beta transform is based on the Mellin transform resp. the inverse Mellin transform, leading to the Fourier-Bessel expansion (2.12.42). More precisely, if one applies the beta transform to the perturbed Dirichlet series and applies the modular relation to the x-part, then one arrives at (2.12.42). On the other hand, if one applies the gamma transform to $\Gamma(s)\varphi(s)$ and applies a certain transformation formula to the x-part, then one arrives at the Hardy integral, leading to the K-Bessel function expression (Table 3.1).*

Here the first and the third columns are the transforms which are inverses to each other while the second column means that the generalized Bessel functions appear in the final expressions. Asai (1970) gives the Fourier–Bessel expansion for number fields and the Hilbert modular group to be given in Sect. 3.7. Cf. Table 3.4 for the Fourier–Whittaker expansion.

The Hardy transform is indeed the main source for deducing many of the formulas of Soni type Soni (1966), Oberhettinger and Soni (1972). For instance, when this procedure is applied to the Hecke–Dirichlet series $\pi^{-s} \times \sum \frac{r_2(n)}{n^s}$, one obtains a well-known formula of Dixon and Ferrar (1931) and when our Dirichlet series is $\zeta^2(s)$, Soni's formula (Soni 1966, Eq. (4)) is derived.

Table 3.1 Hardy and beta transforms and Bessel function expressions

Transf.	Hardy	K-Bessel expr.	Beta
Voronoĭ	–	$2K_0(2\sqrt{x})$	–
Hardy	(3.1.21)	–	–
Mellin	(3.1.3) with $\nu = 0$	–	–
Hecke	(3.1.4)	–	–
Siegel	(3.1.4)	–	–
Rademacher	Gen. of (3.1.4)	–	–
Koshlyakov	–	–	(2.5.18)
Maass	–	$K_{ir}(z)$, $K_\nu(z)$	–
Bellman 50	$W_{\frac{3}{2}}(x)$	$W_1(x) = 2K_0(2\sqrt{x})$	(2.5.18)
Bellman 56	$V_a(x, y)$	$W_1(x) = 2K_0(2\sqrt{x})$	(2.5.18)
Berndt III	(3.1.2)	Fourier-Bessel exp.	(2.5.18)
Berndt IV	–	Fourier-Bessel exp.	(2.5.18)
Asai70	(3.1.2) with $p = 1$	Fourier-Bessel exp.	–
Ka-matsu	–	–	(2.5.18)
KTTY	(3.1.21)	–	–

3.2 Ramified Functional Equation and Fourier–Whittaker Expansion: Statement of Results

Recall Hecke's revolution Hecke (1936), Hecke (1983) (cf. Weil's words in Preface) in the theory of automorphic forms (holomorphic) in pursuit of a generalization of Hamburger's theorem Hamburger (1921). Maass continued Hecke's work to introduce non-holomorphic forms.

Both Maass (1949) and Maass (1964) are devoted to a generalization of Hecke's work but their contents are rather different. Maass (1949) is devoted to the study of Eisenstein–Maass series (3.2.6) while Maass lecture notes Maass (1964) contains results on the theory of Hecke–Eisenstein series (3.4.1).

Maass theory has created a new research horizon and there are enormous amount of works related to it. However, it seems only recently shown by Wang NL et al. (2023) that Maass theory of real-analytic automorphic forms and the Epstein zeta-functions are intimately related and the latter may be accommodated in the former. The link is the Fourier–Whittaker expansion which is perceived as the Chowla–Selberg integral formula and has been accommodated in the Fourier–Bessel expansion for zeta-functions satisfying the Hecke-type functional equation. Another link is the Epstein-type Eisenstein series (3.6.2) which gives rise to the Fourier expansion for real-analytic Eisenstein series by multiplying by y^s. The underlying unifying principle is that *the ramified functional equation is not in RHB correspondence with the (Bochner) modular relation (Fourier series) but with the Fourier–Whittaker expansion.*

Our objective is to unify the theory of zeta-functions allied to Epstein zeta-function with positive definite quadratic forms and those associated to real-analytic automorphic forms (including holomorphic modular forms, etc.) *in the framework of the RHB correspondence between the ramified functional equation and the Fourier–Whittaker expansion, cf. Definition* 2.0.1 *and* Kanemitsu and Tsukada (2014).

For this we show that the following theorem, Theorem 3.2.1, governs both the theory of Epstein zeta-functions and that of zeta-functions associated with real-analytic Maass forms as symbolized by the extended G-function hierarchy (3.2.4). It is a rather special case of Kanemitsu and Tsukada (2014, pp. 115–119, 269, Theorem 4.3, Theorem 10.1). Hopefully, this will liquidate the situation about Maass forms as described in Kaneko (2021).

First we state the definition of the ramified functional equation in parallel to Hecke's one in Definition 2.12.1.

Definition 3.2.1 We state a special case of the setting on Kanemitsu and Tsukada (2014, pp. 115–119) with $H = I = 2$, i.e. that part of the Dirichlet series, for the statement of Theorem 3.2.1. Suppose $\varphi(s) = \sum_{k=1}^{\infty} \frac{\alpha_k}{\lambda_k^s}$ and $\varphi'(s) = \sum_{k=1}^{\infty} \frac{\alpha_k'}{\lambda_k'^s}$ resp. $\psi(s) = \sum_{k=1}^{\infty} \frac{\beta_k}{\mu_k^s}$ and $\psi'(s) = \sum_{k=1}^{\infty} \frac{\beta_k'}{\mu_k'^s}$ are absolutely convergent for $\sigma > \sigma_a^*$ resp. $\sigma < r - \sigma_a^*$, where r is a real number. We say that $\varphi, \varphi', \psi, \psi'$ satisfy the ramified functional Eq. (3.2.2) if there exists a meromorphic function $\chi(s)$ which is convex in the sense that $\lim_{|v|\to\infty} \chi(u + iv) = 0$ in any finite interval and has two expressions in (3.2.2).

Further suppose that all the poles of $\chi(s)$ lie in a compact set $\mathcal{S}$ with a boundary $\mathcal{C}$ contained in the strip $r - \sigma_a^* < \sigma < \sigma_a^*$. Let

$$\mathrm{P}(z) = \frac{1}{2\pi i} \int_{\mathcal{C}} \chi(w)\, z^{-w} \mathrm{d}w \tag{3.2.1}$$

be the residual function (the sum of all residues) and let

$$X(z) = \frac{1}{2\pi i} \int_{L_1} \chi(w)\, z^{-w} \mathrm{d}w$$

be the key-function, where L_1 is an indented Bromwich path such that all poles of $\Gamma(-w)\Gamma(1 - a - b - w)$ lie on the right resp. all poles of $\Gamma(s + w)\Gamma(b + w)$ and $\mathcal{S}$ lie to the left of it.

Theorem 3.2.1 *Let $\chi(s)$ be a meromorphic function satisfying the ramified functional equation (in the sense of Definition 3.2.1)*

$$\chi(s) = \begin{cases} G_{2,2}^{2,2}\left(2 \,\middle|\, \begin{matrix} 1+\mu-s, \frac{1}{2}+\varkappa \\ \mu, -\mu \end{matrix}\right) \sum_{k=1}^{\infty} \frac{\alpha_k}{\lambda_k^s} \\ -G_{2,2}^{2,2}\left(2 \,\middle|\, \begin{matrix} 1+\mu-s, \frac{1}{2}-\varkappa \\ \mu, -\mu \end{matrix}\right) \sum_{k=1}^{\infty} \frac{\alpha_k'}{\lambda_k'^{\,s}}, \\ G_{2,2}^{2,2}\left(2 \,\middle|\, \begin{matrix} 1+\mu-s, \frac{1}{2}+\varkappa \\ \mu, -\mu \end{matrix}\right) \sum_{k=1}^{\infty} \frac{\beta_k}{\mu_k^{r-s}} \\ -G_{2,2}^{2,2}\left(2 \,\middle|\, \begin{matrix} 1+\mu-s, \frac{1}{2}-\varkappa \\ \mu, -\mu \end{matrix}\right) \sum_{k=1}^{\infty} \frac{\beta_k'}{\mu_k'^{\,r-s}}, \end{cases} \tag{3.2.2}$$

where the first resp. the second row is valid for $\sigma > \sigma_a^*$ resp. $\sigma < r - \sigma_a^*$. Then the functional Eq. (3.2.2) and the modular relation are equivalent:

$$X(z) = \begin{cases} \sum_{k=1}^{\infty} \alpha_k G_{1,2}^{2,1}\left(2z\lambda_k \,\middle|\, \begin{matrix} \frac{1}{2}+\varkappa \\ \mu, -\mu \end{matrix}\right) e^{-z\lambda_k} \\ +\sum_{k=1}^{\infty} \alpha_k' G_{1,2}^{2,1}\left(2z\lambda_k' \,\middle|\, \begin{matrix} \frac{1}{2}-\varkappa \\ \mu, -\mu \end{matrix}\right) e^{-z\lambda_k'}, \\ z^{-r} \sum_{k=1}^{\infty} \beta_k G_{1,2}^{2,1}\left(\frac{2\mu_k}{z} \,\middle|\, \begin{matrix} \frac{1}{2}+\varkappa \\ \mu, -\mu \end{matrix}\right) e^{-\frac{\mu_k}{z}} \\ +z^{-r} \sum_{k=1}^{\infty} \beta_k' G_{1,2}^{2,1}\left(\frac{2\mu_k'}{z} \,\middle|\, \begin{matrix} \frac{1}{2}-\varkappa \\ \mu, -\mu \end{matrix}\right) e^{-\frac{\mu_k'}{z}} + \mathrm{P}(z). \end{cases} \tag{3.2.3}$$

Proof is given in Tsukada (2007), Kanemitsu and Tsukada (2014, pp. 217–220) of the most general case of ramified functional equations, which shows that the functional Eq. (3.2.2) implies the modular relation (3.2.3). Since this is an *unprocessed modular relation*, i.e. without the processing gamma factor, the proof of equivalence hinges on the Mellin transform pair in Lemma 3.3.1. RFES in Definition 2.1.1 is a prototype of the ramified functional equation.

The G-function hierarchy extended from Kuzumaki (2011) reads

$$G_{0,1}^{1,0}\left(z \,\middle|\, \begin{matrix} - \\ \mu \end{matrix}\right) = \sqrt{\frac{2}{\pi}} z^{\mu+\frac{1}{2}} K_{\pm\frac{1}{2}}(z) = z^{\mu} e^{-z} \tag{3.2.4}$$

$\searrow$

$$G_{0,2}^{2,0}\left(z^2 \,\middle|\, \begin{matrix} - \\ \frac{\mu}{2}, -\frac{\mu}{2} \end{matrix}\right) = 2K_{\mu}(2z) = \sqrt{\pi} e^{-2z} (4z)^{\mu} \Psi\left(\mu + \frac{1}{2},\ 2\mu+1;\ 4z\right)$$
$$= \sqrt{\frac{\pi}{z}} W_{0,\mu}(4z)$$

$$= 2\sqrt{\pi}e^{2z}G^{2,0}_{1,2}\left(4z\,\middle|\,\begin{matrix}\frac{1}{2}\\ \mu, \ -\mu\end{matrix}\right)$$

$$= \frac{2\sqrt{\pi}e^{-2z}}{\Gamma\left(\mu+\frac{1}{2}\right)\Gamma\left(-\mu+\frac{1}{2}\right)}G^{2,1}_{1,2}\left(4z\,\middle|\,\begin{matrix}\frac{1}{2}\\ \mu, \ -\mu\end{matrix}\right)$$

$\searrow$

$$W_{\varkappa,\mu}(2z) = z^{\mu+\frac{1}{2}}W\left(z;\mu+\frac{1}{2}+\varkappa,\mu+\frac{1}{2}-\varkappa\right)$$

$$= \frac{(2z)^{\frac{1}{2}}}{\Gamma\left(\frac{1}{2}+\mu-\varkappa\right)\Gamma\left(\frac{1}{2}-\mu+\varkappa\right)}G^{2,1}_{1,2}\left(2z\,\middle|\,\begin{matrix}\frac{1}{2}+\varkappa\\ \mu,-\mu\end{matrix}\right)e^{-z}$$

$\searrow$

The Main formula H,

where the arrow $\searrow$ indicates the uplift. The second equality may be replaced by

$$G^{2,0}_{0,2}\left(z^2\,\middle|\,\begin{matrix}-\\ \mu,-\mu\end{matrix}\right) = 2K_{2\mu}(2z).$$

For notation we refer to Sect. 3.1, cf. also Sect. 2.13. We make a convention that the various factors of the G-function in (3.2.4) are included in the coefficients α_k, etc., e.g. the penultimate one is multiplied by $\frac{\sqrt{2}}{\Gamma(\frac{1}{2}-\varkappa+\mu)\Gamma(\frac{1}{2}-\varkappa-\mu)}$, etc. The residual function $\mathrm{P}(z)$, (3.2.1), thus may differ at each occurrence and is to be found accordingly.

It is indicated in Kanemitsu and Tsukada (2014, pp. 269–270) that Maass (1964, Theorem 35) can be interpreted as a ramified modular relation. We prove that under Lemma 3.3.1 (to be proved in Sect. 3.1), the following corollary to Theorem 3.2.1 substantiates this statement.

Corollary 3.2.1 *The modular relation (Fourier–Whittaker expansion)*

$$X(z) = \begin{cases} \sum_{k=1}^{\infty}\alpha_k W(z\lambda_k;a,b) + \sum_{k=1}^{\infty}\alpha'_k W\left(z\lambda'_k;b,a\right), \\ z^{-r}\sum_{k=1}^{\infty}\beta_k W\left(\frac{\mu_k}{z};a,b\right) + z^{-r}\sum_{k=1}^{\infty}\beta'_k W\left(\frac{\mu'_k}{z};b,a\right) + \mathrm{P}(z) \end{cases} \tag{3.2.5}$$

and the ramified functional equation

$$\chi(s) = \begin{cases} \Gamma(s;a,b)\sum_{k=1}^{\infty}\frac{\alpha_k}{\lambda_k^s} + \Gamma(s;b,a)\sum_{k=1}^{\infty}\frac{\alpha'_k}{{\lambda'_k}^s}, \\ \Gamma(r-s;a,b)\sum_{k=1}^{\infty}\frac{\beta_k}{\mu_k^{r-s}} + \Gamma(r-s;b,a)\sum_{k=1}^{\infty}\frac{\beta'_k}{{\mu'_k}^{r-s}} \end{cases}$$

are equivalent. Further they are in RBH correspondence: (non-holomorphic) modular form ↔ zeta-function. Here

$$a = \frac{1}{2} + \mu + \varkappa, \quad b = \frac{1}{2} + \mu - \varkappa.$$

Proof To show the last statement we are to show that (3.2.5) is the Fourier expansion of the relevant non-holomorphic modular form. This is done in Maass (1964) and the statement will be given in Lemma 3.4.1 with a simplified proof in Sect. 3.5. □

Exercise 3.2.1 *Using* (3.3.5) *and* (3.3.6), *express the modified Whittaker function and its Mellin transform in terms of G-functions.*

Solution. For example, the Dirichlet series part is

$$\chi(s) = \begin{cases} G_{2,2}^{2,2}\left(\frac{1}{2}\,\middle|\,\begin{matrix} 1, a+b \\ s, b \end{matrix}\right)\varphi(s) + G_{2,2}^{2,2}\left(\frac{1}{2}\,\middle|\,\begin{matrix} 1, a+b \\ s, a \end{matrix}\right)\varphi'(s), \\ G_{2,2}^{2,2}\left(\frac{1}{2}\,\middle|\,\begin{matrix} 1, a+b \\ r-s, b \end{matrix}\right)\psi(r-s) + G_{2,2}^{2,2}\left(\frac{1}{2}\,\middle|\,\begin{matrix} 1, a+b \\ r-s, a \end{matrix}\right)\psi'(r-s). \end{cases}$$

□

We note that the index $\varkappa$ comes from the multiplier system (3.4.15). Section 3.3 is devoted to elaboration of this corollary.

Corollary 3.2.2 *In the case* $\varkappa = 0$, $\varphi(s) = \varphi'(s)$ *and* $\psi(s) = \psi'(s)$, *Theorem 3.2.1 reads The unramified functional equation*

$$\chi(s) = \begin{cases} G_{2,2}^{2,2}\left(2\,\middle|\,\begin{matrix} 1+\mu-s, \frac{1}{2} \\ \mu, -\mu \end{matrix}\right)\sum_{k=1}^{\infty}\frac{\alpha_k}{\lambda_k^s} \\ G_{2,2}^{2,2}\left(2\,\middle|\,\begin{matrix} 1+\mu-s, \frac{1}{2}+\varkappa \\ \mu, -\mu \end{matrix}\right)\sum_{k=1}^{\infty}\frac{\beta_k}{\mu_k^{r-s}} \end{cases}$$

where the first resp. the second row is valid for $\sigma > \sigma_a^*$ *resp.* $\sigma < r - \sigma_a^*$, *and the modular relations are equivalent:*

$$X(z) = \begin{cases} \sum_{k=1}^{\infty}\alpha_k G_{1,2}^{2,1}\left(2z\lambda_k\,\middle|\,\begin{matrix} \frac{1}{2} \\ \mu, -\mu \end{matrix}\right)e^{-z\lambda_k} \\ z^{-r}\sum_{k=1}^{\infty}\beta_k G_{1,2}^{2,1}\left(\frac{2\mu_k}{z}\,\middle|\,\begin{matrix} \frac{1}{2} \\ \mu, -\mu \end{matrix}\right)e^{-\frac{\mu_k}{z}} + \mathrm{P}(z). \end{cases}$$

Section 3.6 is devoted to concretization of this corollary.

Let Γ denote the Fuchsian group of the first kind, typically $\Gamma = \Gamma(1) = \mathrm{PSL}_2(\mathbb{Z})$ acting on $\mathcal{H}$ through linear fractional transformations (2.12.9). In most of the literature on spectral theory of automorphic forms, the **Eisenstein–Maass series** $E(s,\tau)$ is used.

Γ being assumed to have only one cusp at ∞, $E(s,\tau)$ is defined for $\sigma > 1$ by Motohashi (1997), Venkov (1982), (1990, p. 20), etc. as follows:

$$E(s,\tau) = \sum_{\gamma\in\Gamma_\infty\backslash\Gamma} y^s(\gamma\tau), \quad y(\tau) = \mathrm{Im}\,\tau, \tag{3.2.6}$$

where Γ_∞ is the stabilizer of ∞. $E(s,\tau)$ was first introduced by Maass (1949, p. 162, (106)) as the Eisenstein series of Stufe N (or of level N) for $\Gamma = \Gamma(1)$ by (3.2.9). For simplicity, we mostly assume that the Fuchsian group is the full modular group $\Gamma(1)$, (2.12.8), and denote it by Γ in what follows.

It follows that

$$E(s,\tau) = y^s \sum_{\substack{c,d=-\infty \\ (c,d)=1}}^{\infty}{}' |c\tau + d|^{-2s} = \frac{y^s}{\zeta(2s)} \zeta_{\mathbb{Z}^2}(s,\tau). \tag{3.2.7}$$

The first equality is due to the following. If $\gamma = \begin{pmatrix} a & b \\ c & d \end{pmatrix} \in \Gamma$, $\det\gamma = ad - bc = 1$, then the y-part is given by (2.12.12).

Note that the condition $(c,d) = 1$ is involved. The second equality follows from

Lemma 3.2.1 (Relative primality principle) *Suppose $h(m,n) \neq 0$ save for the case $m = n = 0$ that it is a homogeneous function $\ell h(m,n) = h(\ell m, \ell n)$ and that the series*

$$H(s) = \sum_{m,n=-\infty}^{\infty}{}' h(m,n)^{-s}$$

is absolutely convergent for $\sigma > 1$. Then

$$\frac{1}{\zeta(s)} \sum_{m,n=-\infty}^{\infty}{}' h(m,n)^{-s} = \sum_{\substack{c,d=-\infty \\ (c,d)=1}}^{\infty}{}' h(c,d)^{-s}$$

Proof Let $\mu(n)$ be the Möbius function defined by $\frac{1}{\zeta(s)} = \sum_{n=1}^\infty \frac{\mu(n)}{n^s}$, $\sigma > 1$. The factor $\frac{1}{\zeta(2s)} = \sum_{\ell=1}^\infty \frac{\mu(\ell)}{\ell^{2s}}$ has the *effect of restricting the summation variables m, n to those which are relatively prime*. This is one form of the *relative primality principle* (Li et al. 2017, p. 75).

Thus

$$\frac{1}{\zeta(s)}H(s) = \sum_{\ell=1,m,n=-\infty}^{\infty}{}' \mu(\ell)\,(h(\ell m, \ell n))^{-s} \tag{3.2.8}$$
$$= \sum_{m,n=-\infty}^{\infty}{}' \sum_{\ell|(m,n)} \mu(\ell) h(m,n)^{-s}$$
$$= \sum_{\substack{c,d=-\infty \\ (c,d)=1}}^{\infty}{}' h(c,d)^{-s},$$

where the last equality follows from the Möbius inversion formula to the effect that $\sum_{d|n}\mu(d) = 1$ $(resp.\ \sum_{d|n}\mu(d) = 0)$ if $n = 1$ $(resp.\ n \neq 1)$. □

For a positive integer N and integers μ, ν, Maass (1949, p. 162, (106)) introduces the **Eisenstein series of Stufe** N (in slightly different notation)

$$E(s, \tau; (\mu, \nu), N) = \sum_{\substack{m\equiv\mu \bmod N \\ n\equiv\nu \bmod N}}{}' y^s |m\tau + n|^{-2s}. \tag{3.2.9}$$

Equation (3.2.7) is a special case of this with $N = 1$ and with the factor $\frac{1}{\zeta(2s)}$. The Fourier expansion is given as Maass (1949, p. 163, (112)) (which corresponds to the q-expansion). This is a basis of analytic continuation and the functional equation as stated in Chakraborty (2010), Liu (2026), etc. Cf. (3.4.2) for the kth Hecke–Eisenstein series.

In the same year, Part I of Chowla and Selberg (1949) appeared and its complete form Selberg and Chowla (1967) was published in 1967 both of which contain the Chowla–Selberg integral formula. But Maass theory and the Chowla–Selberg integral formula have been thought of as independent of each other.

We note, however, that (3.2.7) implies that *the non-holomorphic Eisenstein–Maass series is the* y^s *times the (EES)* $\frac{1}{\zeta(2s)}\zeta_{\mathbb{Z}^2}(s, \tau)$.

One year after publication of Maass (1949), Bellman published Bellman (1950) and derived many illustrative non-holomorphic automorphic functions by using the Hardy transform Wang et al. (2022) in which it is revealed that *Maass method of using the DE is one of the ways leading to the Fourier–Whittaker expansion.*

Cf. Siegel (1943), Siegel (1961), Terras (1985), Kanemitsu and Tsukada (2014), etc.

In Sect. 3.6, the rich theory of Epstein zeta-functions is naturally translated to that of the non-holomorphic Eisenstein series (as has been done in Wang et al. (2022)) and vice versa. For Epstein zeta-functions with characters, we refer to Kitaoka (1971) and references therein.

3.3 Maass Forms and Rudiments

As mentioned above, Maass lecture notes (Maass 1964, pp. 160–217) contain some important results on the non-holomorphic Eisenstein series (3.4.1). One missing point here is that *in the real-analytic case* $a = b = s$, $G(\tau, \bar{\tau}; s, s)$ *is the Epstein zeta-function analytic in* s.

Some of the main results are stated in Sect. 3.2: Maass (1964, p. 228, Theorem 35) as Corollary 3.2.5 and Maass (1964, p. 240) as Corollary 3.2.2. One of our aims in this section is to prove Lemma 3.3.1 (which implies Corollary 3.2.5). Before the proof we assemble basic facts about Maass theory.

Let $\{\Gamma, a, b, 1\}$ denote the space of functions $f(\tau, \bar{\tau})$ real-analytic on $\mathcal{H}$ satisfying a DE, automorphy and growth condition as in Maass (1964, p. 186). The following is a corollary to Maass (1964, p. 208, Theorem 33).

Corollary 3.3.1 *(Maass 1964, p. 215, Theorem 34) Let* Γ *be the full modular group* $\Gamma(1)$, *(2.12.8). Then the space* $\{\Gamma, a, a, 1\}$ *is of rank 1 over* $\mathbb{C}$ *for* $a \geq 0$ *and it is generated by the completed Eisenstein series, i.e.* (3.4.1) *multiplied by the suitable gamma factor.*

Research on (3.4.1) has been done by several authors including Knopp (1989), Pasles and Pribitkin (2001).

The relation does not seem to be studied between two non-holomorphic Eisenstein series (3.4.1) and (3.6.2). *Our aim is to clarify the relation and show that* (3.4.1) *entails the case of the Epstein zeta-function which gives rise to the Fourier expansion for the Eisenstein–Maass series.*

Maass theory depends essentially on the Whittaker function as a generalization of the *K-Bessel* function (3.1.1). Here we employ the G-function expressions.

In addition to (3.1.13), we may also use (Erdélyi et al. 1953, I, p. 216, (6)), which reads

$$G_{1,2}^{2,0}\left(x \,\middle|\, \begin{matrix} a \\ b,\ c \end{matrix}\right) = x^{\frac{1}{2}(b+c-1)}\, e^{-\frac{1}{2}x}\, W_{\kappa,\mu}(x),$$

where $\kappa = \frac{1}{2}(b + c + 1) - a,\ \mu = \frac{1}{2}(b - c)$.

Maass introduced the gamma factor as the Mellin transform of a Whittaker function (Maass 1964, p. 218, (1))

$$\Gamma(s; a, b) = \int_0^{\infty} x^s W(x; a, b)\, \frac{\mathrm{d}x}{x}, \tag{3.3.1}$$

where W is the modified Whittaker function (Maass 1964, p. 182, (1))

$$W(\varepsilon y; a, b) = y^{-\frac{1}{2}(a+b)} W_{\varkappa,\mu}(2y), \quad \varkappa = \varepsilon\frac{1}{2}(a - b), \mu = \frac{1}{2}(a + b - 1), \tag{3.3.2}$$

where $y > 0$ and $\varepsilon^2 = 1$.

In Chap. 4, we use the same symbol (4.3) in a different meaning, but there will be no confusion.

Maass notes (Maass 1964, p. 213) that

$$W(-x; a, b) = W(x; b, a). \tag{3.3.3}$$

This is a consequence of the formula (Erdélyi et al. 1953, I, p. 265, (8))

$$W_{\varkappa,\mu}(x) = W_{\varkappa,-\mu}(x), \tag{3.3.4}$$

which in turn is equivalent to the transformation formula (Erdélyi et al. 1953, I, p. 257, (6))

$$\Psi(a, c; x) = x^{1-c}\Psi(a - c + 1, 2 - c; x).$$

This is a counterpart of Kummer's transformation (Erdélyi et al. 1953, I, p. 253, (7)) and such is (3.3.4). The symmetry property (3.1.5) for the K-Bessel function is a special case of (3.3.4).

We note that (3.3.3) *necessitates the division of the Fourier series in Lemma* 3.4.1 *into two parts, positive and negative, yielding the ramified modular relation entailing the ramified functional equation in Corollary* 3.2.1.

In Wang et al. (2022, Lemma 3) the G-function expression for $\Gamma(s; a, b)$ resp. $W(z; a, b)$ is

$$\Gamma(s; a, b) = \frac{2^{\frac{1}{2}(a+b)}}{\Gamma(b)\Gamma(1-a)} G^{2,2}_{2,2}\left(2 \,\middle|\, \begin{matrix} 1-s, 1-b \\ 0, 1-a-b \end{matrix}\right) \tag{3.3.5}$$

resp.

$$W(z; a, b) = \frac{2^{\frac{1}{2}(a+b)}}{\Gamma(b)\Gamma(1-a)} G^{2,1}_{1,2}\left(2z \,\middle|\, \begin{matrix} 1-b \\ 0, 1-a-b \end{matrix}\right) e^{-z}. \tag{3.3.6}$$

These are not in the form which can be accommodated in the hierarchy (3.2.4). It is (3.3.8) that completes the hierarchy (Prudnikov et al. 1986, p. 714, Formula 3).

Lemma 3.3.1 *In view of the expressions*

$$\Gamma\left(s; \mu + \frac{1}{2} + \varkappa, \mu + \frac{1}{2} - \varkappa\right) \tag{3.3.7}$$
$$= \frac{\sqrt{2}}{\Gamma\left(\frac{1}{2} - \varkappa + \mu\right)\Gamma\left(\frac{1}{2} - \varkappa - \mu\right)} G^{2,2}_{2,2}\left(2 \,\middle|\, \begin{matrix} 1+\mu-s, \frac{1}{2}+\varkappa \\ \mu, -\mu \end{matrix}\right),$$

and

$$W\left(z; \mu + \frac{1}{2} + \varkappa, \mu + \frac{1}{2} - \varkappa\right) \tag{3.3.8}$$
$$= \frac{\sqrt{2}z^{-\mu}}{\Gamma\left(\frac{1}{2} - \varkappa + \mu\right)\Gamma\left(\frac{1}{2} - \varkappa - \mu\right)} G_{1,2}^{2,1}\left(2z \,\middle|\, \begin{matrix} \frac{1}{2} + \varkappa \\ \mu, -\mu \end{matrix}\right) e^{-z}.$$

Equation (3.3.1) *is a special case of the heaven-sent Mellin transform pair* $e^{-x}G_{p,q}^{m,n}(ax|\cdot) \leftrightarrow G_{p+1,q}^{m,n+1}(a|\cdot)$ (Erdélyi et al. 1954, I, p. 338, (16)):

$$\int_0^\infty x^{z-1}e^{-x}G_{p,q}^{m,n}\left(ax \,\middle|\, \begin{matrix} a_1, \cdots, a_p \\ b_1, \cdots, b_q \end{matrix}\right) \mathrm{d}x$$
$$= G_{p+1,q}^{m,n+1}\left(a \,\middle|\, \begin{matrix} 1-s, a_1, \cdots, a_p \\ b_1, \cdots, b_q \end{matrix}\right).$$

Proof It suffices to prove the expressions (3.3.7) and (3.3.8). Maass derived Maass (1964, p. 222) that

$$\Gamma(s; a, b) = 2^{\frac{a-b}{2}} \frac{\Gamma(s)\Gamma(s+1-a-b)}{\Gamma(s+1-a)} F\left(b, 1-a; 1+s-a; \frac{1}{2}\right), \tag{3.3.9}$$

where $F = {}_2F_1$ is the **Gauss hypergeometric function** (Erdélyi et al. 1953, I, p. 182). This and

$$G_{2,2}^{2,2}\left(\frac{1}{2} \,\middle|\, \begin{matrix} a, b \\ c, d \end{matrix}\right) = \Gamma(1-a+b)\Gamma(1-b+d)\Gamma(1-a+d)$$
$$\times z^c F\left(1-a+c, 1-b+c; 2-a-b+c+d; \frac{1}{2}\right)$$

(Wolfram 1998-2021, p. 82) imply (3.3.7).

(3.3.9) follows from (Erdélyi et al. 1954, p. 337)

$$\int_0^\infty x^s e^{-\alpha x} W_{\varkappa,\mu}(\beta x) \frac{\mathrm{d}x}{x} = \frac{\beta^{\mu+\frac{1}{2}}}{\left(\alpha + \frac{1}{2}\beta\right)^{\mu+s+\frac{1}{2}}} \frac{\Gamma\left(\mu+s+\frac{1}{2}\right)\Gamma\left(-\mu+s+\frac{1}{2}\right)}{\Gamma\left(s-\varkappa+1\right)}$$
$$\times F\left(\mu+s+\frac{1}{2}, \mu-\varkappa+\frac{1}{2}; s-\varkappa+1; \frac{2\alpha-\beta}{2\alpha+\beta}\right),$$

and (Erdélyi et al. 1953, p. 64, (22))

$$F\left(a, b; c; z\right) = (1-z)^{-b} F\left(c-a, b; c; \frac{z}{z-1}\right).$$

□

The following lemma reduces the ramified to unramified functional equation in Corollary 3.2.2. We appeal to the special case of (3.3.2) with $a = b$ which appears in (3.2.4).

Lemma 3.3.2 (Maass 1964, p. 220, (6)) *The special case of* (3.3.2) *with* $a = b$ *reads*

$$W(x; a, a) = x^{-a} W_{0,a-\frac{1}{2}}(2x) = \sqrt{\frac{2}{\pi}} x^{\frac{1}{2}-a} K_{a-\frac{1}{2}}(x). \tag{3.3.10}$$

Then we have (Maass 1964, p. 220, (6))

$$\Gamma(s; a, a) = \frac{2^{s-a-1}}{\sqrt{\pi}} \Gamma\left(\frac{s}{2}\right) \Gamma\left(\frac{s}{2} - a + \frac{1}{2}\right). \tag{3.3.11}$$

Proof The Mellin transform of (3.3.10) and (3.1.14) leads to (3.3.11). □

Berndt (1969, p. 324) contains the assertion

Proposition 3.3.1 *For* $\mathrm{Re}\, z > 0$, *the functional equation*

$$\Gamma\left(s + \frac{\nu}{2}\right) \Gamma\left(s - \frac{\nu}{2}\right) \varphi(s) = \Gamma\left(r - s + \frac{\nu}{2}\right) \Gamma\left(r - s - \frac{\nu}{2}\right) \psi(r - s)$$

and the modular relation

$$2 \sum_{k=1}^{\infty} \alpha_k K_\nu(2\sqrt{\lambda_k z}) = 2z^{-r} \sum_{k=1}^{\infty} \beta_k K_\nu\left(2\sqrt{\frac{\mu_k}{z}}\right) + \mathrm{P}(z)$$

are equivalent.

Corollary 3.3.2 *Corollary 3.2.1 with* $b = a$ *boils down to an analogue of Proposition 3.3.1: The functional equation*

$$\Gamma(2s + \nu; a, a)\varphi(2s + \nu) = \Gamma(r - 2s - \nu; a, a)\psi(r - 2s - \nu)$$

and the modular relation

$$\sum_{k=1}^{\infty} \alpha_k K_\nu(2\sqrt{\lambda_k z}) = z^{-r} \sum_{k=1}^{\infty} \beta_k K_\nu\left(2\sqrt{\frac{\mu_k}{z}}\right) + \mathrm{P}(z)$$

are equivalent.

Equations (3.1.13), (3.1.12), (Erdélyi et al. 1953, I, p. 216, (4)–(6)), and (3.3.8) lead to the G-function hierarchy (3.2.4).

The following lemma is for Sect. 2.12.

Lemma 3.3.3 (Maass 1964, p. 220, (6)) *The special case of* (3.3.2) *with* $b = 0$ *reads*

$$W(x; a, 0) = 2^{\frac{a}{2}} \Gamma(z).$$

Proof Proof is given on Maass (1964, p. 220). This also follows as a special case of (3.3.9) with $b = 0$. This does not seem to follow from other expressions above, e.g. from (3.1.11). □

3.4 Examples of Real-Analytic Automorphic Forms

We assemble some concrete examples scattered around in literature with comments.

On Maass (1964, p. 169), the **Hecke–Eisenstein series** (HES)

$$G(\tau, \bar{\tau}; a, b) = \sum_{(m,n)\neq(0,0)} \frac{1}{(m\tau + n)^a (m\bar{\tau} + n)^b}, \quad \mathrm{Re}(a+b) > 2 \tag{3.4.1}$$

is defined for $\tau \in \mathcal{H}$, where $k := a - b$ is an even integer (in Maass, k is replaced by $2k$). The special case with $a = 2$ was studied by Hecke (1927) and explicated in Schoenberg (1974, pp. 63–68). It is a prototype of a non-holomorphic modular form for $\{\Gamma(1), a, b, 1\}$. Cf. Example 3.4.1. A slightly more general case than (3.4.1) was developed by Miyake (1989, pp. 268–293): For a positive integer N and integers μ, ν, Hecke–Eisenstein series with congruence conditions is defined Miyake (1989, p. 289, (7.2.52)) by

$$E_k(s, \tau; \mu, \nu; N) = \sum_{\substack{m\equiv\mu \bmod N \\ n\equiv\nu \bmod N}} (m\tau + n)^{-k} |m\tau + n|^{-2s}. \tag{3.4.2}$$

Hence

$$E_k(s, \tau; 1, 1; 1) = G(\tau, \bar{\tau}; s + k, s)$$

and both are non-holomorphic modular forms for $\{\Gamma(1), s + k, s, 1\}$. The result is stated for the Eisenstein series with characters (Miyake 1989, p. 274, (7.2.1)) (where it is referred to as the Eisenstein series with parameter s)

$$E_k(s, \tau; \chi, \psi) = \sum_{m,n=-\infty}^{\infty}{}' \chi(m)\psi(n)(m\tau + n)^{-k} |m\tau + n|^{-2s} \tag{3.4.3}$$

for Dirichlet characters χ, ψ mod N. The main result is Miyake (1989, p. 284, Theorem 7.2.9) giving rise to the Fourier expansion of $E_k(s, \tau; \chi, \psi)$ in terms of confluent hypergeometric functions. Cf. Example 3.4.1. There is no mention of Hecke (1927) nor Maass (1964), although they are very closely connected.

A generalization of (3.2.6), the kth Eisenstein–Maass series (3.4.4), was studied by Noda (2007) and Noda (2009), the latter of which is almost verbatim to the former save for the use of the Mellin–Barnes integral for the confluent hypergeometric function, which appears in the Fourier coefficients.

For non-negative even integer k, the kth Eisenstein–Maass series is defined by

$$E_k(s, \tau) = \sum_{\gamma\in\Gamma_\infty\backslash\Gamma} (c\tau + d)^{-k} y^s(\gamma\tau), \tag{3.4.4}$$

so that

$$E_0(s,\tau) = E(s,\tau).$$

We also have

$$E_k(s,\tau) = \frac{y^s}{\zeta(2s+k)} E_k(s,\tau,1,1,1) \tag{3.4.5}$$

by Lemma 3.2.1 as in (3.2.7). Noda's main results Noda (2007) depend on the expression of Fourier coefficients in terms of the confluent hypergeometric functions, which in turn depends on Miyake's results through (3.4.5).

Similar to (3.4.5), $E_k(s,\tau)$ is also the product of (3.4.1) by $\frac{y^s}{\zeta(2s+k)}$:

$$E_k(s,\tau) = \frac{y^s}{\zeta(2s+k)} G(\tau,\bar{\tau},s+k,s). \tag{3.4.6}$$

We refer to $E_k(s,\tau)$ as the *Hecke–Maass series* or the kth Eisenstein–Maass series.

It turns out that both Eisenstein series (3.2.6) and (3.4.3) have Fourier expansions in terms of the x-part on which their functional equations are based. Maass' results (Maass 1964, p. 212) (Whittaker function (3.3.2)) and Miyake's results (Miyake 1989, Sect. 7.2) (confluent hypergeometric function (3.1.10)) are directly translated into the case of Hecke–Maass series. Of course, these two functions W and Ψ are nearly the same, cf. (3.1.11).

Research on the special case (Maass 1964, p. 209)

$$f(\tau,\bar{\tau};a,b) = \sum_{n--\infty}^{\infty} \frac{1}{(\tau+n)^a(\bar{\tau}+n)^b}, \quad \mathrm{Re}(a+b) > 2 \tag{3.4.7}$$

of (3.4.1) or rather its generalization

$$f(\tau,\bar{\tau};a,b,\mu) = \sum_{n--\infty}^{\infty} \frac{e^{-2\pi i\mu n}}{(\tau+n)^a(\bar{\tau}+n)^b} \tag{3.4.8}$$

has been done by several authors including Knopp (1989), Pasles (2000a), Pasles and Pribitkin (2001), Pribitkin (1999), Pribitkin (2000). It is stated Pasles and Pribitkin (2001) that (3.4.8) first appeared in Knopp (1989) as a result of John Hawkins. Pasles and Pribitkin (2001) develops more general summation formulas with one more factor in the denominator, containing information on other works.

In Pasles (1999), Pasles (2000b), RHB correspondence for nonanalytic automorphic integrals on the Hecke group is studied.

Exercise 3.4.1 *Confirm that* (3.4.1) *resp.* (3.4.7) *is a generalization of EES (1.1.7) resp. the lattice zeta-function (2.1.70).*

Solution. Indeed,

$$G(\tau, \bar{\tau}; a, a) = \zeta_{\mathbb{Z}^2}(a, \tau), \quad f(\tau, \bar{\tau}; a, a) = \phi_1(2a, \tau).$$

□

We will establish the Fourier–Whittaker expansion for the Hecke–Maass functions in Lemma 3.4.1, thus identifying the index $\varkappa$ and rendering the RHB correspondence visible with the ramified functional equation in Corollary 3.2.1. In the process, we may slightly simplify Maass' treatment to express the Fourier coefficient in terms of the Whittaker function. We may deduce results on $E_k(s, \tau)$ from Maass' results (Example 3.4.1) on (3.4.1) and conversely. For the Eisenstein–Maass series and the Epstein-type Eisenstein series, we state Fourier–Bessel expansions for several related zeta-functions, by accommodating them in the Fourier–Whittaker expansion by Example 3.4.1.

Let Γ be the horocyclic group (Maass 1964, p. 295). As on Maass (1964, p. 186), let $\{\Gamma, a, b, v\}$ denote the space of functions $f(\tau, \bar{\tau})$ real-analytic on $\mathcal{H}$ satisfying automorphy

$$f(\tau, \bar{\tau})|_{\gamma} = v(\gamma) f(\gamma\tau, \gamma\bar{\tau}), \gamma \in \Gamma, \tag{3.4.9}$$

v being a multiplier system for Γ of dimension $-r$, the DE (with the operator in (3.4.12))

$$\Omega_{ab} f(\tau, \bar{\tau}) = 0, \tag{3.4.10}$$

and the growth condition

$$f(\tau, \bar{\tau})|_{\alpha^{-1}} = O(y^K), \quad K > 0, \quad y \to \infty, \tag{3.4.11}$$

uniformly in x provided that $\alpha^{-1}\infty$ is a parabolic cusp of Γ.

In the above

$$\Omega_{ab} = \Delta + i(a-b)y\frac{\partial}{\partial x} - (a+b)y\frac{\partial}{\partial y} = \Delta + iry\frac{\partial}{\partial x} - qy\frac{\partial}{\partial y}, \tag{3.4.12}$$

say, where

$$r = a - b, \quad q = a + b \tag{3.4.13}$$

and where

$$\Delta = -y^2\left(\frac{\partial^2}{\partial x^2} + \frac{\partial^2}{\partial y^2}\right) \tag{3.4.14}$$

is the **non-Euclidean Laplace operator** (or hyperbolic Laplacian). As on (Maass 1964, p. 179) let $\{a, b\}$ denote the space of functions $f(\tau, \bar{\tau})$ real-analytic on $\mathcal{H}$ satisfying the DE (3.4.10). Therefore, the space $\{\Gamma, a, b, v\}$ consists of functions $f(\tau, \bar{\tau}) \in \{a, b\}$ satisfying (3.4.9) and (3.4.11). The Hecke–Eisenstein series (3.4.1) belongs to $\{\Gamma, a, b, 1\}$ and satisfies the invariance property $G(\tau, \bar{\tau}, a, b)|_{\gamma_{ab}} = G(\tau, \bar{\tau}, a, b)$, where Maass (1964, p. 169)

$$G(\tau,\bar\tau,a,b)|_{\gamma_{ab}} = (c\tau+d)^{-a}(c\bar\tau+d)^{-b}G(\tau,\bar\tau,a,b), \quad \gamma=\begin{pmatrix} * & * \\ c & d\end{pmatrix}\in\Gamma.$$

Maass' main result (Maass 1964, pp. 228–229, Theorem 35) is stated for the **Hecke group** $\mathfrak{G}(\lambda)$, (Hecke 1936, p. 671), (Maass 1964, p. 226), generated by

$$T^\lambda = \begin{pmatrix} 1 & \lambda \\ 0 & 1\end{pmatrix}, \quad S=\begin{pmatrix} 0 & -1 \\ 1 & 0\end{pmatrix},$$

where $\lambda>0$ but only for $\lambda\ge 2$ or $\lambda = 2\cos\frac{\pi}{\ell}, \ell=3,5,\cdots$, $\mathfrak{G}(\lambda)$ is a discontinuous group (including the full modular group (2.12.8)).

Then (3.4.9) reads for generators of $\mathfrak{G}(\lambda)$

$$\begin{aligned} f(\tau+\lambda,\bar\tau+\lambda) &= e^{2\pi i\varkappa} f(\tau,\bar\tau), \\ f\left(-\frac{1}{\tau},-\frac{1}{\bar\tau}\right) &= \gamma(-i\tau)^a(-i\bar\tau)^b f(\tau,\bar\tau), \end{aligned} \tag{3.4.15}$$

where $\lambda>0, 0\le\varkappa<1, \gamma=\pm1$ and $\lambda=1, \varkappa=0, a=b=0$ does not occur and (3.4.11) reads

$$f(\tau,\bar\tau)=O(y^{K_1}), \quad y\to\infty, \quad f(\tau,\bar\tau)=O(y^{K_2}), \quad y\to+0, \quad K_j>0, \tag{3.4.16}$$

uniformly in x.

Definition 3.4.1 We refer to the function which satisfies automorphy (3.4.15) and is a solution of (3.4.10) with growth condition (3.4.16) as a **Hecke–Maass function**.

Lemma 3.4.1 *The Hecke–Maass function* $f(\tau,\bar\tau)\in[\mathfrak{G}(\lambda),a,b,v]$, v *being described by* (3.4.15), *admits the Fourier–Whittaker expansion*

$$f(\tau,\bar\tau)=a_0u(y,q)+b_0+\sum_{n+\varkappa\ne0} a_{n+\varkappa}W\left(\frac{2\pi(n+\varkappa)}{\lambda}y;a,b\right)e^{\frac{2\pi i(n+\varkappa)}{\lambda}x}, \tag{3.4.17}$$

where W is a modified Whittaker function (3.3.2) *and so* (3.4.17) *is an expansion in the Whittaker function.* $\varepsilon^2=1$, *and (with q as in* (3.4.13)*)*

$$u(y,q)=\frac{y^{1-q}-1}{1-q}. \tag{3.4.18}$$

Proof Proof is given on Maass (1964, p. 187) for a horocyclic group, which can lead to the proof of the expansion on Maass (1964, p. 230) for a Hecke group $\mathfrak{G}(\lambda)$. The first formula on Maass (1964, p. 187) for a Hecke–Maass function $f(\tau,\bar\tau)$ reads, with N replaced by λ

$$f(\tau, \bar{\tau}) = \sum_{n=-\infty}^{\infty} \alpha_{n+\varkappa}(y) e^{\frac{2\pi i(n+\varkappa)}{\lambda} x},$$

where

$$\alpha_{n+\varkappa}(y) = \frac{1}{\lambda} \int_0^{\lambda} f(\tau, \bar{\tau}) e^{-\frac{2\pi i(n+\varkappa)}{\lambda} x} \, \mathrm{d}x. \tag{3.4.19}$$

(3.5.10) with $p = 0, \nu' = 0$ reads

$$\begin{aligned} \Omega_{ab}(e^{i\varepsilon x} g(y)) &= -e^{i\varepsilon x} \left(y^2 \frac{\mathrm{d}^2}{\mathrm{d}y^2} + qy \frac{\mathrm{d}}{\mathrm{d}y} - \varepsilon^2 y^2 + \varepsilon r y \right) g(y) \\ &= e^{i\varepsilon} \left(\Omega_{ab} + \varepsilon^2 y^2 - \varepsilon r y \right) g(y). \end{aligned} \tag{3.4.20}$$

Applying (3.4.20) on (3.4.19) for $n + \varkappa \neq 0$, we see that $\alpha_{n+\varkappa}(y) e^{\frac{2\pi i(n+\varkappa)}{\lambda} x}$ satisfies the condition in Lemma 3.5.1 and so (3.5.12) applies. For $n + \varkappa = 0$, we use the first equality of (3.4.20) and remark that the two independent solutions are given by $1, u(y, q)$, where $u(y, q)$ is defined by (3.4.18). This completes the proof. □

Example 3.4.1 By Lemma 3.4.1, the Hecke–Eisenstein series (3.4.1) admits the Fourier–Whittaker expansion with $\lambda = 1$, $\varkappa = 0$. We determine the coefficients a_n as follows:

$$G(\tau, \bar{\tau}; a, b) = \varphi_k(y, q) + 2(-1)^{\frac{k}{2}} (\sqrt{2}\pi)^q \sum_{n \neq 0} \frac{\sigma_{q-1}(n)}{\Gamma\left(\frac{q}{2} + \varepsilon \frac{r}{2}\right)} W\left(2\pi n y; a, b\right) e^{2\pi i n x},$$

where $\varepsilon = \operatorname{sgn} n$ and $q = a + b, r = a - b$ and

$$\varphi_k(y, q) = 2\zeta(q) + 4(-1)^{\frac{k}{2}} (2y)^{1-q} \pi \frac{\Gamma(q-1)}{\Gamma\left(\frac{q+k}{2}\right) \Gamma\left(\frac{q-k}{2}\right)} \zeta(q-1). \tag{3.4.21}$$

Hence, in particular,

$$\begin{aligned} & G(\tau, \bar{\tau}; s+k, s) \\ &= 2\zeta(2s+k) + 4(-1)^{\frac{k}{2}} (2y)^{1-2s-k} \pi \frac{\Gamma(2s+k-1)}{\Gamma(s+k)\Gamma(s)} \zeta(2s+k-1) \\ &+ 2(-1)^{\frac{k}{2}} (\sqrt{2}\pi)^{2s+k} \sum_{n=1}^{\infty} \frac{\sigma_{2s+k-1}(n)}{\Gamma(s+k)} W\left(2\pi n y; s+k, s\right) e^{2\pi i n x} \\ &+ 2(-1)^{\frac{k}{2}} (\sqrt{2}\pi)^{2s+k} \sum_{n=1}^{\infty} \frac{\sigma_{2s+k-1}(n)}{\Gamma(s)} W\left(-2\pi n y; s+k, s\right) e^{-2\pi i n x}. \end{aligned} \tag{3.4.22}$$

By (3.1.11), (3.4.22) amounts to

$$
\begin{aligned}
&G(\tau, \bar{\tau}; s+k, s) \\
&= 2\zeta(2s+k) + 4(-1)^{\frac{k}{2}}(2y)^{1-2s-k}\pi\frac{\Gamma(2s+k-1)}{\Gamma(s+k)\Gamma(s)}\zeta(2s+k-1) \\
&+ 2\sum_{n\neq 0}\sigma_{2s+k-1}(n)a_n(y,s)e^{2\pi inx},
\end{aligned} \tag{3.4.23}
$$

where

$$
a_n(y,s) = \begin{cases} (-1)^{\frac{k}{2}}(2\pi)^{2s+k}e^{-2\pi ny}\frac{1}{\Gamma(s+k)}\Psi(s, 2s+k; 4\pi ny) & n > 0 \\ (-1)^{\frac{k}{2}}(2\pi)^{2s+k}e^{-2\pi ny}\frac{1}{\Gamma(s)}\Psi(s+k, 2s+k; 4\pi ny) & n < 0. \end{cases}
$$

Equation (3.4.23) is in conformity with Doi–Miyake (Miyake 1989, p. 284, Theorem 7.2.9) for $E_k(s, \tau; 1, 1)$ defined by (3.4.3). In view of (3.4.6) the result of Noda (2007) on $E_k(s, \tau)$ follows with the correction to the effect that it is to be multiplied by 2.

Exercise 3.4.2 *Confirm* (3.4.21).

Solution. Since $\varphi_k(y, q) = a_0 u(y, q) + b_0$ in (3.4.17), we may deduce

$$
\begin{aligned}
&\varphi_k(y,q) = 2\zeta(q) \\
&+ (-1)^{\frac{k}{2}}2^{3-q}\pi\frac{\Gamma(q-1)}{\Gamma\left(\frac{q+k}{2}\right)\Gamma\left(\frac{q-k}{2}\right)}\zeta(q-1) \\
&+ (-1)^{\frac{k}{2}}2^{3-q}\pi\frac{\Gamma(q-1)}{\Gamma\left(\frac{q+k}{2}\right)\Gamma\left(\frac{q-k}{2}\right)}\zeta(q-1)(1-q)u(y,q).
\end{aligned}
$$

□

Remark 3.4.1 It is stated on Maass (1964, p. 183) that the solution of the Whittaker DE (3.1.8) of polynomial growth is a constant multiple of the Whittaker function (Erdélyi et al. 1953, I, p. 278). Equation (3.3.2) makes the results of Doi–Miyake and Maass inconsistent, i.e. the variable of the Whittaker function in Maass results is multiplied by $\frac{1}{2}$! (Table 3.2)

Table 3.2 Spaces of real-analytic functions

Space	Condition(s)
$\{a, b\}$	Real-analytic solutions of (3.4.10)
$\{\Gamma, a, b, v\} \subset \{a, b\}$	(3.4.9), (3.4.11)
$\{\mathfrak{G}, a, b, v\} \subset \{\Gamma, a, b, v\}$	(3.4.15), (3.4.16)

3.5 Analysis of the Operator Ω_{ab}

We give slight simplification of Maass' procedure and accommodate the case of eigen-functions. Here the discontinuous group is the Hecke group. The operator Ω_{an} in (3.4.12) is introduced as follows. The operator

$$\Omega'_{ab} = \Delta + iry\frac{\partial}{\partial x} - \nu + \frac{5}{4}r^2 \tag{3.5.1}$$

in Maass (1964, p. 175, 1') is modified so as to incorporate the effect of multiplication of $f(\tau)$ by a power of y, say $g(\tau) = y^{\frac{q}{2}} f(\tau)$. Indeed, (3.5.9) reads for $q = 0$ resp. $\nu' = 0$

$$\Omega'_{ab}\left(f(\tau)g(y)\right) = -y^2 f(\tau)\frac{\mathrm{d}^2 g}{\mathrm{d}y^2} - 2y^2\frac{\partial f}{\partial y}\frac{\mathrm{d}g}{\mathrm{d}y} + g(y)\Omega'_{ab} f(\tau),$$

resp.

$$\Omega_{ab}(f(\tau)g(y)) = -y^2 f(\tau)\frac{\mathrm{d}^2 g}{\mathrm{d}y^2} - y\left(2y\frac{\partial f}{\partial y} + qf(\tau)\right)\frac{\mathrm{d}g}{\mathrm{d}y} + g(y)\Omega_{ab} f(\tau),$$

which gives for $g(y) = y^{\frac{q}{2}}$,

$$\Omega'_{ab}(f(\tau)y^{\frac{q}{2}}) = -y^{\frac{q}{2}}q\frac{\partial f}{\partial y} + y^{\frac{q}{2}}\left(\Omega'_{ab} - \frac{q}{2}\left(\frac{q}{2} - 1\right)\right)f(\tau), \tag{3.5.2}$$

resp.

$$\Omega_{ab}(f(\tau)y^{\frac{q}{2}}) = -y^{\frac{q}{2}}\frac{q}{2}\left(2y\frac{\partial f}{\partial y} + qf(\tau)\right) + y^{\frac{q}{2}}\left(\Omega_{ab} - \frac{q}{2}\left(\frac{q}{2} - 1\right)\right)f(\tau). \tag{3.5.3}$$

By Lemma 3.5.1, the term $2y\frac{\partial f}{\partial y} + qf(\tau)$ is to vanish, and we are led to the definition

$$\Omega^{\nu'}_{ab} = \Delta + i(a-b)y\frac{\partial}{\partial x} - (a+b)y\frac{\partial}{\partial y} + \nu' = \Delta + iry\frac{\partial}{\partial x} - qy\frac{\partial}{\partial y} + \nu' \tag{3.5.4}$$

$$= \Omega_{ab} + \nu' = \Omega'_{ab} - q\left(y\frac{\partial}{\partial y} + \frac{1}{2}\left(\frac{q}{2} - 1\right)\right),$$

where

$$\nu' = -\nu + \frac{5}{4}r^2 - \frac{q}{2}\left(\frac{q}{2} - 1\right). \tag{3.5.5}$$

Equation (3.5.4) is Maass (1964, p. 176, 1") (in which $\frac{\partial^2}{\partial^2 y}$ is missing). q is chosen so that the constant term $\nu' = 0$, i.e. (Maass 1964, p. 177)

$$q = 1 + \sqrt{5r^2 - 4\nu + 1}, \tag{3.5.6}$$

in which case, $\Omega^0_{ab} = \Omega_{ab}$.

However, as can be seen in the proof of Lemma 3.5.1, in order to eliminate the $\frac{\mathrm{d}}{\mathrm{d}y}$, one must multiply $g(y)$ by $y^{-\frac{q}{2}}$. But as mentioned above, the operator arose after multiplying $f(\tau)$ by $y^{\frac{q}{2}}$. This suggests possible streamlining of the process and for this we are to go back to (3.5.1).

Exercise 3.5.1 *Check the following process. Applying Ω'_{ab} to $e^{i\varepsilon x}g(y)$*

$$\left(y^2\frac{\mathrm{d}^2}{\mathrm{d}y^2} + \nu - \frac{5}{4}r^2 + \varepsilon r y - \varepsilon^2 y^2\right) g\left(y\right) = 0,$$

or by $\varepsilon^2 = 1$

$$\left(y^2\frac{\mathrm{d}^2}{\mathrm{d}y^2} + \nu - \frac{5}{4}r^2 + \frac{1}{2}\varepsilon r y - \frac{1}{4}y^2\right) g\left(\frac{1}{2}y\right) = 0. \tag{3.5.7}$$

Comparing this with (3.1.8), *we derive that*

$$\varkappa = \frac{\varepsilon r}{2}, \quad \mu = \sqrt{5r^2 - 4\nu + 1} = q - 1 \tag{3.5.8}$$

the latter because of (3.5.6).

For our intended application to the Hecke–Eisenstein series in Example 3.4.1, we are to treat (3.5.4) at this stage before going to the definition of Ω_{ab}. Since our main objective is to study (3.4.6) and $G(\tau, \bar{\tau}, s+k, s)$ is in $\{a, b\}$ with $a = s+k$, $b = s$, it follows that $r = k$, $q = 2s + k$.

The space of real-analytic functions on $\mathcal{H}$ which satisfy the DE $\Omega^{\nu'}_{ab} f(\tau) = 0$ is the space of functions whose eigen-values are $-\nu'$.

Lemma 3.5.1 *Let $f(\tau) = f(x, y)$ and $g(y)$ be differentiable functions. Then*

$$\begin{aligned}&\Omega^{\nu'}_{ab}(f(\tau)g(y)) \\ &= -y^2 f(\tau)\frac{\mathrm{d}^2 g}{\mathrm{d}y^2} - y\left(2y\frac{\partial f}{\partial y} + q f(\tau)\right)\frac{\mathrm{d}g}{\mathrm{d}y} + \nu' f(\tau)g(y) + g(y)\Omega_{ab} f(\tau).\end{aligned} \tag{3.5.9}$$

In the case $f(\tau)g(y) \in \{a, b\}$ with $f(\tau) = e^{i\varepsilon x}y^p$ and $\varepsilon^2 = 1$, (3.5.9) *amounts to the DE*

$$\begin{aligned}&\Omega^{\nu'}_{ab}(f(\tau)g(y)) \qquad (3.5.10)\\&= -f(\tau)\left(y^2\frac{\mathrm{d}^2}{\mathrm{d}y^2} + y(2p+q)e^{i\varepsilon x}\frac{\mathrm{d}}{\mathrm{d}y} - \varepsilon^2 y^2 + \varepsilon r y + p(p+q-1) - \nu'\right)g(y)\\&= 0.\end{aligned}$$

If further, $g(y)$ is of polynomial growth as $y \to \infty$, then the solution of (3.5.10) *with $p = -\frac{q}{2}$ of polynomial growth is constant times $g(y)$:*

$$g(y) = y^{-\frac{q}{2}} W_{\frac{1}{2}\varepsilon r, \frac{1}{2}\sqrt{(q-1)^2-4\nu'}}(2y), \qquad (3.5.11)$$

which in the case $\nu' = 0$ reduces to

$$g(y) = y^{-\frac{q}{2}} W_{\frac{1}{2}\varepsilon r, \frac{1}{2}(q-1)}(2y) = W(\varepsilon y; a, b). \qquad (3.5.12)$$

Cf. (3.3.2).

Proof For $f(\tau) = e^{i\varepsilon x} y^p$, $p = -\frac{q}{2}$, $\frac{\mathrm{d}g}{\mathrm{d}y}$ term in (3.5.10) vanishes. Hence writing

$$g(y) = t^{-\frac{q}{2}} w(y), \qquad (3.5.13)$$

we see that (3.5.10) reads

$$\left(y^2\frac{\mathrm{d}^2}{\mathrm{d}y^2} - \frac{q}{2}\left(\frac{q}{2} - 1\right) - \nu' + \varepsilon r y - y^2\right) w(y) = 0. \qquad (3.5.14)$$

For (3.5.14) to coincide with (3.1.8), we are to make the change of variables $y \leftrightarrow \frac{1}{2}y$. Then (3.5.14) reads

$$\left(y^2\frac{\mathrm{d}^2}{\mathrm{d}y^2} - \frac{q}{2}\left(\frac{q}{2} - 1\right) - \nu' + \frac{1}{2}\varepsilon r y - \frac{1}{4}y^2\right) w\left(\frac{1}{2}y\right) = 0. \qquad (3.5.15)$$

Since $g(y)$ is of polynomial growth by assumption, so is $w(y)$, whence $w(y)$ must be the Whittaker function $W_{\frac{1}{2}\varepsilon r, \frac{1}{2}\sqrt{(q-1)^2-4\nu'}}(2y)$. Thus we conclude (3.5.11). □

Remark 3.5.1 Substituting (3.5.5) in (3.5.15), we deduce (3.5.7).

Corollary 3.5.1 *If $\Omega_{ab} f = 0$ and $f(\tau) = y^{-\frac{q}{2}} h(x)$, then* (3.5.9) *reads*

$$\Omega_{ab}(f(\tau)g(y)) = -y^2 f(\tau)\frac{\mathrm{d}^2 g}{\mathrm{d}y^2}.$$

In particular,

$$\Omega_{ab}(f(\tau)y^s) = -s(s-1)f(\tau)y^s, \qquad (3.5.16)$$

i.e. $f(\tau)y^s$ is an eigen-function of Ω_{ab} with eigen-value $-\nu = -s(s-1)$.

In this case, (3.1.8) *holds true with*

$$\mu = \frac{1}{2}\sqrt{(q-1)^2 - 4\nu'} = q - 1, \tag{3.5.17}$$

the second equality in (3.5.8).

In view of Lemma 3.5.1, when applying the operator Ω_{ab} to Hecke–Maass function $f(\tau)$, we may think of it as $f(\tau) = e^{i\omega x}h(y)$.

Remark 3.5.2 Maass forms are often called **Maass wave forms** in view of the statement that the N forms $g_k(\tau)$ satisfy the wave equation (Maass (1949), p. 143):

$$\left(\Delta - \left(r^2 + \frac{1}{4}\right)\right) g_k(\tau) = 0,$$

i.e. they are eigen-functions of the hyperbolic Laplacian. Cf. Example 3.6.1.

We quote from Bellman (1950): "Maass introduced the concept of non-analytic automorphic functions. These are functions of two real variables, $f(x_1, x_2)$, possessing functional equations of the type

$$f(x_1, x_2 + c) = f(x_1, x_2),\ f(x_1' + ix_2') = R(x_1, x_2) + s(x_1, x_2) f(x_1, x_2), \tag{3.5.18}$$

where $x_1' + ix_2' = 1/(x_1 + ix_2)$. These functions are connected with the zeta function of a quadratic number field, and their automorphic properties are deduced not as a direct consequence of the functional equations of the related Dirichlet series, nor as a consequence of the transformations of the associated theta functions, but, in a most interesting manner, as a result of the uniqueness of the solution of the partial differential equation satisfied by $f(x_1, x_2)$."

This elucidating description applies to Maass (1964) rather than Maass (1949).

3.6 From Epstein Type to Hecke Eisenstein Series and Back

We also introduce the completed real-analytic Eisenstein series $E^*(z, s)$ which are somewhat more natural than $E(s, z)$ in view of (3.2.7) and (3.6.1) and state some results on Eisenstein series including it. In particular, Theorem 3.6.1 provides CSIF for five Dirichlet series.

Definition 3.6.1 The **real-analytic Eisenstein series** (RAES) is defined by Liu (2026)

$$E^*(z, s) = \frac{\Gamma(s)}{2\pi^s} \sum_{\substack{m,n \in \mathbb{Z} \\ (m,n) \neq (0,0)}} \frac{y^s}{|mz + n|^{2s}} = \frac{1}{2} y^s \pi^{-s} \Gamma(s) \zeta_{\mathbb{Z}^2}(s, z). \tag{3.6.1}$$

It is also called the (completed) non-holomorphic Eisenstein series in Definition 2.12.2 or also as the completed Epstein-like zeta-function, cf. Siegel (1961), Terras (1985).

By (3.2.7),

$$E^*(z,s) = \frac{1}{2}\xi(2s)E(s,z) = \frac{1}{2}\zeta(2s)\pi^{-s}\Gamma(s)E(s,z),$$

where $\xi(s) = \xi_{\mathbb{Q}}(s)$ is the completed Riemann zeta-function (2.1.9) (Table 3.3).

Here A is the constant in Hecke's functional Eq. (2.12.2) and $\mathfrak{C}$ is an absolute ideal class $\in I/P$ of an imaginary quadratic field K, Q_1 is defined in (1.1.8), and

$$\zeta(s,\mathfrak{C}) = \frac{1}{w}Z(s,Q), \quad Z(s,Q) = \sum_{m,n}{}' \frac{1}{Q(m,n)^s} = a^{-s}\zeta_{\mathbb{Z}^2}(s,\tau) \tag{3.6.2}$$

and (1.1.9). Cf. Definition 1.1.1 and Li et al. (2022, Sect. 2.1). Cf. Table 4.2.
Epstein zeta = (1.1.6)
Epstein-type Eisenstein = (3.2.6)
real-analytic Eisenstein = (3.6.1)
kth Eisenstein–Maass = (3.4.4).

Theorem 3.6.1 (Chowla–Selberg integral formula) (i) *All five Dirichlet series in Table 3.3 have the Fourier–Bessel expansion, For the class zeta-function, it is (2.5.45).*

Table 3.3 Epstein zeta-functions and non-holomorphic Eisenstein series

Zeta	Symbol	A
Class zeta	$\zeta_K(s,\mathfrak{C}) = \sum_{\mathfrak{C}\ni\mathfrak{a}\neq 0} \frac{1}{(N\mathfrak{a})^s}$	$\frac{2\pi}{\sqrt{\lvert\Delta\rvert}}$
Epstein zeta	$Z(s,Q) = \sum'_{m,n} \frac{1}{Q(m,n)^s}$	$\frac{2\pi}{\sqrt{\lvert\Delta\rvert}}$
Epstein-type Eisenstein	$\zeta_{\mathbb{Z}^2}(s,\tau) = Z(s,Q_1)$, (1.1.8)	$\frac{\pi}{y}$
Eisenstein–Maass (3.2.6)	$E(s,\tau) = E_0(s,\tau) = \frac{y^s}{\zeta(2s)}\zeta_{\mathbb{Z}^2}(s,\tau)$	$\frac{\pi}{y}$
Real-analytic Eisenstein	$E^*(s,z) = \frac{1}{2}\left(\frac{\pi}{y}\right)^{-s}\Gamma(s)\zeta_{\mathbb{Z}^2}(s,z)$	1
Hecke Eisenstein (3.4.1)	$G(\tau,\bar{\tau};a,b)$	–
kth Eisenstein–Maass	$E_k(s,z) = \frac{y^s}{\zeta(2s+k)}G(\tau,\bar{\tau},s+k,s)$	–

For the Epstein zeta-function, it is Theorem 2.5.1, (i) as well as (3.6.3) $\times a^{-s}$. *For the Epstein-type Eisenstein series*

$$\zeta_{\mathbb{Z}^2}(s,z) = 2\zeta(2s) + \frac{2\sqrt{\pi}\Gamma(s-\frac{1}{2})}{\Gamma(s)}\zeta(2s-1)y^{-2s+1} + \frac{8\pi^s y^{\frac{1}{2}-s}}{\Gamma(s)}\sum_{n=1}^{\infty} n^{s-\frac{1}{2}}\sigma_{1-2s}(n)\cos 2\pi nx K_{s-\frac{1}{2}}(2\pi ny). \tag{3.6.3}$$

For the Eisenstein–Maass series resp. the real-analytic Eisenstein series

$$E(s,z) = 2y^s + \frac{2\sqrt{\pi}\Gamma(s-\frac{1}{2})}{\Gamma(s)}\frac{\zeta(2s-1)}{\zeta(2s)}y^{-s+1} + \frac{8y^{\frac{1}{2}}\pi^s}{\Gamma(s)\zeta(2s)}\sum_{n=1}^{\infty} n^{s-\frac{1}{2}}\sigma_{1-2s}(n)\cos 2\pi nx K_{s-\frac{1}{2}}(2\pi yn) \tag{3.6.4}$$

resp.

$$E^*(s,z) = \pi^{-s}\Gamma(s)\zeta(2s)y^s + \pi^{\frac{1}{2}-s}\Gamma\left(s-\frac{1}{2}\right)\zeta(2s-1)y^{1-s} + 4\sqrt{y}\sum_{n=1}^{\infty} n^{s-\frac{1}{2}}\sigma_{1-2s}(n)\cos 2\pi nx K_{s-\frac{1}{2}}(2\pi yn), \tag{3.6.5}$$

where $z = x + iy$ $(y > 0)$
(ii) *Save for* $E^*(s,z)$ *they satisfy Hecke's functional Eq.* (2.12.2) *with* $A = \frac{2\pi}{\sqrt{|\Delta|}}$ *for* $\zeta_K(s)$ *(*K *being an imaginary quadratic field with discriminant* Δ*) and* $Z(s,Q)$ *resp.* $A = \frac{\pi}{y}$ *for others.*

This is immediate from Corollary 3.2.2, Lemma 3.3.2, and Example 3.4.1 in view of

$$\zeta_{\mathbb{Z}^2}(s,\tau) = G(\tau,\bar{\tau};s,s). \tag{3.6.6}$$

The theory of $E^*(s,z)$ is developed in Liu (2026) and we refer to the main result as Corollary 3.6.1. The second term on the right of (3.6.5) is by the functional Eq. (2.1.9).

We give a proof of (3.6.3) in the proof of Theorem 3.6.2. Proof of (3.6.4) is also given in Chakraborty (2010), which we state in the proof of Example 3.6.1 (Table 3.4).

Here
Kober = Kober (1934)
Deuring = Deuring (1933), Deuring (1937)
Taylor = Taylor (1940), Taylor (1945)
Bateman–Grosswald = Bateman and Grosswald (1964)
Selberg–Chowla = Chowla and Selberg (1949), Selberg and Chowla (1967)

Table 3.4 Method of proof of the Fourier–Whittaker expansion

Author	Expr. for K_ν	Proof
Kober	Hardy transform	Transf. formula for ϑ
Deuring	Basset, Mellin	Euler–Maclaurin\Poisson
Taylor	Mellin int.via ${}_2F_1$	Funct. eqn. ζ, ${}_2F_1$
Bateman–Grosswald	Basset integral	Poisson sum. formula
Selberg–Chowla	Hardy transform	Transf. formula for ϑ
Berndt	Mellin integral	Beta transform
Terras (1985)	Mellin, Hankel transf.	n-Poisson sum. formula
Maass, Chakraborty	Basset, Hardy transf.	DE of the aut. form
Kuzumaki	Mellin integral	Beta transform

Berndt = Berndt (1971)
Maass, Chakraborty = Maass (1964), Chakraborty (2010)
Kuzumaki = Kuzumaki (2011).

Definition 3.6.2 Corresponding to Definition 3.4.1 (i.e. to $\{\Gamma(1), 0, 0, 1\}$), we state the definition of **Maass forms (of weight** 0). A smooth function $f(\tau) \neq 0$ on $\mathcal{H}$ is called a Maass form if it satisfies (i) automorphy $f(\gamma\tau) = f(\tau)$, (ii) eigen-function of the automorphic Laplacian $\Delta f = \lambda f$, and (iii) of polynomial growth in y.

The following corollary assembles all essential results in Liu (2026).

Corollary 3.6.1 (i) *The real analytic Eisenstein series $E^*(z, s)$ has the meromorphic continuation to the whole s-plane $\mathbb{C}$ in view of its Fourier expansion = Chowla–Selberg integral formula* (3.6.5).

It is analytic except at $s = 1$ and $s = 0$, where it has simple poles with residue $z = 1/2$ and $-1/2$, respectively.

(ii) *The functional equation*

$$E^*(z, s) = E^*(z, 1 - s) \tag{3.6.7}$$

holds.

(iii) *We have*

$$E^*(x + iy, s) \ll y^{\max(\sigma, 1-\sigma)}, \quad y \to \infty. \tag{3.6.8}$$

It follows that the Eisenstein series $E^*(z, s)$ is a Maass form with Laplace eigen-value $s(1 - s)$ in the sense of Definition 3.6.2.

Exercise 3.6.1 *Prove Corollary 3.6.1.*

Solution. Near $s = 1$, the second term on the right of (3.6.5) is

$$\frac{\frac{1}{2}}{s-1} + \cdots,$$

while at $s = 0$, the first term on the right of (3.6.5) is

$$\frac{-\frac{1}{2}}{s-1} + \cdots .$$

Near $s = \frac{1}{2}$, the sum of the first and the second terms is

$$\sqrt{y}\left(\frac{-\frac{1}{2}}{s-1} - \frac{-\frac{1}{2}}{s-1} + \cdots .\right),$$

and so it is a regular point. Other poles of the gamma function are cancelled by the zeros of the zeta-function.
(iii) will be proved in the proof of Example 3.6.1. □

Similar results hold for the relatives of $E^*(z, s)$ and we state them as Examples 3.6.1 and 3.6.2.

In Chakraborty (2010) and Motohashi (1997), the same Eisenstein series is introduced, which is $E^*(z, s)$ (3.6.1).

Exercise 3.6.2 *The K-Bessel function $K_{i\nu}$ and the modified Bessel function of the first kind $I_{i\nu}$ constitute the fundamental system of solutions to the* **modified Bessel differential equation (BDE)** *(cf.* (3.1.8)*).*

$$\frac{\mathrm{d}^2 w}{\mathrm{d}z^2} + \frac{1}{z}\frac{\mathrm{d}w}{\mathrm{d}z} - \left(1 + \frac{\nu^2}{z^2}\right) = 0. \tag{3.6.9}$$

Prove the following. (3.6.9) *reduces to the special case of* (3.1.8).

The asymptotic formulas for $K_{i\nu}$ and $I_{i\nu}$ are valid

$$I_{i\nu}(y) = \frac{e^y}{\sqrt{2\pi y}}\left(1 + O\left(\frac{1 + |\nu|^2}{y}\right)\right) \tag{3.6.10}$$

$$K_{i\nu}(y) = e^{-y}\sqrt{\frac{\pi}{2y}}\left(1 + O\left(\frac{1 + |\nu|^2}{y}\right)\right) \tag{3.6.11}$$

for $y \to \infty$.

Solution. The first assertion follows from the general principle that the first derivative term in a second-order linear DE may be eliminated by multiplying a power z^a.

Consider the second-order linear DE in y

$$w'' + \frac{2s}{y}w' + cw = 0. \tag{3.6.12}$$

Putting

$$w = y^a W,$$

we see that (3.6.12) reduces to

$$W'' + \frac{2s+2a}{y}W' + \frac{a(a-1)+2as+c}{y^2}W = 0.$$

Choosing $a = -s$, (3.6.12) amounts to

$$w'' + \frac{s(1-s)+c}{y^2}w = 0,$$

i.e. the first derivative term eliminated.

The asymptotic formulas are stated on Whittaker and Watson (1927, pp. 372–372) as a consequence of that of $J_\nu(y)$ through

$$I_\nu(z) = e^{-\frac{\pi}{2}i\nu}J_\nu(ze^{\frac{\pi}{2}i\nu}), \quad K_\nu(z) = \frac{\frac{1}{2}\pi}{\sin \pi\nu}\left(I_{-\nu}(z) - I_\nu(z)\right).$$

□

Exercise 3.6.3 *Prove the following. The Bessel DE of order ν*

$$z^2y'' + zy' + (z^2 - \nu^2)y = 0,$$

where ν is an arbitrary (real or complex) number, has the fundamental system consisting of $J_\nu(z)$ and $J_{-\nu}(z)$ for non-integral ν, where

$$J_\nu(z) = \sum_{n=0}^{\infty} \frac{(-1)^n\left(\frac{1}{2}z\right)^{2n+\nu}}{n!\Gamma(\nu+n+1)}$$

and is called the (ordinary) **Bessel function of the first kind** *of order ν or also called the J-***Bessel function**. *The expression* (Erdélyi et al. 1953, II, p. 83, (36))

$$J_\nu(x) = \frac{1}{2\pi i}\int_{(c)} \frac{\Gamma(-z)}{\Gamma(1+\nu+z)}\left(\frac{x}{2}\right)^{\nu+2z} dz = \frac{1}{2}G^{0,1}_{2,0}\left(\frac{x}{2}\,\middle|\,\begin{matrix}1, 1+\nu\\ -\end{matrix}\right) \tag{3.6.13}$$

is often used.

For a proof, cf., e.g. (Watson 1944, pp. 38–40).

The main result of Chakraborty (2010) is part of the following theorem. For more details, cf. Liu (2026). Theorem 3.6.2 and Example 3.6.1 are closely related and the proofs depend on the Fourier expansion. The difference lies in the method for the determination of the Fourier coefficients. In the former, it depends on integral transform formulas (solutions of DE) while in the latter, on DE (Maass). Since from here to the end of this section, it is appropriate to quote Hecke's principle its concise version as stated on Bellman (1961, p. 7): *A periodic function should always be expanded in Fourier series.*

Thus we see that Maass theory is in the stream of Hecke's principle with the method of determination of Fourier coefficients by DE (in contrast to the conventional method of using integral transforms as solutions of DE). Cf. Bellman's description toward the end of Sect. 3.5.

Theorem 3.6.2 *Let the asymptotic formulas* (3.6.10) *and* (3.6.11) *and the limit condition*

$$\lim_{z\to 0} z^{\nu} K_{\nu}(2az) = \frac{1}{2}a^{-\nu}\Gamma(\nu), \ \operatorname{Re}\nu \geq 0, \ \nu \neq 0 \tag{3.6.14}$$

hold for the K-Bessel function as a solution to the modified BDE (3.6.9).

Assume further a Fourier transform formula (3.6.21) (cf. (3.6.24)) *in the form*

$$\int_{-\infty}^{\infty} e^{-\pi\alpha x^2} e(rx)dx = \sqrt{\frac{1}{\alpha}} \exp\left(-\frac{\pi}{\alpha}r^2\right), \quad \alpha > 0. \tag{3.6.15}$$

Then the **formula of Basset**

$$K_{\nu}(z) = \frac{1}{\sqrt{\pi}} (2z)^s \, \Gamma\left(\nu + \frac{1}{2}\right) \int_0^{\infty} \frac{\cos t}{(t^2+z^2)^{2\nu+1}} dt, \tag{3.6.16}$$
$$\operatorname{Re}\nu \geq -\frac{1}{2}, \ |\arg z| < \frac{\pi}{2},$$

and the Hecke integral (for a generalization, cf. Proposition 3.1.1; cf. also (3.1.2))

$$K_{\nu}(z) = \frac{1}{2}\int_0^{\infty} \exp\left(-\frac{z}{2}\left(t+\frac{1}{t}\right)\right) t^{\nu}\frac{dt}{t}, \quad \operatorname{Re}\nu > -\frac{1}{2}, \quad |\arg z| < \frac{\pi}{4} \tag{3.6.17}$$

follow, and at the same time we obtain the Chowla–Selberg integral formula (3.6.5).

Proof We shall prove the Fourier–Bessel expansion (3.6.3) for EES from which other CSIF follow, esp. (3.6.5) follows from (3.6.1).

Proof is by giving closed form for the Fourier coefficients in the spirit of Maass and is similar to the second proof of Chowla and Selberg. The partial sum over $m = 0, n \in \mathbb{Z}$ is (3.6.2) is $2\zeta(2a)$, so that we may write

$$\zeta_{\mathbb{Z}^2}(s, z) = 2\zeta(2s) + 2S, \quad S = S(s, z) = \sum_{n=1}^{\infty}\sum_{m=-\infty}^{\infty} |m + nz|^{-2s} \tag{3.6.18}$$

and

$$S(s, z) = \sum_{\ell=-\infty}^{\infty} c_{\ell}(y) e^{2\pi i \ell x}, \tag{3.6.19}$$

say, where

$$c_\ell = c_\ell(y) = \int_0^1 S(s, x+iy)e^{2\pi i\ell x}\,\mathrm{d}x = \sum_{n=1}^{\infty}\sum_{m=-\infty}^{\infty}\int_0^1 |m+nz|^{-2s} + e^{2\pi i\ell x}\,\mathrm{d}x.$$

on substituting the defining series, changing the order of integration and summation (which is legitimate in view of absolute convergence).

$$c_\ell = \sum_{n=1}^{\infty}\sum_{m=-\infty}^{\infty}\int_0^1 \frac{e^{2\pi i\ell x}}{\{(m+xn)^2+y^2n^2\}^s}\,\mathrm{d}x.$$

Now classifying the inner sum modulo n and making the substitution $x \mapsto x - m/n$, we have

$$c_\ell = \int_{-\infty}^{\infty} \frac{e^{2\pi i\ell x}}{(x^2+y^2)^s} dx \times \sum_{n=1}^{\infty} n^{-2s} \sum_{m \bmod n} e^{2\pi i \frac{\ell m}{n}}$$

whose double sum is $\sum_{n|\ell} n^{1-2s}$, meaning $\sigma_{1-2s}(\ell)$ or $\zeta(1-2s)$ according to $\ell \neq 0$ or $\ell = 0$.
Hence

$$c_\ell = \begin{cases} \sigma_{1-2s}(\ell) \displaystyle\int_{-\infty}^{\infty} \frac{e^{2\pi i\ell x}}{(x^2+y^2)^s} dx, & \ell \neq 0 \\ \zeta(2s-1) \displaystyle\int_{-\infty}^{\infty} \frac{1}{(x^2+y^2)^s} dx, & \ell = 0. \end{cases} \tag{3.6.20}$$

Now the resulting integrals are expressed in closed form in Excrcise 3.6.4. We note that the $\ell = 0$ case becomes

$$\pi^{\frac{1}{2}-s}\Gamma\left(s-\frac{1}{2}\right)\zeta(2s-1)y^{1-s},$$

which, by FER, reduces to the second term on the right-hand side of (3.6.3). Since the $\ell \neq 0$ case of $2S_2$ gives the K-Bessel series in (3.6.3), this complete the proof. □

Exercise 3.6.4 *Prove that*

$$\begin{aligned} &\left(\frac{y}{\pi}\right)^s \Gamma(s) \int_{-\infty}^{\infty} \frac{e^{2\pi i\ell x}}{(x^2+y^2)^s}\,\mathrm{d}x \\ &= \begin{cases} 2|\ell|^{s-1/2}\sqrt{y}K_{s-1/2}(2\pi|\ell|y), & \text{if } \ell \neq 0, \\ \pi^{-s+1/2}\Gamma(s-1/2)y^{1-s}, & \text{if } \ell = 0. \end{cases} \end{aligned} \tag{3.6.21}$$

We note that (3.6.21) *in the case* $\ell \neq 0$ *is the formula of Basset* (3.6.16). *This is a continuous version of the gamma transform. For discrete version, cf. Exercise 3.6.7.*

Solution. Substituting

$$\Gamma(s)\left(\frac{y}{\pi(x^2+y^2)}\right)^s = \int_0^\infty t^s \exp\left(-\pi t\frac{x^2+y^2}{y}\right)\frac{dt}{t}$$

we write the left-hand side of (3.6.21) as

$$\left(\frac{y}{\pi}\right)^s \Gamma(s)\int_{-\infty}^\infty \frac{e^{2\pi i\ell x}}{(x^2+y^2)^s}\,dx = \int_0^\infty \int_{-\infty}^\infty t^s \exp\left(-\pi t\frac{x^2+y^2}{y}\right)e^{2\pi i\ell x}\,dx\frac{dt}{t}, \tag{3.6.22}$$

on changing the order of integration, which is justified by absolute convergence.

Now we have

$$\int_{-\infty}^\infty e^{-t\pi x^2/y}e^{2\pi i\ell x}\,dx = \begin{cases}\sqrt{y/t}\ \exp(-y\pi\ell^2/t), & \text{if } \ell \neq= 0,\\ \sqrt{y/t}, & \text{if } \ell = 0,\end{cases}$$

where the $\ell \neq 0$ case is (3.6.24) and the $\ell = 0$ case is the probability integral (3.6.26).

We substitute this for the inner integral in (3.6.22) to transform (3.6.22) into

$$\left(\frac{y}{\pi}\right)^s \Gamma(s)\int_{-\infty}^\infty \frac{e^{2\pi i\ell x}}{(x^2+y^2)^s}\,dx = \sqrt{y}\int_0^\infty \exp\left(-\pi y\left(\frac{\ell^2}{t}+t\right)\right)t^{s-\frac{1}{2}}\frac{dt}{t}. \tag{3.6.23}$$

By the substitution $t \mapsto |\ell|t$ and the use of (3.1.2) we deduce (3.6.21). We may also prove (3.1.2) in a rounding but instructive way in the solution to Example 3.6.1.

For $\ell = 0$, we may also evaluate the integral $\int_{-\infty}^\infty \frac{1}{(x^2+y^2)^s}\,dx$ on the right of (3.6.21) directly as follows. By the change of variable $x = y\sqrt{u}$, we immediately find that it is

$$\int_{-\infty}^\infty \frac{1}{(x^2+y^2)^s}\,dx = \int_{-\infty}^\infty \frac{u^{1/2-1}}{(1+u)^s}\,du = y^{1-2s}B(1/2, s-1/2),$$

where $B(\alpha, \beta)$ is the beta function. Hence

$$B(1/2, s-1/2) = \frac{\Gamma(1/2)\Gamma(s-1/2)}{\Gamma(s)} = \sqrt{\pi}\frac{\Gamma(s-1/2)}{\Gamma(s)}.$$

and the $\ell = 0$ case follows. □

Exercise 3.6.5 *The Fourier transform formula*

$$\int_{-\infty}^\infty e^{-\pi\alpha x^2+2\pi i r x}\,dx = \sqrt{\frac{1}{\alpha}}\exp\left(-\frac{\pi}{\alpha}r^2\right), \quad \alpha > 0 \tag{3.6.24}$$

plays an important role in several contexts. Prove it in the form

$$\int_0^\infty e^{-x^2} \cos 2ax \, dx = \frac{\sqrt{\pi}}{2} e^{-a^2}, \quad a > 0, \tag{3.6.25}$$

employing the value of the **probability integral**

$$\int_{-\infty}^\infty e^{-x^2} \, dx = \sqrt{\pi}. \tag{3.6.26}$$

Solution. We integrate the function $f(z) = e^{-z^2}$ along the sides of the rectangle with vertices at $(N, 0)$, $(N, N + ia)$, $(-N, -N + ia)$, $(-N, 0)$, $N > 0$. Then by the Cauchy integral theorem, we obtain

$$\frac{1}{2\pi i} \int_{-N}^N f(x) \, dx = \frac{1}{2\pi i} \int_{-N}^N f(x + ia) \, dx + O(e^{-N^2}). \tag{3.6.27}$$

The integral on the right-hand side is

$$\frac{e^{a^2}}{2\pi i} \int_{-N}^N e^{-x^2 - 2iax} \, dx,$$

which approaches as $N \to \infty$

$$\frac{2e^{a^2}}{2\pi i} \int_0^\infty e^{-x^2} \cos 2ax \, dx. \tag{3.6.28}$$

Hence letting $N \to \infty$ in (3.6.27), this must be equal to

$$\frac{1}{2\pi i} \int_{-\infty}^\infty e^{-x^2} \, dx = \frac{1}{2\pi i} \sqrt{\pi},$$

by (3.6.26). Hence Eq. (3.6.28) to be $\sqrt{\pi}$ yields (3.6.25). Writing the left-hand side of (3.6.25) as $\frac{1}{2} \int_{-\infty}^\infty e^{-\pi\alpha x^2 + 2\pi i r x} dx$, making the change of variable $\sqrt{\pi\alpha} x \leftrightarrow x$ and writing $a = \frac{\sqrt{\pi\alpha} r}{2}$, (3.6.24) follows (Table 3.5). □

Here [Z-W] = Zhang and Williams (1995) and Q_1 is defined by (1.1.8).

We now give the first proof of Chowla and Selberg. We need the **Poisson summation formula** which is as fundamental as the functional equation for zeta-functions and so is very useful in many respects, cf., e.g. (Rademacher 1973, pp. 71–79).

Lemma 3.6.1 *If $f \in C^2(\mathbb{R})$ and $f \in L^1 \cap L^2$,*

$$\sum_{n=-\infty}^\infty f(n) = \sum_{n=-\infty}^\infty \hat{f}(n), \tag{3.6.29}$$

where

Table 3.5 Epstein zeta-functions and non-holomorphic Eisenstein series

Author(s)	Epstein	Non-holomorphic
Chakraborty (2010)	–	$\frac{1}{\xi(2s)}E^*(z,s)=E(s,z)$
Kuzumaki (2011)	$\zeta_{\mathbb{Z}^2}(s;\alpha)=Z(s,Q_1)$	$\frac{y^s}{2\zeta(2s)}Z(s,Q_1)$
Z-W	$Z(s,Q)=\sum'_{m,n}\frac{1}{Q(m,n)^s}$	$\frac{y^s}{2\zeta(2s)}Z(s,Q)$
Motohashi (1997)	–	$\frac{1}{\xi(2s)}E^*(z,s)=E(s,z)$

$$\hat{f}(n)=\int_{-\infty}^{\infty}e^{-2\pi int}f(t)\,\mathrm{d}t$$

is the Fourier transform of f and the left-hand side is to mean the mean at discontinuities.

For Hamburger's version of the Poisson summation formula, cf. (4.5.14).

Exercise 3.6.6 *Use Theorem 3.6.1 and Exercise 3.6.5 to prove the theta-transformation formula, (2.1.64)*

$$\sum_{n=-\infty}^{\infty}e^{-\pi(n+\alpha)^2/x}=x^{1/2}\sum_{n=-\infty}^{\infty}e^{-\pi n^2x+2\pi in\alpha} \tag{3.6.30}$$

for Re $x>0$, $|\arg x^{1/2}|<\frac{\pi}{4}$. *Follow the method of Rademacher to prove* (3.6.30) *and determine the value* (3.6.26).

Equation (3.6.30) *is (2.1.64) and implies*

$$\vartheta_3\left(-\frac{1}{\tau},0\right)=\sqrt{-i\tau}\vartheta_3(\tau,0) \tag{3.6.31}$$

(Chan 2020, p. 70, (6.30)).

Solution. Equation (3.6.29) reads

$$\sum_{n=-\infty}^{\infty}e^{-\pi(n+\alpha)^2/x}=\sum_{n=-\infty}^{\infty}\int_{-\infty}^{\infty}e^{-\pi(t+\alpha)^2/x-2\pi int}\,\mathrm{d}t. \tag{3.6.32}$$

Changing the variable $t+\alpha=xu$ in the right-hand side integral of (3.6.32), we see that the right-hand side reduces to

$$x\sum_{n=-\infty}^{\infty}e^{2\pi in\alpha}\int_{-\infty}^{\infty}e^{-\pi xu^2+2\pi inxu}\,\mathrm{d}u,$$

whose inner integral is, by (3.6.24), equal to

$$x^{-1/2}e^{-\pi n^2 x}.$$

Substituting this in (3.6.32) completes the proof.

Rademacher's method rests on expressing the exponent in (3.6.32) as $-\frac{\pi}{x}(t+\alpha+ikx)^2+2\pi k\alpha-\pi k^2 x$, thus expressing the integral in as

$$e^{2\pi k\alpha-\pi k^2 x}\int_L e^{\frac{\pi}{x}z^2}\,\mathrm{d}z, \tag{3.6.33}$$

where L is the line

$$L:\operatorname{Im} z=\operatorname{Im}\alpha+kx,\quad -\infty<\operatorname{Re} z<\infty.$$

The path may be shifted to the real line and the integral (3.6.33) becomes

$$e^{2\pi k\alpha-\pi k^2 x}\frac{C\sqrt{x}}{\sqrt{\pi}},$$

where C is (3.6.26). Hence (3.6.32) becomes

$$\sum_{n=-\infty}^{\infty} e^{-\pi(n+\alpha)^2/x}=\frac{K\sqrt{x}}{\sqrt{\pi}}\sum_{k=-\infty}^{\infty} e^{2\pi k\alpha-\pi k^2 x}. \tag{3.6.34}$$

Putting $x=1$ and $\alpha=0$, this gives $C=\sqrt{\pi}$ and (3.6.34) leads to (3.6.30). □

Exercise 3.6.7 *By the first method of* Selberg and Chowla (1967) *prove* (3.6.3). *Cf. Exercise 2.5.4. For continuous version of gamma transform, cf. Exercise 3.6.4.*

Solution. We start from (3.6.18) and transform $S=S(s,z)$. By the gamma transform,

$$\begin{aligned}\pi^{-s}\Gamma(s)(2S)&=\pi^{-s}\Gamma(s)\sum_{m=-\infty}^{\infty}{}'\sum_{n=-\infty}^{\infty}\left((mx+n)^2+(my)^2\right)^{-s} \qquad (3.6.35)\\&=\sum_{m=-\infty}^{\infty}{}'\int_0^{\infty} t^s e^{-\pi(my)^2 t}\sum_{n=-\infty}^{\infty} e^{-\pi(n+mx)^2 t}\,\frac{\mathrm{d}t}{t}\end{aligned}$$

by interchange of summation and integration. We apply (3.6.30) in the form

$$\sum_{n=-\infty}^{\infty} e^{-\pi(n+\alpha)^2 t}=t^{-\frac{1}{2}}\left(1+2\sum_{n=1}^{\infty} e^{-\pi n^2 t^{-1}}\cos 2\pi n\alpha\right).$$

Then

$$\pi^{-s}\Gamma(s)(2S) = \int_0^\infty t^{s-\frac{1}{2}} \sum_{m=-\infty}^{\infty}{}' e^{-\pi(my)^2 t} \sum_{n=-\infty}^{\infty} e^{-\pi n^2 x^{-1}} \cos 2\pi nmx \, \frac{\mathrm{d}t}{t} \tag{3.6.36}$$

$$= 2\sum_{m=1}^{\infty} \int_0^\infty t^{s-\frac{1}{2}} e^{-\pi(my)^2 t} \frac{\mathrm{d}t}{t}$$

$$+ 4\sum_{m=1}^{\infty}\sum_{n=1}^{\infty} \cos 2\pi nmx \int_0^\infty t^{s-\frac{1}{2}} e^{-\pi\left((my)^2 t + n^2 t^{-1}\right)} \frac{\mathrm{d}t}{t}.$$

The first term on the right of (3.6.36) is $2\sqrt{\pi}\Gamma(s-\frac{1}{2})\zeta(2s-1)y^{-2s+1}$ while the integral in the second term becomes by substitution $t = \frac{n}{ym}u$

$$\left(\frac{n}{ym}\right)^{s-\frac{1}{2}} \int_0^\infty u^{s-\frac{1}{2}} e^{-\pi mny(u+u^{-1})} \frac{\mathrm{d}u}{u}.$$

Hence, putting $mn = \ell$, the second term becomes

$$4y^{\frac{1}{2}-s} \sum_{\ell=1}^{\infty} \ell^{s-\frac{1}{2}} \sigma_{1-2s}(\ell) \cos 2\pi \ell x \int_0^\infty u^{s-\frac{1}{2}} e^{-\pi \ell y(u+u^{-1})} \frac{\mathrm{d}u}{u}$$

whose inner integral is $2K_{s-\frac{1}{2}}(2\pi\ell y)$ by (3.6.17). Hence (3.6.18) leads to

$$\Gamma(s)\zeta_{\mathbb{Z}^2}(s,z) = 2\Gamma(s)\zeta(2s) + 2\pi^s\sqrt{\pi}\Gamma(s-\frac{1}{2})\zeta(2s-1)y^{-2s+1}$$

$$+ 8\pi^s y^{\frac{1}{2}-s} \sum_{\ell=1}^{\infty} \ell^{s-\frac{1}{2}} \sigma_{1-2s}(\ell) \cos 2\pi\ell x \, K_{s-\frac{1}{2}}(2\pi\ell y),$$

whence (3.6.3). □

During elucidation of the Kuznetsov sum formula Agarwal et al. (2023), it has become clear that there are many instances in the theory of automorphic L-functions where known results are not mentioned but proved by another method. One typical example is the case of CSIF, Theorem 2.5.1, cf. Chowla and Selberg (1949), Selberg and Chowla (1967).

Although holomorphy and Fourier expansion of $E(s,z)$ may be proved directly as in Theorem 3.6.2, we may deduce the same results from (1.1.9): (EZF≈EES) and (EMS≈EES).

Continuing Exercise 2.12.2, we state the following

Proposition 3.6.1 *Suppose $w = u + iv$ and $z = x + iy$ are related by $\gamma z = w$, $\gamma \in \Gamma(1)$. We write $f(x,y) = F(u,v) = F(\gamma(x,y))$. Then we have the invariance under the action of $\Gamma(1)$ of the differential form*

$$-f_y \mathrm{d}x + f_x \mathrm{d}y = -F_v \mathrm{d}u + F_u \mathrm{d}v. \tag{3.6.37}$$

This entails invariance of the non-Euclidean Laplacian (3.4.14)

$$\Delta f(x, y) = -v^2(F_{uu} + F_{vv}) = \Delta F(u, v) \tag{3.6.38}$$

as well as the invariance

$$y\frac{\partial f}{\partial n} = v\frac{\partial F}{\partial n}. \tag{3.6.39}$$

of the non-Euclidean outer-normal derivative

$$\frac{\partial}{\partial n} f = \boldsymbol{n} \cdot (f_x, f_y) = -\frac{\mathrm{d}y}{|\mathrm{d}z|}\frac{\partial}{\partial x} f + \frac{\mathrm{d}x}{|\mathrm{d}z|}\frac{\partial}{\partial y} f, \tag{3.6.40}$$

where $|\mathrm{d}z|$ *is the Euclidean arc length element and* $\boldsymbol{n}$ *is the outer normal.*

Proof By the chain rule and the Cauchy–Riemann equation, we have successively

$$\begin{aligned} -f_y\mathrm{d}x + f_x\mathrm{d}y &= -(F_u u_y + F_v v_y)\mathrm{d}x + (F_u u_x + F_v v_x)\mathrm{d}y \\ &= (F_u v_x - F_v u_x)\mathrm{d}x + (F_u v_y + F_v u_y)\mathrm{d}y, \end{aligned}$$

which becomes

$$\begin{aligned} &= -F_v(u_x\mathrm{d}x + u_y\mathrm{d}y) + F_y(v_x\mathrm{d}x + v_y\mathrm{d}y) \\ &= -F_v\mathrm{d}u + F_u\mathrm{d}v. \end{aligned}$$

This proves (3.6.37)

It remains to apply the change of variable formula (2.12.19) to (3.6.37). Differentiating (3.6.37), we deduce that the left-hand side is

$$\mathrm{d}(-f_y\mathrm{d}x + f_x\mathrm{d}y) = (f_{xx} + f_{yy})\mathrm{d}x\mathrm{d}y = -\Delta f\frac{\mathrm{d}x\mathrm{d}y}{y^2}$$

and the right-hand-side is

$$\mathrm{d}(-F_v\mathrm{d}u + F_u\mathrm{d}v) = (F_{uu} + F_{vv})\mathrm{d}u\mathrm{d}v = -\Delta F\frac{\mathrm{d}u\mathrm{d}v}{v^2},$$

whence (3.6.38) follows on account of (2.12.20).

By the second operator $|\mathrm{d}z| = \frac{v}{y}|\mathrm{d}w|$ in (2.12.19), (3.6.40) is equal to

$$\frac{|\mathrm{d}z|}{|\mathrm{d}w|}\left(-\frac{\mathrm{d}v}{|\mathrm{d}w|}F_u + \frac{\mathrm{d}u}{|\mathrm{d}w|}F_v\right) = \frac{v}{y}\frac{\partial}{\partial n}F,$$

whence (3.6.39) follows. □

From here to the end of this section, we provide examples of Maass's method of using the Fourier series and DE.

Example 3.6.1 We shall substantiate the relations (1.1.9): (EZF≈EES) and (EMS≈EES).

We shall first show that EMS has the Fourier–Bessel expansion as a Maass form in Definition 3.6.2 (with $i\nu = s - \frac{1}{2} = \pm i\sqrt{\lambda - 1/4}$):

$$E(s,z) = \sum_{0\neq n\in\mathbb{Z}} a_n\sqrt{y}K_{i\nu}(2\pi|n|y)e(nx) + a_0 y^{1/2+i\nu} + b_0 y^{1/2-i\nu}. \tag{3.6.41}$$

This is the contracted case of Lemma 3.4.1 and holds true for every Maass form.

Then comparing (3.6.41) and (3.6.20), we determine the Fourier coefficients a_n, thereby proving (3.6.4) again. Alternatively appealing to condition (3.6.14) resp. (3.6.15), we deduce both (3.6.4) and Basset's formula (3.6.16), clarifying the close relationship between the Fourier–Bessel expansion for Maass forms and the theory of the K-Bessel function. Another proof of Basset's formula is given in Exercise 3.6.4 based on the continuous HGT.

Proof We proceed as in the proof of Maass that first proving $E(s,z)$ to be an eigen-function of the non-Euclidean Laplace operator Δ and then check that the Fourier coefficient satisfies the Bessel DE. We state the proof of (3.6.41) for $E(s,z)$ but it works for every Maass form. It will turn out that it is simpler to work with $E(s,z)$ than with $\zeta_{\mathbb{Z}^2}(s,z)$.

$E(s,z)$ is 1-periodic in x and smooth, so that it has the convergent Fourier expansion

$$E(s,z) = \sum_{n=-\infty}^{\infty} c_n(y)e^{2\pi inx}, \tag{3.6.42}$$

say. By invariance (3.6.38) of Δ, we have

$$\Delta y(\gamma z)^s = \Delta y(z)^s = s(1-s)y^s.$$

Hence $E(s,z)$ is a Laplace eigen-function with eigen-value $\lambda = s(1-s)$, so that the summand also satisfies the same condition

$$-y^2c_n''(y) + 4\pi^2n^2y^2c_n(y) = -y^2(c_n''(y) + (2\pi in)^2c_n(y)) = \lambda c_n(y). \tag{3.6.43}$$

If $n \neq 0$, then putting $w_n(y) = c_n(4\pi|n|y)$, we find that $w_n''(y)$

$$w_n''(y) + \left(-\frac{1}{4} + \frac{\lambda}{y^2}\right)w_n(y) = 0.$$

By the passage preceding (3.1.12), this has two independent solutions $W_{0,\mu}$ and $M_{0,\mu}$ with $\mu = i\nu = s - \frac{1}{2}$ which reduce to the Bessel functions and we have a general solution

$$c_n(y) = a_n\sqrt{y}K_{i\nu}(2\pi|n|y) + b_n\sqrt{y}I_{i\nu}(2\pi|n|y). \tag{3.6.44}$$

Recall (3.6.23) in the form for $\ell \neq 0$

$$\left(\frac{y}{\pi}\right)^s \Gamma(s) S_2 = \sigma_{1-2s}(\ell)\sqrt{y}\int_0^\infty \exp\left(-\pi y\left(\frac{\ell^2}{t}+t\right)\right) t^{s-\frac{1}{2}}\frac{\mathrm{d}t}{t},$$

which is an integral function in s and of rapid decay in $y \to \infty$ (indeed, it is essentially the Hardy–Hecke transform).

Thus we see that (3.6.19) is of polynomial growth in y as $y \to \infty$ and so the term containing I is ruled out and (3.6.44) amounts to

$$c_n(y) = a_n\sqrt{y}K_{i\nu}(2\pi|n|y).$$

For $n = 0$, (3.6.43) takes the form (3.6.47) and Exercise 3.6.8 gives the solution (3.6.49) for $c_0(y)$. Hence (3.6.41) follows and *a fortiori* the Fourier–Bessel expansion (3.6.41) for the EES up to unknown coefficients a_n. Since the term corresponding to $2y^s$ is determined as $\frac{y^s}{\zeta(2s)}2S_1$ in the proof of Theorem 3.6.2, the a_n are those which are contained in $2S_2$.

Then comparing (3.6.41) and (3.6.21), we have for $n \neq 0$,

$$a_n\sqrt{y}K_{s-1/2}(2\pi|n|y) = c_n(s) = 2S_2 = 2\sigma_{1-2s}(n)\left(\frac{y}{\pi}\right)^s \Gamma(s)\int_{-\infty}^{\infty}\frac{e^{2\pi i\ell x}}{(x^2+y^2)^s}\,\mathrm{d}x \tag{3.6.45}$$

$$= 2\sigma_{1-2s}(n)2|n|^{s-1/2}\sqrt{y}K_{s-1/2}(2\pi|n|y),$$

whence

$$a_n = 4\sigma_{1-2s}(n)|n|^{s-1/2}. \tag{3.6.46}$$

Substituting (3.6.46) in (3.6.45), we deduce

$$2|n|^{s-1/2}\sqrt{y}K_{s-1/2}(2\pi|n|y) = \left(\frac{y}{\pi}\right)^s \Gamma(s)\int_{-\infty}^{\infty}\frac{e^{2\pi i\ell x}}{(x^2+y^2)^s}\,\mathrm{d}x,$$

which leads to Basset's formula (3.6.16) after simple change of variable $z = 2\pi|n|y$, $t = 2\pi|n|x$ and *a fortiori* (3.6.17).

We may also deduce (3.6.46) by appealing not to (3.6.21) but to (3.6.14). Substituting (3.6.23) in (3.6.45), we have

$$a_n\sqrt{y}K_{s-1/2}(2\pi|n|y) = 2\sigma_{1-2s}(n)\sqrt{y}\int_0^\infty \exp\left(-\pi y\left(\frac{n^2}{t}+t\right)\right) t^{s-\frac{1}{2}}\frac{\mathrm{d}t}{t}$$

$$= 2\sigma_{1-2s}(n)\sqrt{y}(\pi y)^{\frac{1}{2}-s}\int_0^\infty \exp\left(\left(\frac{\pi|n|y^2}{u}-u\right)\right) u^{s-\frac{1}{2}}\frac{\mathrm{d}u}{u}$$

by the change of variable.

Multiplying by y^{s-1} and letting $y \to 0$ for $\sigma > 1$, we have the following for $n \neq 0$ by (3.6.14): the left-hand side becomes

$$a_n y^{s-\frac{1}{2}} K_{s-1/2}(2\pi|n|y) \to \frac{1}{2} a_n (2\pi|n|)^{1/2-s} \Gamma\left(s - \frac{1}{2}\right)$$

while the right-hand side becomes

$$2\sigma_{1-2s}(n)\pi^{\frac{1}{2}-s} \int_0^\infty t^{s-\frac{1}{2}} \frac{\mathrm{d}t}{t} = 2\sigma_{1-2s}(n)\pi^{\frac{1}{2}-s} \Gamma\left(s - \frac{1}{2}\right),$$

whence (3.6.46).

Regarding a_0, (3.6.20) and (3.6.21) with $\ell = 0$ establish that

$$a_0 = \frac{2\sqrt{\pi}\Gamma(s-\frac{1}{2})}{\Gamma(s)} \frac{\zeta(2s-1)}{\zeta(2s)},$$

which is the coefficient of the second term on the right of (3.6.4).

This completes the proof. □

We note that (3.6.14) is part of Watson's formula (2.12.50), so that part of FER is involved.

Exercise 3.6.8 *Solve the differential equation*

$$c_0''(y) + \lambda y^{-2} c_0(y) = 0, \tag{3.6.47}$$

Solution. We recall the method of reduction of order of a second-order differential equation

$$y'' + p(t)y' + q(t)y = 0. \tag{3.6.48}$$

With $\varphi(t)$ one solution of (3.6.48), we write

$$y(t) = \varphi(t)v(t).$$

Then (3.6.48) reduces to the first-order linear DE in v':

$$\varphi v'' + \left(2\varphi' + p(t)\varphi\right) v' = 0.$$

This is solved as

$$v'(t) = c \frac{1}{\varphi^2(t)} \exp\left(-\int p(t)\, \mathrm{d}t\right).$$

Hence (3.6.47) is solvable as

$$c_0(y) = \varphi(y)\left(c_1 + c_2\int \frac{1}{\varphi^2(y)}dy\right).$$

To find a particular solution $\varphi(y)$ we set $\varphi(y) = y^s$. Then (3.6.48) reads $(s(s-1)+\lambda)\varphi(y) = 0$ whose characteristic equation $s^2 - s + \lambda = 0$ has two roots $i\nu = s - \frac{1}{2} = \pm i\sqrt{\lambda - 1/4}$. Hence

$$\begin{aligned} c_0(y) &= y^s\left(c_1 + c_2\int y^{-2s}dy\right) \\ &= a_0 y^{1-s} + b_0 y^s \\ &= a_0 y^{1/2-i\nu} + b_0 y^{1/2+i\nu}. \end{aligned} \tag{3.6.49}$$

□

Example 3.6.2 In this example, we deduce (3.6.3) again by the method of Example 3.6.1, i.e. we shall find a pseudo-Maass Fourier expansion for EES corresponding to (3.6.41):

$$\zeta_{\mathbb{Z}^2}(s, z) = \sum_{0\neq n\in\mathbb{Z}} f_n y^{\frac{1}{2}-s} K_{s-\frac{1}{2}}(2\pi|n|y)e(nx) + f_0 + g_0 y^{1/2-2s}.$$

For $\zeta_{\mathbb{Z}^2}(s, z)$ has the convergent Fourier expansion as in (3.6.42)

$$\zeta_{\mathbb{Z}^2}(s, z) = \sum_{n=-\infty}^{\infty} d_n(y)e^{2\pi inx},$$

say. By Exercise 3.6.9, it satisfies the DE (3.6.52) and so does the Fourier coefficient c_n:

$$\left(\frac{\partial^2}{\partial^2 x} + \frac{\partial^2}{\partial^2 y} + \frac{2s}{y}\frac{\partial}{\partial y}\right)c_n(y) = 0. \tag{3.6.50}$$

Then by the method of eliminating the first derivative stated in the solution to Exercise 3.6.2, we see that (3.6.50) leads to

$$d_n''(y) + \frac{2s}{y}d_n'(y) - 4\pi^2 n^2 d_n(y) = 0$$

corresponding to (3.6.43). This is transformed into

$$w'' + \frac{s(1-s)+c}{y^2}w = 0, \quad c = 4\pi^2 n^2 y^2$$

by the change of variable

$$w(y) = y^{-s}d_n(y) \tag{3.6.51}$$

and finally into the modified Bessel DE as in Example 3.6.1.

The rest remains the same and we conclude that

$$d_n(y) = y^{-s}\sqrt{y}K_{s-\frac{1}{2}}(2\pi yn).$$

We confirm that under (3.6.51), $y^s\zeta_{\mathbb{Z}^2}(s, z)$ becomes a Maass form. Cf. Lemma 3.5.1 and its corollary.

Exercise 3.6.9 *Suppose the expression (1.1.8). Then show that Q_1^{-s} satisfies the DE*

$$\left(\frac{\partial^2}{\partial^2 x} + \frac{\partial^2}{\partial^2 y} + \frac{2s}{y}\frac{\partial}{\partial y}\right) Q_1^{-s} = 0. \tag{3.6.52}$$

Solution. We have

$$\frac{\partial}{\partial y} Q_1^{-s} = -sQ_1^{-s-1}(2yn^2), \frac{\partial^2}{\partial^2 y} Q_1^{-s} = -(2n^2)sQ_1^{-s-1}(1 + (-s-1)Q_1^{-1}(2n^2y^2))$$

and

$$\begin{aligned}\frac{\partial}{\partial x} Q_1^{-s} &= -sQ_1^{-s-1}(2xn^2 + 2mn)\\ \frac{\partial^2}{\partial^2 x} Q_1^{-s} &= -(2n^2)sQ_1^{-s-1}(1 + (-s-1)Q_1^{-1}(2n^2x^2 + 4mnx + 2m^2)).\end{aligned}$$

Hence it follows that

$$\left(\frac{\partial^2}{\partial^2 x} + \frac{\partial^2}{\partial^2 y}\right) Q_1^{-s} = 4n^2 s^2 Q_1^{-s-1}$$

and the right-hand side may be written as $-\frac{2s}{y}\frac{\partial}{\partial y} Q_1^{-s}$. This proves (3.6.52). □

Exercise 3.6.10 *Complete the proof of Example 3.6.2 proceeding almost verbatim to that of Example 3.6.1.*

Once the (unramified) functional equation is known, then all are unified in the framework of "The Fourier-Bessel expansion" (Kanemitsu and Tsukada 2014, Chap. 4)

As another example we take up the Poincaré series, holomorphic and non-holomorphic, (3.9.1) (Table 3.6).

Table 3.6 Holomorphic, non-holomorphic Maass forms

Name	Holom.	Non-holom.
Eisenstein	$\zeta_{\mathbb{Z}^2}(s,\alpha)=Z(s,Q_1)$	$E(s,z)=\frac{y^s}{2\zeta(s)}\zeta_{\mathbb{Z}^2}(s,z)$
Epstein zeta	$Z(s,Q)=\sum'_{m,n}\frac{1}{Q(m,n)^s}$	$\frac{y^s}{\zeta(2s)}Z(s,Q_1)$
Poincaré ser.	$U_{am}(z)$, (3.8.5)	$E_{am}(z\|\psi)$

3.7 Fourier–Bessel Expansion and Kronecker Limit Formulas for Non-holomorphic Eisenstein Series of Asai

Asai (1970) gives the Kronecker limit formula for the non-holomorphic Eisenstein series of the Hilbert modular group $\mathrm{SL}_2(\mathfrak{O})$, $\mathfrak{O}=\mathfrak{O}_K$ being the ring of integers (assumed to be a PID) in an algebraic number field F. The method is in the spirit of Exercise 3.6.7 using the Hardy–Hecke transform, which is described in Proposition 3.1.1. We shall give a version according to Liu et al. (2025).

For algebraic preliminaries in this section, cf. the beginning of 2.5. Let K be an algebraic number field of degree $n=r_1+2r_2$, where r_1 resp. $2r_2$ indicates the real resp. imaginary conjugates. For every $\theta\in F$, $\theta\to\theta^{(j)}$ denotes the conjugation map which sends the imaginary ones to $\theta^{(j)}=\theta^{(j+r_1)}$ for $r_1+1\le j\le r_1+r_2$.

Definition 3.7.1 Let $\mathcal{H}_q$ denote the quarternion upper half-plane ($z=\begin{pmatrix}x & -y\\ y & x\end{pmatrix}$, $y>0$).
Let $\mathcal{H}_n=\mathcal{H}^{r_1}\times\mathcal{H}_q{}^{r_2}$ be the upper half-space corresponding to F.
$\mathcal{H}_n\ni \boldsymbol{z}=(z_j)$, where

$$\begin{cases} z_j=x_j+iy_j, & e_j=1,\quad 1\le j\le r_1\\ z_j=\begin{pmatrix}x_j & -y_j\\ y_j & x_j\end{pmatrix}, & e_j=2,\quad r_1+1\le j\le r_1+r_2,\end{cases}\tag{3.7.1}$$

$$\boldsymbol{y}(\boldsymbol{z})=(y(z_j)),\quad y_j=y(z_j)>0,\quad N\boldsymbol{y}(\boldsymbol{z})=\prod_{j=1}^{r_1+r_e}y(z_j)^{e_j},$$

$$\boldsymbol{x}(\boldsymbol{z})=(x(z_j)),\quad x_j=x(z_j).$$

In the case where K is purely imaginary, the number of complex conjugates is viewed as r_1 above.
The group acting on $\mathcal{H}_n$ is $G=\mathrm{SL}_2(\mathbb{R})^{r_1}\times\mathrm{SL}_2(\mathbb{C})^{r_2}$. For $G\ni\sigma=(\sigma_j)$, $\sigma_j=\begin{pmatrix}\alpha_j & \beta_j\\ \gamma_j & \delta_j\end{pmatrix}$, the action is defined by $\sigma\boldsymbol{z}=(\sigma_j z_j)$, where

$$\sigma_j z_j = \frac{\alpha_j z_j + \beta_j}{\gamma_j z_j + \delta_j}, \tag{3.7.2}$$

and where for $r_1 + 1 \le j \le r_1 + r_2$, α_j, etc. are embedded as $\begin{pmatrix} \alpha_j & \\ & \alpha_j \end{pmatrix}$ and $\alpha_j z_j$ are matrix products.
$\Gamma = \mathrm{SL}_2(\mathfrak{O}) \subset G$ is its discontinuous subgroup under the identification $\sigma_j = \sigma^{(j)} = \begin{pmatrix} \alpha^{(j)} & \beta^{(j)} \\ \gamma^{(j)} & \delta^{(j)} \end{pmatrix}$ for $\begin{pmatrix} \alpha & \beta \\ \gamma & \delta \end{pmatrix} \in \Gamma$; $y(\sigma^{(j)} z_j) = y(\gamma^{(j)}, \delta^{(j)}; z_j)$ holds, so that

$$\boldsymbol{y}(\sigma \boldsymbol{z}) = \boldsymbol{y}(\boldsymbol{\gamma}, \boldsymbol{\delta}; \boldsymbol{z}); \tag{3.7.3}$$

Γ_0 is the subgroup of Γ given by

$$\Gamma_0 = \{\sigma = \begin{pmatrix} \alpha & \beta \\ 0 & \delta \end{pmatrix} \in \Gamma\}, \tag{3.7.4}$$

cf. (2.12.10);
Δ is the absolute value of the discriminant of K: $\Delta = N\mathfrak{O}^* = |N\omega|^{-1}$, where $\mathfrak{O}^* = (\omega)$ is the inverse different.

We introduce generalizations of real-analytic Eisenstein series (3.2.6) and EES (1.1.7) in the following definition. In the spirit of Exercise 2.5.4, we stick to the Epstein-type Eisenstein series associated with $F\ \zeta_{\mathfrak{O}^2}(s, z)$ and we deviate from Asai and give some new notation. For every $\mu, \nu \in F$ with $(\mu, \nu) \neq (0, 0)$ and every $\boldsymbol{z} = (z_j) \in \mathcal{H}_n$, we define

$$\tilde{y}(\mu^{(j)}, \nu^{(j)}; z_j) = \frac{1}{|\mu^{(j)} + \nu^{(j)} x_j|^2 + |\nu^{(j)} y_j|^2} \tag{3.7.5}$$

and

$$\tilde{\boldsymbol{y}}(\mu, \nu; \boldsymbol{z}) = (\tilde{y}(\mu^{(j)}, \nu^{(j)}; z_j)), \quad N\tilde{\boldsymbol{y}}(\mu, \nu; \boldsymbol{z}) = \prod_{j=1}^{r_1+r_2} \tilde{y}(\mu^{(j)}, \nu^{(j)}; z_j)^{e_j}. \tag{3.7.6}$$

Asai introduced

$$y(\mu^{(j)}, \nu^{(j)}; z_j) = y_j \tilde{y}(\mu^{(j)}, \nu^{(j)}; z_j) = \frac{y_j}{|\mu^{(j)} + \nu^{(j)} x_j|^2 + |\nu^{(j)} y_j|^2}$$

and

$$\boldsymbol{y}(\mu, \nu; \boldsymbol{z}) = (y(\mu^{(j)}, \nu^{(j)}; z_j)), \quad N\boldsymbol{y}(\mu, \nu; \boldsymbol{z}) = \prod_{j=1}^{r_1+r_2} y(\mu^{(j)}, \nu^{(j)}; z_j)^{e_j}.$$

It follows that

$$N\boldsymbol{y}(\mu,\nu;\boldsymbol{z}) = N\boldsymbol{y}(\boldsymbol{z})N\tilde{\boldsymbol{y}}(\mu,\nu;\boldsymbol{z}) = \prod_{j=1}^{r_1+r_2} y_j^{e_j} \prod_{j=1}^{r_1+r_2} \tilde{y}(\mu^{(j)},\nu^{(j)};z_j)^{e_j}. \tag{3.7.7}$$

Definition 3.7.2 A generalization of the Eisenstein–Maass series (3.2.6) is

$$E(s,\boldsymbol{z}) = \sum_{\sigma\in\Gamma_0\backslash\Gamma} N\boldsymbol{y}(\sigma\boldsymbol{z})^s, \tag{3.7.8}$$

and that of (EES) is

$$\tilde{\zeta}_{\mathfrak{O}^2}(s,\boldsymbol{z}) = \sum_{\mu,\nu}{}' N\tilde{\boldsymbol{y}}(\mu,\nu;\boldsymbol{z})^s = \sum_{\mu,\nu}{}' \prod_{j=1}^{r_1+r_2} \left(\frac{1}{|\mu^{(j)}+\nu^{(j)}x_j|^2+|\nu^{(j)}y_j|^2}\right)^s, \tag{3.7.9}$$

where the prime on the summation sign means that (μ,ν) runs over all non-associated pairs $(\mu,\nu)\in\mathfrak{O}^2$ except $(0,0)$, and where two pairs (μ,ν) and (μ_1,ν_1) are called associated if both relations $\mu x = \varepsilon\mu$ and $\nu_1 = \varepsilon\nu$ hold with the same unit ε of $\mathfrak{O}$. The real-analytic Eisenstein series (3.6.1) is generalized to

$$E^*(s,\boldsymbol{z}) = \sum_{\mu,\nu}{}' N\boldsymbol{y}(\mu,\nu;\boldsymbol{z})^s = N\boldsymbol{y}(\boldsymbol{z})^s\tilde{\zeta}_{\mathfrak{O}^2}(s,\boldsymbol{z}). \tag{3.7.10}$$

Then as a reminiscent to (3.2.7),

$$\begin{aligned}\zeta_K(2s)E(s,\boldsymbol{z}) &= E^*(s,\boldsymbol{z}) = N\boldsymbol{y}(\boldsymbol{z})^s\tilde{\zeta}_{\mathfrak{O}^2}(s,\boldsymbol{z}) \\ &= N\boldsymbol{y}(\boldsymbol{z})^s\sum_{\mu,\nu}{}' N\tilde{\boldsymbol{y}}(\mu,\nu;\boldsymbol{z}) = N\boldsymbol{y}(\boldsymbol{z})^s\sum_{\mu,\nu}{}'\prod_{j=1}^{r_1+r_2}\tilde{y}(\mu^{(j)},\nu^{(j)};z_j)^{(e_j)s}.\end{aligned} \tag{3.7.11}$$

Theorem 3.7.1 (Asai 1970, (13)) *The Chowla–Selberg integral formula*

$$\begin{aligned}\tilde{\zeta}_{\mathfrak{O}^2}(s,\boldsymbol{z}) &= \zeta_K(2s) \\ &+ \frac{1}{\sqrt{\Delta}}N\boldsymbol{y}(\boldsymbol{z})^{1-2s}\left(\frac{\sqrt{\pi}\Gamma(s-\frac12)}{\Gamma(s)}\right)^{r_1}\left(\frac{\pi\Gamma(2s-1)}{\Gamma(2s)}\right)^{r_2}\zeta_K(2s-1) \\ &+ \frac{2^{r_1+r_2}}{\sqrt{\Delta}^s}N\boldsymbol{y}(\boldsymbol{z})^{\frac12-s}\left(\frac{\pi^s}{\Gamma(s)}\right)^{r_1}\left(\frac{(2\pi)^{2s}}{\Gamma(2s)}\right)^{r_2}\sum_{\substack{(\mu,\nu)'\\ \mu\nu\neq0}}\left|\frac{N\nu}{N\mu}\right|^{s-\frac12}e^{2\pi i R(\mu\nu\omega\boldsymbol{x})} \\ &\times\prod_{j=1}^{r_1+r_2}K_{e_j(s-\frac12)}\left(2e_j\pi|(\mu\nu\omega)^{(i)}|y_j\right)\end{aligned} \tag{3.7.12}$$

holds true, where

$$R(\mu\nu\omega\boldsymbol{x}) = \sum_{j=1}^{r_1+r_2} e_j \operatorname{Re}\left((\mu\nu\omega)^{(j)} x_j\right) \tag{3.7.13}$$

and $(\mu, \nu)'$ *is a class of pairs* $(\mu, \nu) \in \mathfrak{O} \times \mathfrak{O}$ *under the equivalence relation* $(\mu, \nu) \sim (\mu\varepsilon, \nu\varepsilon^{-1})$ *for a unit* ε.

Proof Proof goes in the spirit of, and parallel to, the first proof of Selberg and Chowla (1967) with some more complications, cf. Exercise 3.6.7. By the dissection (to secure $\nu \neq 0$)

$$\sum_{(\mu,\nu)\neq(0,0)} = \sum_{\substack{0\neq(\mu)\subset\mathfrak{O} \\ \nu=0}} + \sum_{0\neq(\nu)\subset\mathfrak{O}} \sum_{\mu\in\mathfrak{O}},$$

we find that

$$\tilde{\zeta}_{\mathfrak{O}^2}(s, z) = \sum_{0\neq(\mu)\subset\mathfrak{O}} \tilde{y}(\mu, 0, z)^s + S_2 = \zeta_K(2s) + S_2, \tag{3.7.14}$$

where

$$S_2 = \sum_{0\neq(\nu)\subset\mathfrak{O}} \sum_{\mu\in\mathfrak{O}} \prod_{j=1}^{r_1+r_2} \frac{(e_j\pi)^{e_j s}}{\Gamma(e_j s)} I_j(s, z) \tag{3.7.15}$$

and where

$$I_j(s, z) = \int_0^\infty t^{e_j s} e^{-e_j \pi t_j \tilde{y}(\mu^{(i)}, \nu^{(i)}, z_j)^{-1}} \frac{dt_j}{t_j}, \tag{3.7.16}$$

$$\sum_{\mu\in\mathfrak{O}} \prod_j I_j(s, z) = \int_0^\infty \cdots \int_0^\infty \prod_j t_j^{e_j s} e^{-\pi(\sum_j e_j t_j y_j^2 |\nu^{(j)}|^2)} \theta(\boldsymbol{t}) \frac{dt_j}{t_j}.$$

Here $\theta(\boldsymbol{t})$ is Hecke's theta-function

$$\theta(\boldsymbol{t}) = \sum_{\mu\in\mathfrak{O}} e^{-\pi(\sum_j e_j t_j |\mu^{(j)} + \nu^{(j)} x_j|^2)},$$

which satisfies the transformation formula

$$\begin{aligned} \theta(\boldsymbol{t}) &= \sqrt{\Delta}^{-1} \prod_j t_j^{-\frac{e_j}{2}} \sum_{\mu_1\in\mathfrak{O}^*} e^{-\pi \sum_j e_j t_j^{-1} |\mu_1^{(j)}|^2 + 2\pi i R(\mu_1\nu\boldsymbol{x})} \\ &= \sqrt{\Delta}^{-1} \prod_j t_j^{-\frac{e_j}{2}} + \sqrt{\Delta}^{-1} \prod_j t_j^{-\frac{e_j}{2}} \sum_{0\neq\mu_1\in\mathfrak{O}^*} e^{-\pi \sum_j e_j t_j^{-1} |\mu_1^{(j)}|^2 + 2\pi i R(\mu_1\nu\boldsymbol{x})} \end{aligned} \tag{3.7.17}$$

where R is defined in (3.7.13).

Substituting (3.7.17) in the second equality (3.7.16), we find that the first term of (3.7.17) contributes

$$\sqrt{\Delta}^{-1}\int_0^\infty\cdots\int_0^\infty\prod_j t_j^{e_j(s-\frac{1}{2})}e^{-\pi(\sum_j e_j y_j^2|\nu^{(j)}|^2)t_j}\frac{\mathrm{d}t_j}{t_j} \tag{3.7.18}$$
$$=\sqrt{\Delta}^{-1}\prod_j\Gamma\left(e_j\left(s-\frac{1}{2}\right)\right)\pi^{n(\frac{1}{2}-s)}2^{r_2(1-2s)}N\mathbf{y}(z)^{1-2s}|\nu|^{1-2s}.$$

Hence that part of S_2 in (3.7.15) corresponding to (3.7.18) is

$$S_2^{(1)}:=\sum_{0\neq(\nu)\subset\mathfrak{O}}\sum_{\mu_1=0}\prod_{j=1}^{r_1+r_2}\frac{(e_j\pi)^{e_js}}{\Gamma(e_js)}I_j(s,z) \tag{3.7.19}$$
$$=\sqrt{\Delta}^{-1}N\mathbf{y}(z)^{1-2s}\left(\frac{\sqrt{\pi}\Gamma\left(\left(s-\frac{1}{2}\right)\right)}{\Gamma(s)}\right)^{r_1}\left(\frac{2\pi\Gamma\left(2\pi\left(2s-1\right)\right)}{\Gamma\left(2s\right)}\right)^{r_2}\zeta_K(2s-1).$$

Now what remains is that part of S_2 which contains the second term in (3.7.17)

$$S_2^{(2)}:=\sum_{0\neq(\nu)\subset\mathfrak{O}}\sum_{\mu_1\neq0}\prod_{j=1}^{r_1+r_2}\frac{(e_j\pi)^{e_js}}{\Gamma(e_js)}I_j(s,z) \tag{3.7.20}$$
$$=\sqrt{\Delta}^{-1}\prod_j\frac{(e_j\pi)^{e_js}}{\Gamma(e_js)}\sum_{(\mu,\nu)'}e^{2\pi iR(\mu\nu\omega\mathbf{x})}\prod_j H_{\mu,\nu}^{(j)}$$

on expressing the sum $\sum_{0\neq(\nu)\subset\mathfrak{O}}\sum_{\mu_1\neq0}$ as $\sum_{\substack{(\mu,\nu)',\mu\nu\neq0\\ \mu_1=\mu\omega}}$ or $\sum_{(\mu,\nu)'}$ as stated in the theorem. Here

$$H_{\mu,\nu}^{(j)}=\int_0^\infty t_j^{e_j(s-\frac{1}{2})}e^{-e_j\pi\left(y_j^2|\nu^{(j)}|^2t_j+|\mu^{(j)}\omega^{(j)}|^2t_j^{-1}\right)}\frac{\mathrm{d}t_j}{t_j}, \tag{3.7.21}$$

which is the Hecke transform in Proposition 3.1.1. Hence

$$H_{\mu,\nu}^{(j)}=\left|\frac{\mu^{(j)}\omega^{(j)}}{\nu^{(j)}y_j}\right|^{e_j(s-\frac{1}{2})}2K_{e_j(s-\frac{1}{2})}(2|\mu^{(j)}\nu^{(j)}\omega^{(j)}|y_j),$$

so that

$$\prod_j H_{\mu,\nu}^{(j)}=2^{r_1+r_2}\left|\frac{\mu\omega}{\nu N\mathbf{y}}\right|^{s-\frac{1}{2}}\prod_j K_{e_j(s-\frac{1}{2})}(2|\mu^{(j)}\nu^{(j)}\omega^{(j)}|y_j) \tag{3.7.22}$$
$$=2^{r_1+r_2}\Delta^{\frac{1}{2}-s}N\mathbf{y}^{\frac{1}{2}-s}\left|\frac{\mu}{\nu}\right|^{s-\frac{1}{2}}\prod_j K_{e_j(s-\frac{1}{2})}(2|\mu^{(j)}\nu^{(j)}\omega^{(j)}|y_j).$$

Table 3.7 Correspondence between $E(s, z)$ and $E(s, z)$

Data	$E(s, z)$	$E(s, z)$
PID	$\mathfrak{O} = (\omega)$	$\mathbb{Z} = (1)$
Mod. gr.	$\Gamma = \mathrm{SL}_2(\mathfrak{O})$	$\mathrm{SL}_2(\mathbb{Z})$
Γ_0	(3.7.4)	$\{\pm 1\}$
Factor gr.	$\Gamma_0\backslash\Gamma$	$\mathrm{PSL}_2(\mathbb{Z})$
UHS	$\mathcal{H}_n$	$\mathcal{H}$
|discr.|	Δ	1

Hence

$$S_2^{(2)} = 2^{r_1+r_2}\Delta^{-s} N\boldsymbol{y}^{\frac{1}{2}-s} \prod_j \frac{(e_j\pi)^{e_j s}}{\Gamma(e_j s)} \times \sum_{(\mu,\nu)'} \left|\frac{\mu}{\nu}\right|^{s-\frac{1}{2}} e^{2\pi i R(\mu\nu\omega\boldsymbol{x})} \prod_j K_{e_j(s-\frac{1}{2})}(2|\mu^{(j)}\nu^{(j)}\omega^{(j)}|y_j). \tag{3.7.23}$$

Thus, substituting (3.7.19) and (3.7.23) in (3.7.14), we obtain the Chowla–Selberg integral formula for $\tilde{\zeta}_{\mathfrak{O}^2}(s, z)$, whence we complete the proof. □

The special case of this is the Chowla–Selberg integral formula for the Eisenstein–Maass series, cf. Theorem 3.6.1 (Table 3.7).

Corollary 3.7.1 *Suppose $r_2 = 0$, so that there is no quaternion upper half-plane and $\mathcal{H}$ is the product of r_1 copies of the upper half-plane. In the imaginary quadratic case, we regard $r_1 = 1$. Then the Chowla–Selberg integral formula* (3.6.4) *reads*

$$\mathcal{E}(s, z) = \frac{1}{2}\zeta(2s)E(s, z) = y^s\zeta(2s) + y^{1-s}\frac{\sqrt{\pi}\Gamma(s-\frac{1}{2})}{\Gamma(s)}\zeta(2s-1) + \frac{4y^{\frac{1}{2}}\pi^s}{\Gamma(s)}\sum_{n=1}^{\infty} n^{s-\frac{1}{2}}\sigma_{1-2s}(n)\cos 2\pi nx\, K_{s-\frac{1}{2}}(2\pi yn)\,. \tag{3.7.24}$$

Proof We specify the data. $\mathfrak{I} = \mathbb{Z} = \langle 1\rangle$ and so $\omega = 1$, $\Delta = 1$ and $\tilde{\zeta}_{\mathfrak{O}^2}(s, z) = \frac{1}{2}\zeta_{\mathfrak{O}^2}(s, z)$. □

Corollary 3.7.2 *$E(s, z)$ satisfies the functional equation similar to (4.1.9):*

$$A^{-2s}\Gamma^{r_1}(s)\,\Gamma^{r_2}(2s)E(s, z) = A^{-(2-2s)}\Gamma^{r_1}(1-s)\,\Gamma^{r_2}(2-2s)E(1-s, z). \tag{3.7.25}$$

Proof follows in the same way as that of (2.5.22) on using (4.1.9).

3.8 Modular Forms: $SL_2(\mathbb{R})$-Case

In the next section, we will provide Petersson's trace formula as a precursor to the Kuznetsov trace formula, cf., e.g. Agarwal et al. (2023). In this section, as a starter, we supplement some basics on holomorphic automorphic forms w.r.t $SL_2(\mathbb{R})$ to the material presented in Sect. 2.12, cf., e.g. Lang (1975), etc. Although we mainly treat discrete subgroups of

$$SL_2(\mathbb{R}) = \left\{\gamma = \begin{pmatrix} a & b \\ c & d \end{pmatrix} |a, b, c, d \in \mathbb{R}, ad - bc = 1\right\}, \tag{3.8.1}$$

expecting future generalizations, we state some results on discrete subgroups of $G = PGL_2(\mathbb{R})$. (S and G are for special and general; recall the full modular group (2.12.8)). We may let $SL_2(\mathbb{R})$ act on $\bar{\mathbb{C}} = \mathbb{C} \cup \{\infty\}$ by the linear fractional transformation (or Möbius transformation, cf. (2.12.9)):

$$\gamma z = \frac{az + b}{cz + d} \quad \gamma = \begin{pmatrix} a & b \\ c & d \end{pmatrix}, \quad z \in \mathcal{H}.$$

By Exercise 2.12.2, the upper half-plane $\mathcal{H}$ (or the Lobačevskiĭ plane) is stable under the action of $SL_2(\mathbb{R})$.

Let Γ denote the **Fuchsian group of the first kind** (a discrete subgroup of $G = PGL_2(\mathbb{R})$ with $\Gamma\backslash G$ non-compact and its fundamental domain of finite volume) and let $\mathcal{M}_k(\Gamma)$ (M is for modular) be the linear space of all *holomorphic* functions $f : \mathcal{H} \to \mathbb{C}$ satisfying the automorphic property

$$j_\gamma(z)^{-k} f(\gamma z) = f(z), \qquad \gamma \in \Gamma, \tag{3.8.2}$$

where $j_\gamma(z)$ indicates the denominator (2.12.13). More concretely (3.8.2) reads

$$(cz + d)^{-k} f\left(\frac{az + b}{cz + d}\right) = f(z), \quad z \in \mathcal{H}, \quad \gamma \in \Gamma. \tag{3.8.3}$$

The element f of $\mathcal{M}_k(\Gamma)$ is called an **automorphic form** (or a modular form) of weight k with respect to Γ, where $k \geq 0$ is an *even* integer (or some write $2k$ for k).

One motivation for considering condition (3.8.2) can be found on Serre (1973, p. 80).

In view of (2.12.14), (3.8.2) reads

$$\frac{f(\gamma z)}{f(z)} = \left(\frac{d(\gamma z)}{dz}\right)^{-\frac{k}{2}}$$

or

$$f(\gamma z)\, d(\gamma z)^{\frac{k}{2}} = f(z)\, dz^{\frac{k}{2}},$$

i.e. the differential form $f(z)\,\mathrm{d}z^{\frac{k}{2}}$ of degree $\frac{k}{2}$ is invariant under Γ.

For each cusp $\mathfrak{a}$ we fix the element $\sigma_{\mathfrak{a}} \in \mathrm{SL}_2(\mathbb{R})$ such that

$$\sigma_{\mathfrak{a}}\infty = \mathfrak{a}, \quad \sigma_{\mathfrak{a}}^{-1}\Gamma_{\mathfrak{a}}\sigma_{\mathfrak{a}} = \Gamma_{\infty},$$

where $\Gamma_{\mathfrak{a}}$ is the stabilizer of $\mathfrak{a}$ (parabolic subgroup)

$$\Gamma_{\mathfrak{a}} = \{\gamma \in \Gamma | \gamma\mathfrak{a} = \mathfrak{a}\}.$$

Every $f \in \mathcal{M}_k$ has the **Fourier expansion**

$$j_{\sigma_{\mathfrak{a}}}(z)^{-k} f(\sigma_{\mathfrak{a}} z) = \sum_{n=0}^{\infty} \hat{f}_{\mathfrak{a}}(n) e(nz), \tag{3.8.4}$$

which converges absolutely and uniformly on compact subsets. If at every cusp $\mathfrak{a}$, the constant term vanishes

$$\hat{f}_{\mathfrak{a}}(0) = 0,$$

then f is called a **cusp form** (or a Spitzenform). Every cusp form is of exponential decay at its cusps, and, in particular, $y^{k/2} f(z)$ is bounded on $\mathcal{H}$. Let $\mathcal{S}_k(\Gamma)$ ($\mathcal{S}$ is for Spitzenform) denote the space of all cusp forms of weight k which is equipped with the **Petersson scalar product** (or inner product) Petersson (1939)

$$\langle f, g \rangle_k = \int_{\mathcal{F}} y^k f(z)\overline{g(z)}\,\mathrm{d}z = \int_{\mathcal{F}} y^k f(z)\overline{g(z)}\,\frac{\mathrm{d}x\mathrm{d}y}{y^2}$$

which makes sense in $\mathcal{M}_k(\Gamma)$ for $k > 2$. Cf. Exercise 2.12.3.

Theorem 3.8.1 *$\mathcal{S}_k(\Gamma)$ is spanned by holomorphic Poincaré series (in contrast to non-holomorphic ones in* (3.9.1)*):*

$$U_{\mathfrak{a}m}(z) = \sum_{\gamma \in \Gamma_{\alpha}\backslash\Gamma} j_{\sigma_{\mathfrak{a}}^{-1}\gamma}(z)^{-k} e(m\sigma_{\mathfrak{a}}^{-1}\gamma z). \tag{3.8.5}$$

Proof The subspace spanned by the holomorphic Poincaré series is closed and any function orthogonal to the subspace is 0 by Petersson's formula (3.9.2). □

In the case $\mathfrak{a} = \infty$, (3.8.5) reduces to

$$U_m(z, k) = U_{\infty m}(z, k) = \sum_{\gamma \in \Gamma_{\infty}\backslash\Gamma} j_{\gamma}(z)^{-k} e(m\gamma(z)).$$

We shall briefly state the effect of the Hecke operator (3.8.6) on Maass cusp forms in Example 3.6.1 in analogy to Exercise 2.12.4. Apply

$$(T_n f)(z) = n^\lambda \sum_{d|n} \left(\frac{n}{d}\right)^k \sum_{b \bmod d} f\left(\frac{\frac{n}{d}z + b}{d}\right), \tag{3.8.6}$$

to the Fourier expansion

$$f(z) = \sum_{0 \neq m \in \mathbb{Z}} a(m) \sqrt{y} K_{i\nu}(2\pi|m|y) e^{2\pi i m x}. \tag{3.8.7}$$

Then

$$(T_n f)(z)$$
$$= \sqrt{y} n^{\lambda + \frac{1}{2}} \sum_{0<d|n} \sum_{m \neq 0} \left(\frac{n}{d}\right)^k a(m) K_{i\nu}\left(\frac{2\pi|m|ny}{d^2}\right) e\left(\frac{mn}{d^2}x\right) \frac{1}{d} \sum_{b \bmod d} e^{2\pi i \frac{m}{d} b}$$

and as in Exercise 2.12.4,

$$(T_n f)(z) = \sqrt{y} n^{\lambda + \frac{1}{2}} \sum_{d|n} \sum_{d|m, m \neq 0} \left(\frac{n}{d}\right)^k a(m) K_{i\nu}\left(2\pi \frac{|m|n}{d^2} y\right) e^{2\pi i \frac{mn}{d^2} x}.$$

Writing $m = \ell d$, $\ell \neq 0$ and then $\delta = \frac{n}{d}$, we have

$$(T_n f)(z) = \sqrt{y} n^{\lambda + \frac{1}{2}} \sum_{d|n} \sum_{\ell \neq 0} \delta^k a\left(\frac{\ell \delta n}{\delta^2}\right) K_{i\nu}\left(2\pi|\ell|\delta y\right) e^{2\pi i \ell \delta x},$$

whence collecting terms with $\ell\delta = m$, we conclude that

$$(T_n f)(z) = \sum_{m \neq 0} \tilde{t}_n(m) \sqrt{y} K_{i\nu}(2\pi|m|y) e^{2\pi i m x},$$

where correspondingly to (2.12.23)

$$\tilde{t}_n(m) = n^{\lambda + \frac{1}{2}} \sum_{d|(m,n)} d^k a\left(\frac{mn}{d^2}\right). \tag{3.8.8}$$

Choosing $k = 0$, $\lambda = -\frac{1}{2}$, we find that (3.8.6) and (3.8.8) are to read

$$(T_n f)(z) = n^{-\frac{1}{2}} \sum_{d|n} \sum_{b \bmod d} f\left(\frac{\frac{n}{d}z + b}{d}\right)$$

and

$$\tilde{t}_n(m) = \sum_{d|(m,n)} a\left(\frac{mn}{d^2}\right).$$

Corollary 3.8.1 (Corollary to Theorem 2.12.1) *We have the multiplication formula*

$$T_m T_n = \sum_{d|(m,n)} T_{\frac{mn}{d^2}}.$$

Corresponding to Example 2.12.1, we define the L-function associated to the Maass form f in (3.8.7) by

$$L(s, f). = \sum_{m=1}^{\infty} \frac{a(m)}{m^s}, \quad \sigma > \frac{3}{2},$$

where $a(m)$ are Maass–Fourier coefficients (Motohashi 1968, pp. 12–13). Then it has the Euler product corresponding to (2.12.37)

$$L(s, f) = \prod_p \left(1 - \frac{a(p)}{p^s} + \frac{p^{k-1}}{p^{2s}}\right)^{-1}.$$

3.9 The Petersson Trace Formula

In this section, we provide the Petersson trace formula as a prototype of the Kuznetsov trace formula. The unfolding method of Rankin (1939) and Selberg (1940) has been extensively used in spectral theory of automorphic functions.

In pursuit of elucidation of the Kuznetsov sum formula (Agarwal et al., 2023), it has turned out that the key idea is due to Selberg (1940) (Selberg (1956)) and depends on two ways of expressing the inner product of two Poincaré series

$$(P_m(\cdot, s_1), P_n(\cdot, \bar{s}_2)) = \int_{\mathcal{F}} P_m(z, s_1)\overline{P_n(z, \bar{s}_2)}\, dz,$$

where $P_m(z, s)$ is the real-analytic Poincaré series defined on Venkov (1990, p. 36) which is referred to in Kanemitsu et al. (2017). We state the one from Motohashi (1997, p. 4):

$$P_m(z, s) = P_m(z, s, 1) = \sum_{\gamma \in \Gamma_\infty \backslash \Gamma} y(\gamma z)^s e^{2\pi i m \gamma(z)}, \quad z \in \mathcal{H}. \tag{3.9.1}$$

The inner product of Poincaré series is computed in Motohashi (1997, p. 44) for integers $m, n > 0$. Equating the Unfolding formula and the Parseval formula, we may deduce the Kuznetsov sum formula in an accessible way. For partial proofs cf. Agarwal et al. (2023) and Kanemitsu et al. (2017).

As mentioned above, Petersson (1932) already used unfolding method to prove the formula for the Fourier coefficients in (3.8.4)

$$\hat{f}_{\mathfrak{a}}(n) = \frac{(4\pi n)^{k-1}}{(k-2)!} \langle f, U_{\mathfrak{a}n}(\cdot, k)\rangle_k \tag{3.9.2}$$

and obtained

$$j_{\sigma_{\mathfrak{b}}}(z)^{-k} U_{\mathfrak{a}m}(\sigma_{\mathfrak{b}} z, k) = \sum_{n=1}^{\infty} \left(\frac{n}{m}\right)^{\frac{k-1}{2}} \hat{U}_{\mathfrak{a}\mathfrak{b}}(m, n) e(nz) \tag{3.9.3}$$

with

$$\hat{U}_{\mathfrak{a}\mathfrak{b}}(m, n) = \delta_{\mathfrak{a}\mathfrak{b}} \delta_{mn} + 2\pi i^k \sum_c \frac{S(m, n; c)}{c} J_{k-1}\left(\frac{4\pi\sqrt{mn}}{c}\right), \tag{3.9.4}$$

where $S(m, n; c) = S_{\mathfrak{a}\mathfrak{b}}(m, n; c)$ indicates the Kloosterman sum at cusps. We quote only the cusp at ∞ case:

$$S(m, n; c) = \sum_{\substack{d (\mathrm{mod} c) \\ (d, c)=1}} e^{2\pi i \frac{md^* + nd}{c}}, \tag{3.9.5}$$

and where d^* is the inverse of d mod c.

Below we give a rather sketchy treatment of Theorems 3.9.1 and 3.9.2. For details Cf. Agarwal et al. (2023) or Iwaniec (1995-2002).

Let $\{f_{jk}\}_{1 \le j \le \vartheta_k}$ be an orthonormal basis of $\mathcal{M}_k(\Gamma)$ and let

$$f_{jk}(z) = \sum_{n=1}^{\infty} \hat{f}_{\mathfrak{a}jk}(n) e(nz)$$

be the expansion at the cusp $\mathfrak{a}$.

Let

$$U_{\mathfrak{a}m}(z, k) = \frac{(k-2)!}{(4\pi m)^{k-1}} \sum_j \overline{\hat{f}_{\mathfrak{a}jk}(m)} f_{jk}(z)$$

be the expansion of Poincaré series with respect to this basis. We write this as

$$\begin{aligned} U_{\mathfrak{a}m}(\sigma_{\mathfrak{b}} z, k) &= \frac{(k-2)!}{(4\pi m)^{k-1}} \sum_j \overline{\hat{f}_{\mathfrak{a}jk}(m)} f_{jk}(\sigma_{\mathfrak{b}} z) \\ &= \frac{(k-2)!}{(4\pi m)^{k-1}} \sum_j \overline{\hat{f}_{\mathfrak{a}jk}(m)} j_{\sigma_{\mathfrak{b}}}(z)^k \sum_{n=1}^{\infty} \hat{f}_{\mathfrak{b}jk}(n) e(nz). \end{aligned} \tag{3.9.6}$$

Equations (3.9.3) and (3.9.6) multiplied by $j_{\sigma_{\mathfrak{b}}}(z)^{-k}$, we conclude that

$$\frac{(k-2)!}{(4\pi m)^{k-1}}\sum_{n=1}^{\infty}\sum_{j}\overline{\hat{f}_{\mathfrak{a}jk}(m)}\hat{f}_{\mathfrak{b}jk}(n)e(nz)$$
$$= j_{\sigma_{\mathfrak{b}}}(z)^{-k}U_{\mathfrak{a}m}(\sigma_{\mathfrak{b}}z,k)=\sum_{n=1}^{\infty}\left(\frac{n}{m}\right)^{\frac{k-1}{2}}\hat{U}_{\mathfrak{a}\mathfrak{b}}(m,n)e(nz),$$

whence that

$$\frac{(k-2)!}{(4\pi m)^{k-1}}\sum_{j}\overline{\hat{f}_{\mathfrak{a}jk}(m)}\hat{f}_{\mathfrak{b}jk}(z)=\left(\frac{n}{m}\right)^{\frac{k-1}{2}}\hat{U}_{\mathfrak{a}\mathfrak{b}}(m,n)$$

or

$$\frac{(k-2)!}{(4\pi\sqrt{mn})^{k-1}}\sum_{j}\overline{\hat{f}_{\mathfrak{a}jk}(m)}\hat{f}_{\mathfrak{b}jk}(z)=\hat{U}_{\mathfrak{a}\mathfrak{b}}(m,n). \tag{3.9.7}$$

Equations (3.9.7) and (3.9.4), we deduce the **Petersson trace formula**.

Theorem 3.9.1 *Let m, n be positive integers and k a positive even integer. Then*

$$\sum_{c}\frac{S(m,n;c)}{c}J_{k-1}\left(\frac{4\pi\sqrt{mn}}{c}\right) \tag{3.9.8}$$
$$=\delta_{\mathfrak{a}\mathfrak{b}}\delta_{mn}\frac{(-1)^{k-1}}{2\pi}+\frac{(-1)^{k}}{2\pi}\frac{(k-2)!}{(4\pi\sqrt{mn})^{k-1}}\sum_{j}\overline{\hat{f}_{\mathfrak{a}jk}(m)}\hat{f}_{\mathfrak{b}jk}(n).$$

The name is stated only on Motohashi (1997, p. 51) and no other books have this terminology. Some papers refer to the Petersson trace formula and refer to Iwaniec (1997, p. 54, Theorem 3.6) in which there is no mention of this terminology, but stated as Petersson's formulas.

Multiplying (3.9.8) by $2(2k-1)N_f(2k-1)$ and summing over $k=1,2,\cdots$, we conclude

Theorem 3.9.2 (Iwaniec 1995-2002, Theorem 9.6) *Let $\mathfrak{a}$, $\mathfrak{b}$ be cusps of the Fuchsian group of the first kind Γ and let $m, n > 0$ be integers. Then for any test function f satisfying the condition*

$$f(0)=0,\quad f^{(j)}(x)\ll(x+1)^{-j-1},\quad j=0,1,2,$$

we have

$$- \delta_{\mathfrak{a}\mathfrak{b}}\delta_{mn} f^{\infty} + \sum_{c} c^{-1} S_{\mathfrak{a}\mathfrak{b}}(m, n; c) f^{0}\left(\frac{4\pi\sqrt{|mn|}}{c}\right)$$

$$= \sum_{k=1}^{\infty} i^{2k} N_f(2k-1)\bar{\psi}_{\mathfrak{a}j2k}(m)\psi_{\mathfrak{b}j2k}(n),$$

where $\psi_{\mathfrak{a}jk(m)}$ *are the normalized Fourier coefficients*

$$\psi_{\mathfrak{a}jk}(m) = \left(\frac{\pi^{-k}\Gamma(k)}{(4\,m)^{k-1}}\right)^{1/2} \hat{f}_{\mathfrak{a}jk}(m).$$

This is the exact complement to the Kuznetsov sum formula (Agarwal et al. 2023, Theorem 4.2).

Proof of (3.9.2) by unfolding method. We restrict to the case $\mathfrak{a} = \infty$. Then

$$\langle f, U_m(\cdot, k)\rangle_k = \int_{\Gamma\backslash\mathcal{H}} \sum_{n=0}^{\infty} \hat{f}(n) e^{2\pi i m z} \sum_{\gamma\in\Gamma_\infty\backslash\Gamma} \overline{\operatorname{Im}\gamma z} \cdot \overline{e^{2\pi i n z}} y^{k-1}\, \mathrm{d}x\mathrm{d}y.$$

Since $\int_{\Gamma\backslash\mathcal{H}} \sum_{\gamma\in\Gamma_\infty\backslash\Gamma} = \int_{\Gamma_\infty\backslash\mathcal{H}}$ (**unfolding method**) and the series in the integrand are of exponential decay in y, we may freely interchange the summation and integration to deduce

$$\langle f, U_m(\cdot, k)\rangle_k = \sum_{n=0}^{\infty} \hat{f}(n) \int_{\Gamma_\infty\backslash\mathcal{H}} e^{2\pi i(m-n)x}\, \mathrm{d}x y^{\bar{s}+k-2} e^{2\pi(m+n)y}\, \mathrm{d}y$$

$$= \sum_{n=0}^{\infty} \delta_{mn} \int_0^{\infty} y^{\bar{s}+k-2} e^{2\pi(m+n)y}\, \mathrm{d}y$$

by first integrating $\int_0^1 e^{2\pi i(m-n)x}\, \mathrm{d}x$. Hence the only $n = m$ term survives to give

$$\hat{f}(m) \int_0^{\infty} y^{\bar{s}+k-2} e^{2\pi(m+n)y}\, \mathrm{d}y = \frac{\hat{f}(m)}{4\pi m^{\bar{s}+k-1}} \Gamma(\bar{s}+k-1)$$

for $\operatorname{Re}\bar{s}+k-1 > 0$ whence in particular (3.9.2).

References

Abramowitz, M., & Stegun, I. A. (1965). *Handbook of mathematical functions*. New York: Dover.

Agarwal, P., Kanemitsu, S., & Kuzumaki, T. (2023). Automorphic L-functions by way of the Kuznetsov sum formula. *Advances in Computational Mathematics and Modelling, 1*, 47–72.

Asai, T. (1970). On a certain function analogous to log $|\eta(z)|$. *Nagoya Mathematical Journal,40*, 193–221.

Bateman, P. T., & Grosswald, E. (1964). On Epstein's zeta function. *Acta Arithmetica, 9*, 365–373.

Bellman, R. (1956). On a class of functional equations of modular type. *Proceedings of the National Academy of Sciences of the USA,42*, 84–87.

Bellman, R. (1961). *A brief introduction to theta functions*. New York: Holt Rinehart and Winston Inc.

Bellman, R. (1950). Generalized Eisenstein series and non-analytic automorphic functions. *Proceedings of the National Academy of Sciences of the USA, 36*, 356–359.

Berndt, B. C. (1967). Generalized Dirichlet series and Hecke's functional equation. *Proceedings of the Edinburgh Mathematical Society, 15*, 309–313.

Berndt, B. C. (1969). Identities involving the coefficients of a class of Dirichlet series III. *Transactions of the American Mathematical Society, 146*, 323–348.

Berndt, B. C. (1970). Identities involving the coefficients of a class of Dirichlet series IV. *Transactions of the American Mathematical Society, 149*, 179–185.

Berndt, B. C. (1971). Identities involving the coefficients of a class of Dirichlet series VI. *Transactions of the American Mathematical Society, 160*, 157–167.

Chakraborty, K. (2010). On the Chowla-Selberg integral formula for nonholomorphic Eisenstein series. *Integral Transforms and Special Functions, 21*(12), 917–923. https://doi.org/10.1080/10652469.2010.491256

Chan, H. H. (2020). *Theta functions, elliptic functions and* π. Berlin-Boston: de Gruyter.

Chowla, S., & Selberg, A. (1949). On Epstein's zeta-function (I). *Proceedings of the National Academy of Sciences of the USA35*, 371–374; *Collected papers of atle selberg I* (pp. 367–370). Springer (1989). *The collected papers of Sarvadaman Chowla II, CRM* (pp. 719–722) (1999).

de Azevedo Pribitkin, W. (1999). The Fourier coefficients of modular forms and Niebur modular integrals having small positive weight. I. *Acta Arithmetica,91*, 291–309.

de Azevedo Pribitkin, W. (2000). The Fourier coefficients of modular forms and Niebur modular integrals having small positive weight. II. *Acta Arithmetica,93*, 343–358.

Deuring, M. F. (1937). On Epstein's zeta function. *Annals of Mathematics,* 2nd Ser. *38*, 585–593.

Deuring, M. F. (1933). Imaginäre quadratische Zahlkörper mit der Klassenzahl 1. *Mathematische Zeitschrift, 37*(1933), 405–415.

Erdélyi, A., Magnus, W., Oberhettinger, F., & Tricomi, F. G. (Eds.). (1953). *Higher transcendental functions* (Vol. 1–3). New York: McGraw-Hill.

Erdélyi, A., Magnus, W., Oberhettinger, F., & Tricomi, F. G. (Eds.). (1954). *Tables of integral transforms* (Vol.1–2). New York-Toronto-London: McGraw-Hill Book Company Inc.

Hamburger, H. (1921). Über die Riemannschen Funktionalgleichung der ζ-Funktion. *Mathematische Zeitschrift10*, 240–254. (Erste Mitteilung)

Hardy, G. H. (1908). Some multiple integrals. *Quarterly Journal of Mathematics (Oxford)5*(2), 357–375; *Collected papers* (Vol. 5, pp. 434–452) (1972). Comments 453.

Hecke, E. (1923). *Vorlesungen über die Theorie der algebraischen Zahlen*, Akad. Verl., Leipzig; English translate *Lectures on algebraic number theory*, Springer, Berlin etc. 1981.

Hecke, E. (1927). Theorie der Eisensteinschen Reihen höherer Stufe und ihre Anwendung auf Funktionentherie und Arithmetik. *Abh. Math. Sem. Univ. Hamburg,5*, 199–204. Mathematische Werke. *Vandenhoeck und Ruprecht, Göttingen,1959*, 461–486.

Hecke, E. (1936). Über die Bestimmung Dirichletscher Reihen durch ihre Funktionalgleichung. *Mathematische Annalen,112*, 664–669. *Mathematische Werke,* 591–626. Vandenhoeck u. Ruprecht, Göttingen 1959.

Hecke, E. (1983). *Lectures on Dirichlet series, modular functions and quadratic forms*. In B. Schoenberg in coll. with W. Maak, Vandenhoeck u. Ruprecht, Göttingen (first edition: *Dirichlet series*, Planographed Lecture Notes, Princeton IAS, Edwards Brothers, Ann Arbor, 1938).

Hecke, E. (1921). Analytische Funktionen und algebraische Zahlen 1. *Abhandlungen aus dem Mathematischen Seminar der Universität Hamburg, 1*, 102–126.

Iwaniec, H. (1995). *Introduction to the spectral theory of automorphic forms*, Biblioteca de la Revista Matemática Iberoamericana, Madrid. Second edition: *Spectral methods of automorphic forms*, Grad. Stud. Math. Vol. 53, Amer. Math. Soc. and Revista Matemática Iberoamericana, Providence, 2002.

Iwaniec, H. (1997). *Topics in classical automorphic forms. Graduate studies in mathematics* (Vol. 17). Providence: American Mathematical Society.

Kaneko, M. (2017). Book review of [Bringmann et al. (2021)]. *Sugaku,73*, 108–111.

Kanemitsu, S., Li, H.-L., & Agarwal, P. (2017). Note on the Kuznetsov sum formula. *Basic Sciences Journal of Textile Universities, 30*(3), 293–301.

Kanemitsu, S., Tanigawa, Y., Tsukada, H., & Yoshimoto, M. (2004). On Bessel series expressions for some lattice sums II. *Journal of Physics A: Mathematical and General, 37*, 719–734.

Kanemitsu, S., & Tsukada, H. (2014). *Contributions to the theory of zeta-functions: the modular relation supremacy*. Singapore etc: World Scientific.

Katsurada, M., & Matsumoto, K. (1997). Explicit formulas and asymptotic expansions for certain mean square of Hurwitz zeta functions II. In A. Laurinčikas (Ed.), *Analytic and probabilistic methods in number theory. New Trends in Probability and Statistics* (Vol. 4, pp. 119–134). VSP/TEV.

Katsurada, M., & Matsumoto, K. (2002). Explicit formulas and asymptotic expansions for certain mean square of Hurwitz zeta functions III. *Compositio Mathematica, 131*, 239–266.

Kitaoka, Y. (1971). A simple proof of the functional equation of a certain L-function. *Journal of Number Theory, 3*, 155–158.

Knopp, M. I. (1989). On the Fourier coefficients of cusp forms having small positive weight. *Proceedings of Symposia in Pure Mathematics, 49*(2), 111–127.

Kober, H. (1934). Transformationsformeln gewisser Besselscher Reihen, Beziehungen zu Zeta-Funktionem. *Mathematische Zeitschrift, 39*, 609–624.

Koshlyakov, N. S. (1954). Investigation of some questions of analytic theory of the rational and quadratic fields, I-III (Russian). *Izv. Akad. Nauk SSSR, Ser. Mat.18*, 113–144, 213–260, 307–326, Errata: ibid. *19*(1955), 271.

Kuzumaki, T. (2011). Asymptotic expansions for a class of zeta-functions. *Ramanujan Journal, 24*, 331–343.

Lang, S. (1975). $SL_2(\mathbb{R})$. New York, etc: Addison-Wesley.

Li, H. -Y., Kuzumaki, T., & Kanemitsu, S. (2022). On zeta functions and allied theta-functions, to appear *Advances in Applied Analysis and Number Theory,* Chapter 4, 51–97. World Scientific. https://doi.org/10.1142/9789811272608 0004

Li, H.-L., Li, F.-H., Wang, N.-L., & Kanemitsu, S. (2017). *Number theory and its applications II*. London-Singapore-New Jersey: World Scientific.

Liu, J. -Y. (2026). *A quick introduction to Maass forms*, based on Lectures at Postech, 2007, in preparation.

Liu, G. -D., Kuzumaki, T., & Kanemitsu, S. (2025). Integral transforms in number theory, a review, to appear.

Maass, H. (1964). *Lectures on modular functions of one complex variable*. Bombay: Tata Institute of Social Sciences, Revised in 1983.

Maass, H. (1949). Über eine neue Art von nichtanalytischen automorphen Funktionen und die Bestimmung Dirichletscher Reihen durch Funktionalgelichungen. *Mathematische Annalen, 121*, 141–183.

Mellin, H. (1910). Abriss einer einheitlichen Theorie der Gamma- und der hypergeometrischen Funktionen. *Mathematische Annalen, 68*, 305–337.

Miyake, T. (1989). *Modular forms*. New York etc: Springer.

Motohashi, Y. (1968). A new proof of the limit formula of Kronecker. *Proceedings of the Japan Academy Series A Mathematical Sciences, 66*, 614–616.
Motohashi, Y. (1997). *Spectral theory of the Riemann zeta-function*. Cambridge etc: Cambridge University Press.
Noda, T. (2007). A note on the non-holomorphic Eisenstein series. *Ramanujan Journal, 14*, 405–411.
Noda, T. (2009). Laplace-Mellin transform of the non-holomorphic Eisenstein series. *RIMS Kokyuroku, 1665*, 94–99.
Oberhettinger, F. (1974). *Tables of Mellin transforms*. Berlin-Heidelberg-New York: Springer.
Oberhettinger, F., & Soni, K. L. (1972). On some relations which are equivalent to functional equations involving the Riemann zeta-function. *Mathematische Zeitschrift, 127*, 17–34.
Paris, R. B., & Kaminski, D. (2001). *Asymptotics and mellin-barnes integrals*. Cambridge: Cambridge University Press.
Pasles, P. C., de Azevedo Pribitkin, W. (2001). A generalization of the Lipschitz summation formula and some applications. *Proceedings of the American Mathematical Society,129*, 3177–3184.
Pasles, P. C. (1999). Nonanalytic automorphic integrals on the Hecke groups. *Acta Arithmetica, 90*, 155–171.
Pasles, P. C. (2000a). Convergence of Poincarè series with two complex coweigths. *Contemporary Mathematics, 251*, 453–461. AMS, R. I.
Pasles, P. C. (2000). A Hecke correspondence theorem for nonanalytic automorphic integrals. *Journal of Number Theory, 83*, 256–281.
Petersson, H. (1932). Über die Entwicklungskoeffizienten der automorphen Formen. *Acta Mathematica, 58*, 160–215.
Petersson, H. (1939). Über eine Metrisierung der ganzen Modulformen. *Jahresber. Deutsch. Math. Verein., 49*, 49–75.
Prudnikov, A. P., Bychkov, Y. A., & Marichev, D. I. (1986). *Integrals and series. Supplementary chapters*. Moscow: Nauka.
Rademacher, H. (1927). Zur additiven Primzahltheorie algebraischer Zahlkörper III. *Mathematische Zeitschrift, 27*, 321–426.
Rademacher, H. (1973). *Topics in analytic number theory*. Berlin-Heidelberg: Springer.
Rankin, R. A. (1939). Contributions to the theory of Ramanujan's function $\tau(n)$ and similar arithmetical functions, I, II. *Proceedings of the Cambridge Philosophical Society, 36*, 351–372.
Riemann, B. (1855). Zur Theorie der Nobili'schen Farbenringe. *Poggendorff's Ann. Phy. Chem.95*, 28–36. *Collected works of Bernhard Riemann* (2nd ed.). In H. Weber (Ed.) (pp. 55–62). New York: Dover (1953); Anmerkungen (pp. 63–66).
Schoenberg, B. (1974). *Elliptic modular functions. An introduction*. Berlin-Heidelberg-New York: Springer.
Selberg, A., & Chowla, S. (1967). On Epstein's zeta-function. *J. Reine Angew Math.227*, 86–110; *Collected papers of Atle Selberg I* (pp. 521–545). Springer (1989); *The collected papers of Sarvadaman Chowla II. CRM* (pp. 1101–1125) (1999).
Selberg, A. (1940). Bemerkungen über eine Dirichletsche Reihe, die mit der Theorie der Modulformen naher verbunden ist. *Ark Math Naturvid B, 43*, 47–50.
Selberg, A. (1956). Harmonic analysis and discontinuous groups in weakly symmetric Riemannian spaces with applications to Dirichlet series. *The Journal of the Indian Mathematical Society, 20*, 47–87.
Siegel, C. L. (1943). Contributions to the theory of Dirichlet L-series and the Epstein zeta-functions. *Annals of Mathematics, 44*, 143–172.
Siegel, C. L. (1961). *Lectures on advanced analytic number theory*. Bombay: Tata Institute of Social Sciences.
Soni, K. (1966). Some relations associated with an extension of Koshliakov's formula. *Proceedings of the American Mathematical Society, 17*, 543–551.
Steen, S. (1930). Divisor functions: their differential equations and recurrence formulas. *Proceedings of the London Mathematical Society, 2*(31), 47–80.

Stieltjes, T. J. (1886). Recherche sur quelques séries semi-convergentes. *Ann. École Norm. Sup.,3*(3), 201–258 (1859). Ouevres complètes de Thomas Jan Stieltjes, Tome II, ed. by *Soc. Math. Amsterdam, Nordhohoff, Groningen,1918*, 2–58.

Taylor, P. R. (1940). The functional equation for Epstein's zeta-function. *The Quarterly Journal of Mathematics, 11*, 177–182.

Taylor, P. R. (1945). On the Riemann zeta function. *The Quarterly Journal of Mathematics, 16*, 1–21.

Terras, A. (1985). *Harmonic analysis on symmetric spaces and applications I*. Berlin etc: Springer.

Tsukada, H. (2007). A general modular relation in analytic number theory. In *Number theory: sailing on the sea of number theory. Proceedings 4th China-Japan seminar on number theory 2006* (pp. 214–236). Singapore etc: World Scientific.

Venkov, A. B. (1982). Spectral theory of automorphic functions. *Trudy Mat. Inst. Steklov,153*, 3–171 (1981); English translate in *Proceedings of the Steklov Institute of Mathematics,* (4), 1–163 (1982).

Venkov, A. B. (1990). *Spectral Theory of automorphic function and its applications*. Dordrecht-Boston-London: Kluwer Academic Publishers.

Voronoĭ, G. F. (1904). Sur une fonction transcendente et ses applications à la sommation de qulques séries. *Ann. École Norm. Sup.,21*(3), 459–533. *Sob. Soč. II Izd. Akad. Nauk Ukr. SSR Kiev,1952*, 51–165.

Wang, N. -L., Kuzumaki, T., & Kanemitsu, S. (2020). Unified theory of zeta-functions allied to Epstein zeta-functions and associated with Maass forms. *Mathematics* 2185255. MDPI.

Wang, X. -H., Wang, N. -L., & Kanemitsu, S. (2022). An extension of Maass theory to general Dirichlet series. *Integral Transforms and Special Functions*. https://doi.org/10.1080/10652469.2022.2057483

Watson, G. N. (1944). *A treatise on the theory of Bessel functions* (2nd ed.). Cambridge: Cambridge University Press.

Whittaker, E. T., & Watson, G. N. (1927). *A course in modern analysis* (4th ed.). Cambridge: Cambridge University Press.

Wolfram, S. (1998). *Mathematical functions site*. New York: Wolfram Research Inc.

Zhang, N.-X., & Williams, K. (1995). On the Epstein zeta function. *Tamkang Journal of Mathematics, 26*, 165–176.

Chapter 4
A Unifying Principle for Modular Relations and Summation Formulas

The main figures in this chapter are the Dedekind zeta-function (DZF), functional equation of (DZF), unifying principle, summation formulas, confluent hypergeometric function (CHGF). In this chapter we establish a unifying principle which elucidates the sources of various summation formulas used throughout in mathematics research. Ample examples are given: Lipschitz summation formula, Koshlyakov's generalized Dedekind zeta-function, Plana summation formula, Voronoĭ summation formula, etc. Cf. Lindelöf (1905), in which Chap. 3, pp. 52–86 is devoted to the study of summation formulas.

In the final section we shall elucidate Hamburger et al. results in the light of the *Principle*(=Theorem 4.1.1) and RHB correspondence, i.e. through Fourier-Bessel expansion and the Riesz sum.

4.1 A Unifying Principle in the Theory of Modular Relations

The Voronoĭ summation formula is known to be equivalent to the functional equation for the square of the Riemann zeta-function in case the function in question is the Mellin transform of a suitable function. There are some other famous summation formulas which are treated as independent of the modular relation.

In this section we shall establish a far-reaching principle which furnishes the following. Given a zeta-function $Z(s)$ satisfying a suitable functional equation, one can generalize it to $Z_f(s)$ in the form of an integral involving the Mellin transform $F(s)$ of a certain suitable function $f(x)$ and process it further as $\tilde{Z}_f(s)$. Under the condition that $F(s)$ is expressed as an integral and the order of two integrals are interchangeable, one can obtain a closed form for $\tilde{Z}_f(s)$.

Recall MTP in (2.1.1) and the zeta-function (2.1.4) which we assume is absolutely convergent for $\mathrm{Re}s = \sigma > 1$. The abscissa of absolute convergence can be σ_a^* but

K. Chakraborty et al., *Modular Relations and Parity in Number Theory*,
Infosys Science Foundation Series, https://doi.org/10.1007/978-981-96-6471-9_4

we assume it to be 1 for the sake of simplicity. Suppose it satisfies the functional equation of the form (where we understand the right-hand side member may be a different Dirichlet series)

$$Z(s) = G(s)Z(1-s),$$

where $G(s)$ is a certain gamma factor to be specified in each occasion and the line of reflection is chosen to be $\frac{1}{2}$ instead of a more general $r \in \mathbb{R}$ (cf. *Convention* below).

We have for $c > 1$

$$Z_f(s) = \sum_{n=1}^{\infty} a_n f(\lambda_n) = \frac{1}{2\pi i} \int_{(c)} F(z) Z(z)\,\mathrm{d}z \tag{4.1.1}$$

in the first instance. Here and in what follows we use this type of suggestive notation to mean that the processing factors—f and F in this case—may depend also on the extraneous parameter s. Similarly for (4.1.2), (4.1.3) etc. E.g. HGT is the case

$$f(x) = f(x,s) = e^{-xs}, \quad F(s) = F(z,s) = \Gamma(z)s^{-z}$$

and (4.1.1) reads

$$Z_f(s) = \sum_{n=1}^{\infty} a_n e^{-\lambda s} = \frac{1}{2\pi i} \int_{(c)} \Gamma(z) s^{-z} Z(z)\,\mathrm{d}z$$

which is the inverse of (2.1.5). These may not lead to a new circumstance and we consider the integral in (4.1.1) for $Z_f(s)$ is processed (in that case, the series expression is rather involved). We write $\tilde{Z}_f(s)$ for a processed integral for $Z_f(s)$

$$\tilde{Z}(s) = \tilde{Z}_f(s) = \frac{1}{2\pi i} \int_{(c)} F_f(z) \mathfrak{G}_f(z) Z(z)\,\mathrm{d}z \tag{4.1.2}$$

for some $c > 1$, where $F_f(z)$ is the processed Mellin inversion $F(s)$ (2.1.1) given by (4.1.3) and $\mathfrak{G}_f(z)$ is a certain gamma factor which may be incorporated, not necessarily depending on f. Similarly, $\mathfrak{F}_F(w)$ may be a function which processes f. These gamma factors and $\mathfrak{G}_F(w,z)$ may be superfluous but they are kept for further possible applications. Cf. Table 4.1.

Table 4.1 Gamma factors in respective cases

Section	$\mathfrak{F}_F(w)$	$\mathfrak{G}_F(w,z)$	$\mathfrak{G}_f(z)$	$K(\cdot), \gamma(\cdot)$	$\delta(w,z)$
Section 2	$\Gamma(-w)\Gamma(w+s)$	$\Gamma(w+z)$	$\cos\frac{\pi}{2}z$	(4.2.6)	$(-2\pi\alpha\omega)^{-w}$
Section 3	1	$\Gamma(w-z)$	$\Gamma(z)\cos\frac{\pi}{2}z$	(4.3.8)	$\gamma(z)$
Section 4	$\frac{f(iw)-f(-iw)}{2i}$	$\mathfrak{G}_f(z)$	$\frac{1}{\sin\frac{\pi}{2}z}$	(4.4.7)	1

The following theorem gives the *Principle* for obtaining a closed form for $\tilde{Z}_f(s)$. In subsequent sections, we illustrate this theorem, referring to it as an application of the *Principle*.

Theorem 4.1.1 *Suppose the integration path* (c) *of* (4.1.2) *may be shifted to* $(-d)$, $0 < d < 1$ *(in the first instance) with the resulting residual function* $\mathrm{P}(s)$ *(sum of residues in the horizontal strip* $-d < \sigma < c$*):*

$$\tilde{Z}_f(s) = \frac{1}{2\pi i}\int_{(-d)} F_f(z)\mathfrak{G}_f(z)Z(z)\,\mathrm{d}z + \mathrm{P}(s) = J(s) + \mathrm{P}(s),$$

say, that

$$F_f(z) = \int \mathfrak{G}_F(w, z)\mathfrak{F}_F(w)\delta(w, z)\,\mathrm{d}w, \tag{4.1.3}$$

where the integral for $F_f(z)$ *may be the infinite integral over* $(0, \infty)$*, the contour integral or may indicate the integrand itself and* $\mathfrak{G}_F(w, z)$ *may depend only on* z *and may work as* $\mathfrak{G}_f(z)$. $\delta(w, z) = \delta(w)\gamma(z)$ *and* $\gamma(z) = \gamma(z, s)$ *and* $\delta(w) = \delta(w, s)$ *are simple functions specified at each occasion. Suppose that the order of integration is interchangeable:*

$$\frac{1}{2\pi i}\int_{(-d)} \mathrm{d}z \int \mathrm{d}w = \int \mathrm{d}w \frac{1}{2\pi i}\int_{(-d)} \mathrm{d}z.$$

Then we write

$$I(w) := \frac{1}{2\pi i}\int_{(-d)} \mathfrak{G}_F(w, z)M(z)Z(1-z)\gamma(z)\,\mathrm{d}z = \sum_{n=1}^{\infty} a_n K(w, n), \tag{4.1.4}$$

where

$$K(w, n) = \frac{1}{2\pi i}\int_{(-d)} \mathfrak{G}_F(w, z)M(z){\lambda_n}^{z-1}\gamma(z)\,\mathrm{d}z$$

and where

$$M(z) = G(z)\mathfrak{G}_f(z).$$

Then if both $K(w, n)$ *and*

$$J(s) = \int I(w)\mathfrak{F}_F(w)\delta(w)\,\mathrm{d}w \tag{4.1.5}$$

admit closed form, then (4.1.5) *gives a closed form for* $\tilde{Z}_f(s)$.

For a better glimpse of this theorem, cf. Example 4.1.1 below. For convenience of simpler applications in Sect. 4.5, etc., we extract the case of unprocessed case (4.1.1) as Corollary 4.1.1 (i.e. the integral operator (4.1.3) for F_f is the identity operator). This is *a result of common knowledge* (à la Shannon) but is the underlying principle.

Corollary 4.1.1 *In the same setting as in Theorem 4.1.1, consider the* **unprocessed zeta-function** $Z_f(z)$ *in* (4.1.1). *Then*

$$Z_f(s) = \sum_{n=1}^{\infty} a_n f(n) = \mathrm{P}(s) + J(s), \quad J(s) = \frac{1}{2\pi i} \int_{(-d)} F(z)\, Z(z)\, dz, \tag{4.1.6}$$

where $\frac{1}{2} < d < 1$, *say and where*

$$J(s) = I(s) = \frac{1}{2\pi i} \int_{(-d)} F(z)G(z)Z(1-z)\gamma(z.s)\, dz = \sum_{n=1}^{\infty} a_n K(s,n), \tag{4.1.7}$$

$$K(s,n) = \frac{1}{2\pi i} \int_{(-d)} F(z)G(z){\lambda_n}^{z-1}\gamma(z,s)\, dz.$$

Then if (4.1.7) *is expressed in a closed form, it gives a closed form for* $Z_f(z)$.

For a proof, it is enough to note the specification

$$F_f(z) = \mathfrak{G}_F(w,z), \quad F_f(z)\mathfrak{G}_f(z) = \mathfrak{G}_F(w,z)\mathfrak{G}_f(z) = F(z),$$
$$F(z)M(z) = G(z), \quad \mathfrak{G}_f(z) = 1, \quad \mathfrak{F}_F(w)\delta(w) = 1.$$

In subsequent sections we state consequences of Theorem 4.1.1 on the basis of Table 4.1. Section 4.5 is also an unprocessed case depending on Corollary 4.1.1. We present results in terms of the Dedekind zeta-function $\zeta_\Omega(s)$ of an algebraic number field Ω of degree at most 2, which we let represent the case of the Riemann zeta-function $\zeta(s)$, the zeta-function $\Phi(s)$ associated to a modular form and $\zeta^2(s)$ which is a generating function of the divisor function according as $\Omega = \mathbb{Q}$, Ω an imaginary quadratic field or Ω a real quadratic field. Walfisz (1923, 1925) generalize the divisor problem to the case of the mth power of the Dedekind zeta-function and obtain the closed form for the partial sum for arbitrary degree. A. A. Walfisz' results Walfisz (1961a, b) are mentioned in Remark 2.12.2. These are unified as in Table 4.2 by viewing the zeta-functions as represented by the Dedekind zeta-function.

Table 4.2 Dedekind zeta-function and allied functional equation

Field	Zeta	Covering	Gamma factor	Conj.
Rational	Riemann1859	$\zeta(s)$	$\Gamma\left(\frac{s}{2}\right)$	(1, 0)
Imaginary quadr.	Dedekind	$\zeta_\Omega(s)$,$L(s, f_0)$	$\Gamma(s)$	(0, 1)
Real quadr.	Dedekind	$\zeta_\Omega(s)$,$\zeta^2(s)$	$\Gamma^2\left(\frac{s}{2}\right)$	(2, 0)

We assume facts about the Dedekind zeta-function stated in the beginning of Sect. 2.5 in the special case with slightly different notation.

We let Ω be an algebraic number field of degree $\varkappa \le 2$ with r_1 real resp. $2r_2$ imaginary conjugates and let

$$\zeta_\Omega(s) = \sum_{n=1}^{\infty} \frac{a_n}{n^s}, \quad a_n = \sum_{N\mathfrak{a}=n} 1, \tag{4.1.8}$$

be the Dedekind zeta-function (2.5.1), which is absolutely convergent for $\sigma = \operatorname{Re} s > 1$ on the ground that

$$a_n = O(n^\eta),$$

for every $\eta > 0$.

The functional equation reads

$$A^{-s}\Gamma^{r_1}\left(\frac{s}{2}\right)\Gamma^{r_2}(s)\zeta_\Omega(s) = A^{-(1-s)}\Gamma^{r_1}\left(\frac{1-s}{2}\right)\Gamma^{r_2}(1-s)\zeta_\Omega(1-s), \tag{4.1.9}$$

where our A is the inverse of Koshlyakov's:

$$A = \frac{2^{r_2}\pi^{\frac{\chi}{2}}}{\sqrt{|\Delta|}}.$$

The Dedekind zeta-function has a simple pole at $s = 1$ with residue ρ (2.5.2) with w replaced by W (since w is used as an extraneous parameter here).

It is often the case that we denote the left-hand side of (4.1.9) by $\xi_\Omega(s)$ the **completed zeta-function** and express it as

$$\xi_\Omega(s) = \xi_\Omega(1-s). \tag{4.1.10}$$

Also the **regularized zeta-function** is often considered by cancelling the pole at $s = 1$ while retaining the symmetry:

$$\tilde{\xi}_\Omega(s) = \frac{1}{2}s(s-1)A^{-s}\Gamma^{r_1}\left(\frac{s}{2}\right)\Gamma^{r_2}(s)\zeta_\Omega(s) = \tilde{\xi}_\Omega(1-s) \tag{4.1.11}$$

Putting

$$G(s) = \frac{\Gamma^{r_1}\left(\frac{1-s}{2}\right)\Gamma^{r_2}(1-s)}{\Gamma^{r_1}\left(\frac{s}{2}\right)\Gamma^{r_2}(s)},$$

we express (4.1.9) as

$$\zeta_\Omega(s) = A^{2s-1}G(s)\zeta_\Omega(1-s). \tag{4.1.12}$$

Definition 4.1.1 Under the notation in Definition 2.12.1, we consider the functional equation

$$A^{-s}\Gamma^{r_1}\left(\frac{s}{2}\right)\Gamma^{r_2}(s)\varphi(s) = A^{-(1-s)}\Gamma^{r_1}\left(\frac{1-s}{2}\right)\Gamma^{r_2}(1-s)\psi(1-s). \quad (4.1.13)$$

Here the completion and regularization above also apply. We call the case $(r_1, r_2) = (1, 0)$, $(r_1, r_2) = (0, 1)$, $(r_1, r_2) = (2, 0)$ of (4.1.13) the **Riemann, Hecke, and Voronoǐ type functional equation–(FER), (HTFE), (VFE)**, respectively. For the Hecke type, we understand (4.1.13) to mean (2.12.2) in Definition 2.12.1, with the line of reflection being $\sigma = \frac{r}{2}$, so as to include the weight aspect.

HFE includes the case of the congruence-subgroup L-function in Sect. 2.15. In particular (2.15.2) explains our setting with the constant A and $G(s)$ being the same as that of the imaginary quadratic case and the line of reflection $\frac{k}{2}$:

$$L(s, f_0) = CA^{2s-1}G(s)L(k-s, f_0).$$

In such a case, we understand $CL(k-s, f_0)$ is another Dirichlet series $\psi(k-s)$ in Definition 4.1.1. Cf. Lemma 2.12.1 above. The cusp-form L-function $L(s, f)$ in Example 2.12.1 is a special case.

Convention. Throughout in what follows, we make a convention that we let (4.1.9) represent the general functional Eq. (4.1.13) and for the Hecke type $(r_1, r_2) = (0, 1)$, it is to be understood as in Definition 4.1.1, with all the basic results in Theorem 2.12.3. We state the results with $r = 1$ elsewhere. The transition to the general r is simple while treating different Dirichlet series $\varphi(s)$ and $\psi(s)$ is more involved. In most of applications, we proceed as in Example 4.1.1 below, so that a_n means the Ideal-function in (4.1.8) or coefficients of other allied zeta-functions. We denote the residual function by $\mathrm{P}(s)$ which may be different for different zeta-functions.

The **Hurwitz-Lerch L-function** $\Phi(z, s, \alpha; \chi)$ is defined by

$$\Phi(z, s, \alpha; \chi) = \sum_{n=0}^{\infty} \frac{\chi(n)z^n}{(n+\alpha)^s}$$

which is absolutely and uniformly convergent on any compact subset of

$$\{|z| \le 1\} \times \{\sigma > 1\} \times \{\alpha > -1\}$$

(Morita 1977, p. 30). The special case where χ is the trivial character mod 1 is the **Hurwitz-Lerch zeta-function** $\Phi(z, s, w)$ defined by

$$\Phi(z, s, \alpha) = \sum_{n=0}^{\infty} \frac{z^n}{(n+\alpha)^s} \quad (4.1.14)$$

which is absolutely and uniformly convergent on any compact subset of

$$\{|z| \leq 1\} \times \{\sigma > 1\} \times \{\alpha \in \mathbb{C}, \alpha \neq 0, -1, -2, \cdots\}, \tag{4.1.15}$$

where a suitable branch is chosen of $(n+\alpha)^s$. Cf. Erdélyi et al. (1953, I, pp. 27–31). for $\alpha \in \mathbb{C}$ not being a non-positive integer (often restricted to $0 < \alpha \leq 1$), and wither $|z| < 1$ or $|z| = 1,\ \sigma = \mathrm{Re}\, s > 1$.

Its uniformization is the **Lipschitz-Lerch transcendent** is stated in Srivastava and Choi (2001, pp. 121–123):

$$L(x, s, \alpha) = \Phi(e^{2\pi i x}, s, \alpha) = \sum_{n=0}^{\infty} \frac{e^{2\pi i n x}}{(n+\alpha)^s}, \quad x \in \mathbb{R}. \tag{4.1.16}$$

For intended applications of (4.1.16), we choose

$$\{\mathrm{Im}\, z \geq 0\} \times \{\sigma > 1\} \times \{\alpha \in \mathbb{C}, \mathrm{Re}\, \alpha \geq 0\}, \tag{4.1.17}$$

as the domain of absolute convergence. The convergence condition stated on (Srivastava and Choi 2001, p. 122, (11)) is mainly as a special case, esp. the second condition $x \in \mathbb{Z}$ and $\sigma > 1$ is the convergence condition for the Hurwitz zeta-function. The first condition $x \in \mathbb{R}\backslash\mathbb{Z}, \sigma > 0$ (for uniform convergence not absolute convergence) may be changed to $x \in \mathcal{H}$, the upper half-plane and $\sigma > 0$ corresponds to the condition $|z| < 1$ for $\Phi(z, s, \alpha)$ above. Other special cases of Hurwitz-Lerch zeta-function are the following: The polylogarithm

$$\mathrm{Li}_s(z) = \Phi(z, s, 1) = \sum_{n=1}^{\infty} \frac{z^n}{n^s}$$

whose uniformization $\mathrm{Li}_s\left(e^{2\pi x}\right)$ is the boundary Lerch zeta-function (LZF) (2.1.10); The Hurwitz zeta-function

$$\zeta(s, \alpha) = \Phi(1, s, \alpha) = \sum_{n=1}^{\infty} \frac{z^n}{n^s}.$$

Laurinchikas and Garunkstis (2002) is devoted to the theory of the Lip-schitz-Lerch transcendent, which is called the Lerch zeta-function there and pp. 17–30 contain detailed proofs of the functional equation. Cf. also Srivastava and Choi (2001, p. 121, (1)), in which the Hurwitz-Lerch zeta-function part is verbatim to Erdélyi et al. (1953, I, pp. 27–31).

The distribution property (or also Kubert identity) of $\Phi(z, s, w)$ has been studied in Arai et al. (2015) and the Lipschitz summation formula is studied in Li et al. (2013, pp. 128–132) and recently Li et al. (2020) interpreted it as the functional equation for $L(x, s, \alpha)$. I.e. Theorem 4.2.1 is the same as the functional equation, Corollary 4.2.1.

We define the Lipschitz-Lerch transcendent associated with the field Ω by

$$L_\Omega(x,s,\alpha)=\sum_{n=0}^{\infty}a_n\frac{e^{2\pi inx}}{(n+\alpha)^s}, \tag{4.1.18}$$

where a_0 is a constant, which is determined so that (4.1.18) reduces to (2.1.11) for $\Omega=\mathbb{Q}$. Hence we put $a_0=-2\zeta_\Omega(0)=1$. Koshlykov (1954, II, p. 241) introduced the **Hurwitz type Dedekind zeta-function**

$$\zeta_\Omega(s,w)=L_\Omega(0,s,w)=-2\frac{\zeta_\Omega(0)}{w^s}+\sum_{n=1}^{\infty}\frac{a_n}{(n+w)^s},\quad \sigma>1,\quad 0<w<1, \tag{4.1.19}$$

which will be studied in Sect. 4.3.

Example 4.1.1 We assume the conditions in Theorem 4.4.1. We proceed almost verbatim to Liu et al. (2023) and (4.1.1) reads for $Z(s)=\zeta_\Omega(s)$

$$\zeta_{\Omega,f}(s)=\sum_{n=1}^{\infty}a_nf(n)=\frac{1}{2\pi i}\int_{(c)}F(s)\,\zeta_\Omega(s)\mathrm{d}s,$$

which is processed as $\tilde{Z}_f(s)=\tilde{\zeta}_{\Omega,f}(s)$ in (4.1.2).

Then in view of the estimate

$$\zeta_\Omega(s)=O(|t|^{\chi(1-\sigma)}),\quad \sigma\le 0$$

uniformly in the strip $\sigma_1\le\sigma\le\sigma_2$ and the Stirling formula (2.1.6), we may shift the line of integration to $\sigma=-d$, $0<d<1$, passing through the poles in the strip $-d<\sigma<c$. Hence we have

$$\tilde{Z}_f(s)=J(s)+\mathrm{P}(s),$$

where

$$J(s)=\frac{1}{2\pi i}\int_{(-d)}F_f(z)\,\mathfrak{G}_f(z)\zeta_\Omega(z)\,\mathrm{d}s$$

and $\mathrm{P}(s)$ is the sum of all residues in the strip $-d<\sigma<c$. Then applying (4.1.3), changing the order of integration and applying the functional Eq. (4.1.12), we arrive at (4.1.5) with $I(w)$ in (4.1.4).

Then since $1-d>1$, we may use the Dirichlet series for $\zeta_\Omega(1-\zeta)$ and write

$$I(w)=\sum_{n=1}^{\infty}a_nK(w,n),\quad K(w,n)=\frac{1}{2\pi i}\int_{(-d)}\mathfrak{G}_F(w,z)M(z)n^{z-1}\gamma(z)\,\mathrm{d}z. \tag{4.1.20}$$

The most essential part is justification of the change of order of integration. In Theorems 4.2.1 and 4.4.1, the exponential reduction is lost and integration path(s) are to be taken so that the exponent of $|t|$ is < -1.

In the setting of Corollary 4.1.1, (4.1.20) reads

$$J(s) = I(s) = \sum_{n=1}^{\infty} a_n K(s,n), \quad K(s,n) = \frac{1}{2\pi i} \int_{(-d)} F(z)M(z)n^{z-1}\gamma(z,s)\,\mathrm{d}z \tag{4.1.21}$$

and $\gamma(z,s)$ involves the factor A^{2z-1}.

We note that the beta transform (2.5.18) is a special case of

$$G_{1,1}^{1,1}\left(z \,\middle|\, \begin{matrix} a \\ b \end{matrix}\right) = \Gamma(1-a+b)z^b(1+z)^{a-b-1},$$

which is not stated in Erdélyi et al. (1953). We recall (3.1.16).

In the next section we shall make an essential use of the confluent hypergeometric function (CHGF) $U(a,c;z)$, which is denoted $\Psi(a,c;z)$ and defined by (3.1.10) above. We use the special case of (3.1.13) (Erdélyi et al. 1953, I, p. 256, (5)) as its definition and restate basic knowledge about it (already stated in terms of the Whittaker function):

$$\Gamma(a)\Gamma(1+a-c)\,z^a U(a,c;z) = G_{1,2}^{2,1}\left(z \,\middle|\, \begin{matrix} 1 \\ a, 1+a-c \end{matrix}\right) \tag{4.1.22}$$

valid for $|\arg z| < \frac{3\pi}{2}$. With (4.1.22) its definition, it verifies (3.1.10), which is the most relevant to us (Table 4.3).

CHGF is one-valued and analytic for $-\pi < \arg z < \pi$, and is one of two independent solutions to the confluent hypergeometric DE (3.1.9) satisfying the boundary condition that $w(z)$ is of polynomial order as $z \to \infty$ (cf. e.g. (Erdélyi et al. 1953, p. 278)).

The other independent solution is given by (Erdélyi et al. 1953, p. 256, (4))

$$\begin{aligned} {}_1F_1(a,c;x) &= \frac{\Gamma(c)}{\Gamma(a)} G_{2,1}^{1,1}\left(\frac{1}{-x} \,\middle|\, \begin{matrix} 1, c \\ a \end{matrix}\right) = \frac{\Gamma(c)}{\Gamma(a)} \frac{1}{2\pi i} \int_{(\gamma)} \frac{\Gamma(-z)\Gamma(a+z)}{\Gamma(c+z)} (-x)^{s-1}\,\mathrm{d}z \\ &= \frac{\Gamma(c)}{\Gamma(a)} G_{1,2}^{1,1}\left(-x \,\middle|\, \begin{matrix} 1-a \\ 0, 1-c \end{matrix}\right). \end{aligned}$$

Table 4.3 Fundamental system for the confluent hypergeometric DE

Theory	Polyn. growth	∞
CHF	$U(a,c;x) = \Psi(a,c;x)$	${}_1F_1(a,c;x) = \Phi(a,c;x)$
Whittaker function	$W_{\varkappa,\mu}(x)$	$M_{\varkappa,\mu}(x)$
Bessel function	$K_\nu(x)$	$I_\nu(x)$

This is also denoted $\Phi(a, c; x)$.

Letting $c = 1$, $a = w + 1$ and $z + 1 = s$, we have

$$
{}_1F_1(w+1, 1; x) = \frac{1}{\Gamma(w+1)}\frac{1}{2\pi i}\int_{(\gamma+1)} \frac{\Gamma(1-s)\Gamma(w+s)}{\Gamma(s)}(-x)^{s-1}\, ds. \tag{4.1.23}
$$

The Whittaker function is a main ingredient in the Fourier-Whittaker expansion in Lemma 3.4.1. The Fourier-Whittaker expansion with CHGF is presented in Example 3.4.1. In the same vein, the K-Bessel function takes place in Example 3.6.1.

4.2 General Lipschitz Summation Formula

By (4.1.22), the Mellin transform $F(s)$ of

$$
f(x) = f(x, z, s, w) = \frac{z^x}{(x+w)^s}, \quad \sigma > 1
$$

is

$$
F(a) = w^{a-s}\Gamma(z)\, U(a, a+1-s, -\log z^w). \tag{4.2.1}
$$

The following is (Li et al. 2020, Corollary 3), a corollary to the Ewald expansion. It is a consequence of Li et al. (2020, Theorem 4), which is the modular relation corresponding to the ramified functional equation. We give a new proof depending on the *Principle*.

Theorem 4.2.1 (Lerch's transformation formula) *We have the functional equation*

$$
\begin{aligned}
&L(x, 1-s, \alpha) \\
&= \frac{\Gamma(s)}{(2\pi)^s}\Big(e^{\frac{\pi i}{2}s - 2\pi x\alpha i} L\big(-\alpha, s, x\big) + e^{-\frac{\pi i}{2}s + 2\pi(1-x)\alpha i} L\big(\alpha, s, 1-x\big)\Big),
\end{aligned} \tag{4.2.2}
$$

where $0 < \alpha,\ x < 1$.

Proof Note that (4.2.2) may be written as

$$
L(i\omega, 1-s, \alpha) = e^{2\pi\omega\alpha}\frac{\Gamma(s)}{(2\pi)^s}\Big(e^{\frac{\pi i}{2}s} L(-\alpha, s, i\omega) + e^{-\frac{\pi i}{2}s} L(\alpha, s, -i\omega) - \frac{1}{\omega^s}\Big), \tag{4.2.3}
$$

on putting $x = i\omega$.

Hence recalling Definition (4.1.18) of $L_\Omega(x, s, \alpha)$, we are to consider

$$\frac{1}{2}e^{\frac{\pi is}{2}} f_+(x) = \frac{1}{2}\frac{1}{2\pi i}\int_{(c)} x^{-z} F_+(z)\,\mathrm{d}z,$$
$$\frac{1}{2}e^{-\frac{\pi is}{2}} f_-(x) = \frac{1}{2}\frac{1}{2\pi i}\int_{(c)} x^{-z} F_-(z)\,\mathrm{d}z$$

where $F_\pm(z)$ is the Mellin transform of $f_\pm(x)$ defined by

$$f_+(x) = f(x, -\alpha, s, \omega) = \frac{e^{-2\pi i\alpha x}}{(x+i\omega)^s}, \quad f_-(x) = f(x, \alpha, s, -\omega) = \frac{e^{2\pi i\alpha x}}{(x-i\omega)^s}.$$

Since in (4.2.1)

$$z^w = e^{\pm 2\pi i\alpha \pm i\omega} = e^{-2\pi\omega},$$

we see that

$$F_\pm(z) = (\pm i)^z \omega^{z-s}\Gamma(z)\, U(z, z+1-s, -2\pi\alpha\omega).$$

by (4.2.1). Hence (4.1.22) gives

$$\omega^s\Gamma(s)(-2\pi\alpha e^{\mp\frac{\pi i}{2}})^z F_\pm(z) = \Gamma(s)\,\Gamma(z)\,(-2\pi\alpha\omega)^z U(z, z+1-s; -2\pi\alpha\omega) \tag{4.2.4}$$
$$= \frac{1}{2\pi i}\int_{(\gamma)} \Gamma(-w)\,\Gamma(w+z)\,\Gamma(w+s)\,(-2\pi\alpha\omega)^{-w}\,\mathrm{d}w,$$

so that

$$F_f(z) = \frac{1}{2}(F_+(z) + F_-(z)) = \frac{1}{\omega^s\Gamma(s)}(-2\pi\alpha)^{-z}\cos\frac{\pi}{2}z$$
$$\times\left(\frac{1}{2\pi i}\int_{(\gamma)} \Gamma(-w)\,\Gamma(w+z)\,\Gamma(w+s)\,(-2\pi\alpha\omega)^{-w}\,\mathrm{d}w\right)$$

Hence (4.2.4) implies

$$\frac{1}{2}e^{\frac{\pi is}{2}} f_+(x) + \frac{1}{2}e^{-\frac{\pi is}{2}} f_-(x) = \frac{1}{2}\frac{1}{2\pi i}\int_{(c)} x^{-z}(F_+(z) + F_-(z))\,\mathrm{d}z$$
$$= \frac{1}{\omega^s\Gamma(s)}\frac{1}{2\pi i}\int_{(c)} x^{-z}(-2\pi\alpha)^{-z}\cos\frac{\pi}{2}z\,\mathrm{d}z$$
$$\times\left(\frac{1}{2\pi i}\int_{(\gamma)} \Gamma(-w)\,\Gamma(w+z)\,\Gamma(w+s)\,(-2\pi\alpha\omega)^{-w}\,\mathrm{d}w\right).$$

Letting $x = n$ and summing over $n = 1, 2, \cdots$ after multiplying by a_n, we have

$$\tilde{Z}_f(s) := \frac{1}{2}\left(e^{\frac{\pi i}{2}s} L_\Omega(-\alpha, s, i\omega) + e^{-\frac{\pi i}{2}s} L_\Omega(\alpha, s, -i\omega) - \frac{2}{\omega^s}\right)$$
$$= \frac{1}{2\pi i}\int_{(c)} F_f(z)\zeta_\Omega(z)\,\mathrm{d}z$$

for $c > 1$. We may shift the integration path to $\operatorname{Re} z = -d$ at this stage or simply substitute (4.2.4) to deduce that

$$J(s) = \tilde{Z}_f(s) = \frac{1}{\omega^s\Gamma(s)}\frac{1}{2\pi i}\int_{(c)} \zeta_\Omega(z)(-2\pi\alpha)^{-z}\Gamma(w+z)\cos\frac{\pi}{2}z\,\mathrm{d}z$$
$$\times\left(\frac{1}{2\pi i}\int_{(\gamma)} \Gamma(-w)\Gamma(w+s)(-2\pi\alpha\omega)^{-w}\,\mathrm{d}w\right)$$

where $\gamma < 0$ is chosen so small that $\sigma = \gamma$ not only separates the poles of gamma factors but also satisfies $\gamma + c < -\frac{1}{2}$. Then by the Stirling formula, we may change the order of integration, so that

$$J(s) = \frac{1}{2\pi i}\int_{(\gamma)} I(w)\Gamma(-w)\Gamma(w+s)(-2\pi\alpha\omega)^{-w}\,\mathrm{d}w \tag{4.2.5}$$

where

$$I(w) = \frac{\omega^{-s}}{\Gamma(s)}\frac{1}{2\pi i}\int_{(c)} \zeta_\Omega(z)(-2\pi\alpha)^{-z}\Gamma(w+z)\cos\frac{\pi}{2}z\,\mathrm{d}z.$$

Moving the integration path to $\operatorname{Re} z = -d < 0$ and applying (4.1.12), we obtain

$$I(w) = \frac{\omega^{-s}}{\Gamma(s)}\frac{1}{2\pi i}\int_{(-d)} \Gamma(w+z)G(z)\cos\frac{\pi}{2}z\zeta_\Omega(1-z)A^{2z-1}(-2\pi\alpha)^{-z}\,\mathrm{d}z$$
$$= \frac{1}{2\pi i}\int_{(-d)} \Gamma(w+z)M^{(1)}(z)\zeta_\Omega(1-z)\gamma(z)\,\mathrm{d}z = \sum_{n=1}^{\infty} a_n K(w,n),$$

where

$$K(w,n) = \frac{1}{2\pi i}\int_{(-d)} \Gamma(w+z)M^{(1)}(z)n^{z-1}\gamma(z)\,\mathrm{d}z, \tag{4.2.6}$$
$$\gamma(z) = \frac{\omega^{-s}}{A\Gamma(s)}\left(-2A^{-2}\pi\alpha\right)^{-z}$$

and where

$$M^{(1)}(z) = M_\Omega^{(1)}(z) = G(z)\cos\frac{\pi}{2}z = \frac{1}{\Gamma(z)}M^{(2)}(z),$$

where $M^{(2)}(z)$ is defined by (4.3.7) below. Hence we have

$$
\begin{aligned}
M_{\mathbb{Q}}^{(1)}(z) &= 2^{z-1}\sqrt{\pi}\frac{1}{\Gamma(z)}, \\
M_{\mathbb{Q}(\sqrt{|\Delta|})}^{(1)}(z) &= \frac{\Gamma(1-z)}{\Gamma(z)}\cos\frac{\pi}{2}z = G(z)\cos\frac{\pi}{2}z, \quad \Delta < 0 \\
M_{\mathbb{Q}(\sqrt{\Delta})}^{(1)}(z) &= 2^{2z-1}\frac{\Gamma(1-z)}{\Gamma(z)}\sin\frac{\pi}{2}z, \quad \Delta > 0.
\end{aligned} \tag{4.2.7}
$$

For $\Omega = \mathbb{Q}$, it follows that

$$
\begin{aligned}
I(w) &= \sum_{n=1}^{\infty}\frac{a_n}{n}\frac{\omega^{-s}\sqrt{\pi}}{2A\Gamma(s)}\frac{1}{2\pi i}\int_{(-d)}\frac{\Gamma(w+z)}{\Gamma(z)}\left(-\alpha n^{-1}\right)^{-z}\mathrm{d}z \\
&= \sum_{n=1}^{\infty}\frac{a_n}{n}\frac{\omega^{-s}}{2\Gamma(s)}G_{1,1}^{1,0}\left(-\alpha n^{-1}\,\middle|\,\begin{matrix}0\\ w\end{matrix}\right),
\end{aligned} \tag{4.2.8}
$$

where $a_n = 1$ but we keep them for other use. By Wolfram (1998-2021, p. 26) (Prudnikov et al. 1986, p. 631, 8.4.2, (3))

$$
G_{1,1}^{1,0}\left(x\,\middle|\,\begin{matrix}a\\ b\end{matrix}\right) = \frac{\theta(1-|z|)}{\Gamma(a-b)}(1-x)^{a-b-1}x^{b},
$$

we have

$$
\begin{aligned}
G_{1,1}^{1,0}\left(x\,\middle|\,\begin{matrix}0\\ w\end{matrix}\right) &= x^{w}G_{1,1}^{1,0}\left(x\,\middle|\,\begin{matrix}-w\\ 0\end{matrix}\right) \\
&= \frac{x^{w}}{\Gamma(-w)}(1-x)^{-w-1} = \frac{1}{\Gamma(-w)(1-x)}\left(\frac{1}{x}-1\right)^{-w}.
\end{aligned} \tag{4.2.9}
$$

Substituting (4.2.9) in (4.2.8), we conclude that

$$
I(w) = \sum_{n=1}^{\infty}\frac{a_n}{n}\frac{\omega^{-s}}{2\Gamma(s)\Gamma(-w)}\left(1+\frac{\alpha}{n}\right)^{-1}\left(-1-\frac{n}{\alpha}\right)^{-w}. \tag{4.2.10}
$$

Substituting (4.2.10) in (4.2.5), we have

$$
\begin{aligned}
J(s) &= \sum_{n=1}^{\infty}a_n\frac{\omega^{-s}}{2\Gamma(s)}(n+\alpha)^{-1}\frac{1}{2\pi i}\int_{(\gamma)}\Gamma(w+s)\left(2\pi\omega(n+\alpha)\right)^{-w}\mathrm{d}w \\
&= \sum_{n=1}^{\infty}a_n\frac{\omega^{-s}}{2\Gamma(s)}(n+\alpha)^{-1}G_{0,1}^{1,0}\left(2\pi\omega(n+\alpha)\,\middle|\,\begin{matrix}-\\ s\end{matrix}\right).
\end{aligned} \tag{4.2.11}
$$

Since

$$G_{0,1}^{1,0}\left(x \,\middle|\, \begin{matrix} - \\ s \end{matrix}\right) = x^s G_{0,1}^{1,0}\left(x \,\middle|\, \begin{matrix} - \\ 0 \end{matrix}\right) = x^s e^{-x},$$

Equation (4.2.11) leads to

$$\begin{aligned} J(s) &= \sum_{n=1}^{\infty} a_n \frac{\omega^{-s}}{2\Gamma(s)}(n+\alpha)^{-1}(2\pi\omega(n+\alpha))^s e^{-2\pi\omega(n+\alpha)} \\ &= \frac{e^{-2\pi\omega\alpha}}{2\Gamma(s)} \sum_{n=1}^{\infty} a_n (n+\alpha)^{s-1} e^{-2\pi\omega n} = \frac{e^{-2\pi\omega\alpha}}{2\Gamma(s)} L_{\mathbb{Q}}(i\omega, 1-s, \alpha), \end{aligned} \tag{4.2.12}$$

i.e. (4.2.3). □

Corollary 4.2.1 (Lipschitz summation formula) *For the complex variables* $z = x + iy$, $x > 0$, $s = \sigma + it$, $\sigma > 1$ *and the real parameter* $0 < w \leq 1$, *we have the Lipschitz summation formula*

$$\frac{(2\pi)^s}{\Gamma(s)} \sum_{n=0}^{\infty} (n+w)^{s-1} e^{-2\pi z(n+w)} = \sum_{n=-\infty}^{\infty} \frac{e^{2\pi i n w}}{(z+in)^s}. \tag{4.2.13}$$

Under the condition $0 < w < 1$ *this formula holds in the wider half-plane* $\sigma > 0$.

Indeed, changing s resp. x in (4.2.2) by $1 - s$ resp. wi leads to (4.2.13) and vice versa.

Equation (4.2.13) has sometimes been referred to as the Lipschitz summation formula with good reason (Grosswald (1984), Kanemitsu et al. (2000), Rademacher (1973) etc.). Knopp-Robbins Knopp and Robbins (2001) in their Remark 1 state their view on the Lipschitz summation formula to the effect that it is conceptually simpler than Riemann's original method of using the theta series. However, at least the special case of the Lipschitz summation formula (Corollary 4.2.1) can be readily deduced from the partial fraction expansion for the cotangent function, which is known to be equivalent to the functional equation for the Riemann zeta-function, so that we may say these are all equivalent.

In the case where Ω is a quadratic field, the argument is similar depending on (4.1.23). In view of the form of (4.2.7), it suffices to incorporate $e^{\pm\frac{\pi}{2}z}$. We rewrite $n^{z-1}\gamma(z)$ as

$$n^{z-1}\gamma(z) = -\frac{A^2 i}{2\pi\alpha}(-X)^{z-1}, \quad X = X(n) = \frac{1}{2\pi\alpha}A^2 n i,$$

whence

$$K(w,n) = \frac{A\omega^{-s}}{\pi\alpha\Gamma(s)} \frac{1}{2\pi i} \int_{(-d)} \frac{\Gamma(w+z)\Gamma(1-z)}{\Gamma(z)} \frac{1}{2i}\left((-X)^{z-1} - (-\bar{X})^{z-1}\right) dz, \tag{4.2.14}$$

$$= \frac{A\omega^{-s}\Gamma(w+1)}{\pi\alpha\Gamma(s)}\frac{1}{2i}\left({}_1F_1(w+1,1,X) - {}_1F_1(w+1,1,\bar{X})\right).$$

We state the intermediate results for want of a better treatment.

Proposition 4.2.1

$$\tilde{Z}_f(s) = \frac{A\omega^{-s}}{\pi\alpha\Gamma(s)}\sum_{n=1}^{\infty} a_n \frac{1}{2\pi i}\int_{(-d)} \Gamma(-w)\Gamma(w+1)\Gamma(w+s) \times \frac{1}{2i}\left({}_1F_1(w+1,1,X) - {}_1F_1(w+1,1,\bar{X})\right)\delta(w)\,\mathrm{d}w$$

for $\Delta < 0$ *and*

$$\tilde{Z}_f(s) = \frac{2A\omega^{-s}}{\pi\alpha\Gamma(s)}\sum_{n=1}^{\infty} a_n \frac{1}{2\pi i}\int_{(-d)} \Gamma(-w)\Gamma(w+1)\Gamma(w+s) \times \frac{1}{2}\left({}_1F_1(w+1,1,4X) + {}_1F_1(w+1,1,4\bar{X})\right)\delta(w)\,\mathrm{d}w$$

for $\Delta > 0$*, where* $X = \frac{1}{2\pi\alpha}A^2 ni$ *and* $\delta(w) = (-2\pi\alpha\omega)^{-w}$.

4.3 Koshlyakov's Generalized Dedekind Zeta-Function

On Koshlyakov (1954, p. 244), there are two succeeding formulas for J which are incorrect, with the last one being a unified expression. The final results which are stated in three different cases are correct although they depend on these two equalities which are incorrect. We state the corrected version of these equalities. The penultimate one which follows from Koshlyakov (1954, (23.13)) on applying (4.1.12), should read for $\beta < s+1$, $1 < s \in \mathbb{R}$

$$J = \frac{\zeta_\Omega(0)}{\omega^s} + \frac{1}{2\pi i}\int_{\beta-i\infty}^{\beta+i\infty} \frac{\omega^{\frac{1-s}{2}}A^{z-s}}{2\Gamma(s)}\Gamma\left(\frac{z+s-1}{2}\right)\Gamma\left(\frac{s-z+1}{2}\right)\cos\frac{\pi}{2}\frac{s-z+1}{2} \times G\left(\frac{s-z+1}{2}\right)\zeta_\Omega\left(\frac{z-s+1}{2}\right)\frac{\mathrm{d}s}{\omega^{\frac{z}{2}}} \tag{4.3.1}$$

and the last one

$$J = \frac{\zeta_\Omega(0)}{\omega^s} + \frac{\pi^{1-\frac{\chi}{2}}\omega^{\frac{1-s}{2}}A^{-s}}{2\Gamma(s)} \tag{4.3.2}$$

$$\times \sum_{n=1}^{\infty} \frac{F(n)}{n^{\frac{1-s}{2}}} \frac{1}{2\pi i} \int_{\beta-i\infty}^{\beta+i\infty} 2^{r_1 \frac{s-z-1}{2}+r} \Gamma\left(\frac{z+s-1}{2}\right) \Gamma^{\chi-1}\left(\frac{z-s-1}{2}\right)$$

$$\times \sin^r \frac{\pi}{2} \frac{s-z+1}{2} \cos^{r_2} \frac{\pi}{2} \frac{s-z+1}{2} \frac{\mathrm{d}s}{(A^{-1}\sqrt{n\omega})^z}.$$

It seems, however, the transformation $z = \frac{\sigma - s*1}{2}$ (in (4.3.10) below) is not necessary and that he stuck to this because he wanted to apply (3.1.3).

Koshlyakov (1954, pp. 243–247) establishes that the general Lipschitz summation formula is equivalent to the functional equation for the Dedekind zeta-function as a closed form for Koshlyakov's (Koshlyakov 1954, (23.10)):

$$\tilde{Z}_f(s) := \frac{e^{\frac{\pi i}{2}s}}{2} \zeta_\Omega(s, i\omega) + \frac{e^{-\frac{\pi i}{2}s}}{2} \zeta_\Omega(s, -i\omega) + 2\frac{\zeta_\Omega(0)}{\omega^s} \tag{4.3.3}$$

$$= \frac{e^{\frac{\pi i}{2}s}}{2} \sum_{n=1}^{\infty} \frac{a_n}{(n+i\omega)^s} + \frac{e^{-\frac{\pi i}{2}s}}{2} \sum_{n=1}^{\infty} \frac{a_n}{(n-i\omega)^s}.$$

By (2.5.18), we have

$$\frac{1}{2} e^{\frac{\pi i s}{2}} \frac{1}{(x+i\omega)^s} = \frac{\omega^{-s}}{2} \frac{1}{\left(1+\frac{x}{i\omega}\right)^s} = \frac{\omega^{-s}}{2} \frac{1}{2\pi i} \int_{(c)} \frac{\Gamma(z)\Gamma(s-z)}{\Gamma(s)} e^{\frac{\pi i z}{2}} \omega^z \frac{\mathrm{d}z}{x^z}.$$

Hence

$$\frac{1}{2} e^{\frac{\pi i s}{2}} \frac{1}{(x+i\omega)^s} + \frac{1}{2} e^{-\frac{\pi i s}{2}} \frac{1}{(x-i\omega)^s} = \frac{\omega^{-s}}{\Gamma(s)} \frac{1}{2\pi i} \int_{(c)} \Gamma(z)\Gamma(s-z) \cos\frac{\pi}{2} z \omega^z \frac{\mathrm{d}z}{x^z},$$

so that

$$F_f(w) = \frac{\omega^{-s}}{\Gamma(s)} \Gamma(s-w)\Gamma(w) \cos\frac{\pi}{2} w \frac{1}{\omega^{-w}} = \mathfrak{G}_F(w,z)\mathfrak{F}_F(w)\delta(w,s),$$

$$\mathfrak{G}_F(w,z) = 1, \mathfrak{F}_F(w) = \Gamma(s-w)\Gamma(w) \cos\frac{\pi}{2} w, \delta(w,s) = \frac{\omega^{-s}}{\Gamma(s)} \frac{1}{\omega^{-w}}$$

i.e. without integration. For $c > 1$ we have

$$\tilde{Z}_f(s) = \frac{\omega^{-s}}{\Gamma(s)} \frac{1}{2\pi i} \int_{(c)} \Gamma(s-z)\Gamma(z) \cos\frac{\pi}{2} z \frac{\zeta_\Omega(z)}{\omega^{-z}} \mathrm{d}z. \tag{4.3.4}$$

Moving the integration path to $\operatorname{Re} z = -d < 0$, we pass the simple pole at $z = 0$ with residue $\frac{\zeta_\Omega(0)}{\omega^s}$ (noting that the pole of $\zeta_\Omega(z)$ at $z = 1$ is cancelled by the zero of $\cos\frac{\pi}{2}s$) and

$$\tilde{Z}_f(s) = \frac{\zeta_\Omega(0)}{\omega^s} + \frac{\omega^{-s}}{\Gamma(s)} \frac{1}{2\pi i} \int_{(-d)} \Gamma(s-z)\Gamma(z) \cos\frac{\pi}{2} z A^{2z-1} G(z) \zeta_\Omega(1-z) \frac{dz}{\omega^{-z}} \tag{4.3.5}$$

$$= \frac{\zeta_\Omega(0)}{\omega^s} + J(s),$$

say, where we used (4.1.12) and put

$$\begin{aligned} J(s) = I(s) &= \frac{\omega^{-s}}{A\Gamma(s)} \frac{1}{2\pi i} \int_{(-d)} \Gamma(s-z) M^{(2)}(z) \zeta_\Omega(1-z) \frac{dz}{(A^2\omega)^{-z}} \\ &= \frac{1}{2\pi i} \int_{(-d)} \Gamma(s-z) M^{(2)}(z) \zeta_\Omega(1-z) \gamma(z,s)\, dz \end{aligned} \tag{4.3.6}$$

and

$$M^{(2)}(z) = M_\Omega^{(2)}(z) = \Gamma(z) G(z) \cos\frac{\pi}{2} z. \tag{4.3.7}$$

Substituting the Dirichlet series for $\zeta_\Omega(1-z)$ and changing the order of integration and summation, we have

$$J(s) = \sum_{n=1}^\infty a_n K(s,n), \quad K(s,n) = \frac{1}{2\pi i} \int_{(-d)} \Gamma(s-z) M^{(2)}(z) n^{z-1} \gamma(z,s)\, dz \tag{4.3.8}$$

$$\gamma(z,s) = \frac{\omega^{-s}}{A\Gamma(s)} (A^2\omega)^z.$$

It suffices to transform $M(z)$ in a simple form in three cases using (4.1.20). Since the procedure depends on the use of the reciprocal relation and the duplication formula, we only state the final results.

$$\begin{aligned} &M_{\mathbb{Q}}^{(2)}(z) = 2^{z-1}\sqrt{\pi}, \\ &M_{\mathbb{Q}(\sqrt{|\Delta|})}^{(2)}(z) = \Gamma(1-z) \cos\frac{\pi}{2} z, \quad \Delta < 0 \\ &M_{\mathbb{Q}(\sqrt{\Delta})}^{(2)}(z) = 2^{2z-1} \Gamma(1-z) \sin\frac{\pi}{2} z, \quad \Delta > 0. \end{aligned} \tag{4.3.9}$$

It follows that

Theorem 4.3.1

$$\frac{e^{\frac{\pi i}{2}s}}{2} \zeta_\Omega(s, i\omega) + \frac{e^{-\frac{\pi i}{2}s}}{2} \zeta_\Omega(s, -i\omega) + 2\frac{\zeta_\Omega(0)}{\omega^s} = \tilde{Z}_f(s) \tag{4.3.10}$$

$$= \frac{\zeta_\Omega(0)}{\omega^s} + \frac{\pi^{1-\frac{\chi}{2}} \omega^{-s}}{A\Gamma(s)} \sum_{n=1}^\infty \frac{a_n}{n}$$

$$\times \frac{1}{2\pi i}\int_{(-d)} 2^{r_1(z-1)+r}\Gamma(s-z)\,\Gamma^{\chi-1}(1-z)\sin^r\frac{\pi}{2}z\cos^{r_2}\frac{\pi}{2}z(A^2\omega n)^z\,dz,$$

which leads to (4.3.2).

I. The case $\Omega=\mathbb{Q}$.
Since

$$\begin{aligned}K(s,n)&=\frac{\sqrt{\pi}\,\omega^{-s}}{2nA\Gamma(s)}\frac{1}{2\pi i}\int_{(-d)}\Gamma(s-z)\,(2A^2\omega n)^z\,dz\\&=\frac{\sqrt{\pi}\,\omega^{-s}}{2nA\Gamma(s)}(2A^2\omega n)^s\frac{1}{2\pi i}\int_{(s+d)}\Gamma(u)\,(2A^2\omega n)^{-u}\,du\\&=\frac{(2\pi)^s}{2\Gamma(s)}\frac{1}{n^{1-s}}e^{-2\pi wn}\end{aligned}$$

by $A=\sqrt{\pi}$, (4.3.8) reads

$$\tilde{Z}_f(s)=-\frac{1}{2}\omega^{-s}+\frac{(2A^2)^s}{2\Gamma(s)}\sum_{n=1}^{\infty}\frac{e^{-2\pi wn}}{n^{1-s}}$$

by $\zeta_{\mathbb{Q}}(0)=-\frac{1}{2}$. Hence (both sides being equal to $J(s)-\frac{1}{2}\omega^{-s}$)

$$\frac{e^{\frac{\pi i}{2}s}}{2}\zeta(s,i\omega)+\frac{e^{-\frac{\pi i}{2}s}}{2}\zeta(s,-i\omega)-\frac{1}{2}\omega^{-s}=\frac{(2\pi)^s}{2\Gamma(s)}\sum_{n=1}^{\infty}\frac{e^{-2\pi\omega n}}{n^{1-s}},$$

which amounts to the special case of Theorem 4.2.1 with $w=1$, i.e. RHF.
II. The case $\Omega=\mathbb{Q}(\sqrt{\Delta})$, $\Delta<0$.

In this case, as can be seen from (4.3.9), there is no effect of processing $Z_f(s)$ to form $\tilde{Z}_f(s)$ in (4.3.3) and each term admits the Fourier-Bessel expansion. However, as noticed in deriving (4.3.5), the cosine function cancels the pole of $\zeta_\Omega(z)$ at $z=1$ and so treating each term, we must compute the residue at $z=1$. In place of (4.3.4), we consider

$$\frac{e^{\frac{\pi i}{2}s}}{2}\sum_{n=1}^{\infty}\frac{a_n}{(n+i\omega)^s}=\frac{\omega^{-s}}{\Gamma(s)}\frac{1}{2\pi i}\int_{(c)}\Gamma(s-z)\Gamma(z)e^{\frac{\pi i}{2}z}\frac{\zeta_\Omega(z)}{\omega^{-z}}\,dz.$$

The residue of the integrand at $z=1$ is

$$\frac{\omega^{-s}}{\Gamma(s)}\operatorname*{Res}_{z=1}\Gamma(s-z)\Gamma(z)e^{\frac{\pi i}{2}z}\frac{\zeta_\Omega(z)}{\omega^{-z}}=\frac{\omega^{1-s}\rho i}{s-1},$$

where ρ is defined by (2.5.3). Hence

$$\frac{e^{\frac{\pi i}{2}s}}{2}\sum_{n=1}^{\infty}\frac{a_n}{(n+i\omega)^s}=\frac{\omega^{1-s}R_{r_1,r_2}i}{s-1}+\frac{\zeta_\Omega(0)}{2\omega^s}+\frac{1}{2}\sum_{n=1}^{\infty}a_nK^{(\varepsilon)}(s,n), \tag{4.3.11}$$

where

$$K^{(\varepsilon)}(s,n)=\frac{\omega^{-s}}{2nA\Gamma(s)}\frac{1}{2\pi i}\int_{(-d)}\Gamma(s-z)\Gamma(1-z)\xi^{2z}\,dz. \tag{4.3.12}$$

with $\xi=e^{\frac{\pi}{4}i}A\sqrt{\omega n}=\varepsilon A\sqrt{\omega n}$, say, where

$$\varepsilon=e^{\frac{\pi}{4}i}.$$

Putting $s-z=w$, the integral becomes

$$\xi^{s+1}\frac{1}{2\pi i}\int_{(s+d)}\Gamma(w)\Gamma(w-(s-1))\xi^{s-1-2w}\,dw=2\xi^{s+1}K_{s-1}(2\xi) \tag{4.3.13}$$

by (3.1.3). Substituting (4.3.13) in (4.3.11), we infer that

$$\begin{aligned}&\frac{e^{\frac{\pi i}{2}s}}{2}\sum_{n=1}^{\infty}\frac{a_n}{(n+i\omega)^s}\\&=\frac{\omega^{1-s}R_{r_1,r_2}i}{s-1}+\frac{\zeta_\Omega(0)}{2\omega^s}+A^s\frac{\omega^{\frac{1-s}{2}}}{\Gamma(s)}\varepsilon^{s+1}\sum_{n=1}^{\infty}\frac{a_n}{n^{\frac{1-s}{2}}}K_{s-1}(2\varepsilon A\sqrt{\omega n})\end{aligned} \tag{4.3.14}$$

and the counterpart (with $\bar{\varepsilon}$ in place of ε in (4.3.12)) reads

$$\begin{aligned}&\frac{e^{\frac{-\pi i}{2}s}}{2}\sum_{n=1}^{\infty}\frac{a_n}{(n-i\omega)^s}\\&=-\frac{\omega^{1-s}R_{r_1,r_2}i}{s-1}+\frac{\zeta_\Omega(0)}{2\omega^s}+A^s\frac{\omega^{\frac{1-s}{2}}}{\Gamma(s)}\bar{\varepsilon}^{s+1}\sum_{n=1}^{\infty}\frac{a_n}{n^{\frac{1-s}{2}}}K_{s-1}(2\bar{\varepsilon}A\sqrt{\omega n}).\end{aligned} \tag{4.3.15}$$

Hence the corrected version of Koshlyakov (1954, (23.15)) should read

$$\begin{aligned}\tilde{Z}_f(s)=\frac{\zeta_\Omega(0)}{\omega^s}+A^s\frac{\omega^{\frac{1-s}{2}}}{\Gamma(s)}\Bigg(&\varepsilon^{s+1}\sum_{n=1}^{\infty}\frac{a_n}{n^{\frac{1-s}{2}}}K_{s-1}(2\varepsilon A\sqrt{\omega n})\\&+\bar{\varepsilon}^{s+1}\sum_{n=1}^{\infty}\frac{a_n}{n^{\frac{1-s}{2}}}K_{s-1}(2\bar{\varepsilon}A\sqrt{\omega n})\Bigg),\end{aligned} \tag{4.3.16}$$

where $\zeta_\Omega(0) = -\frac{h}{W}$. Of course, this follows from (4.3.6) by Euler's formula.
III. The case $\Omega = \mathbb{Q}(\sqrt{\Delta})$, $\Delta < 0$.
Equation (4.3.10) reads

$$\tilde{Z}_f(s) = \frac{\zeta_\Omega(0)}{\omega^s} + J(s), \quad J(s) = \sum_{n=1}^{\infty} a_n K_1(s,n),$$

where

$$K_1(s,n) = \frac{1}{2\pi i}\int_{(d)} \Gamma(s-z)\Gamma(1-z)\sin\frac{\pi}{2}z(4A^2\omega n)^z\,\mathrm{d}z.$$

By Euler's identity, we have

$$J(s) = \frac{1}{2i}\sum_{n=1}^{\infty} a_n \left(K_1^{(\varepsilon)}(s,n) - K_1^{(\bar{\varepsilon})}(s,n)\right)$$

where

$$K_1^{(\varepsilon)}(s,n) = \frac{1}{2\pi i}\int_{(d)} \Gamma(s-z)\Gamma(1-z)\xi_1{}^z\,\mathrm{d}z, \quad \xi_1 = 2\varepsilon A\sqrt{\omega n}.$$

Hence, similarly to (4.3.16)

$$\frac{e^{\frac{\pi i}{2}s}}{2}\zeta_\Omega(s, i\omega) + \frac{e^{-\frac{\pi i}{2}s}}{2}\zeta_\Omega(s, -i\omega) = (2A)^s \frac{\omega^{\frac{1-s}{2}}}{\Gamma(s)}$$
$$\times\left(\frac{\varepsilon^{s+1}}{i}\sum_{n=1}^{\infty}\frac{a_n}{n^{\frac{1-s}{2}}}K_{s-1}(4\varepsilon A\sqrt{\omega n}) - \frac{\bar{\varepsilon}^{s+1}}{i}\sum_{n=1}^{\infty}\frac{a_n}{n^{\frac{1-s}{2}}}K_{s-1}(4\bar{\varepsilon}A\sqrt{\omega n})\right),$$

Kanemitsu and Tsukada (2014, pp. 127–134) expounds Koshlyakov's argument toward the proof of the functional equation for (4.3.3). It is shown that each sum may be expressed as the Fourier-Bessel expansion which leads to Koshlyakov's results. Koshlyakov also applied the beta transform and naturally these lead to the same result. The difference is that there is no need to consider the sum of two terms on the left of (4.3.3) since each term admits the expression by the general theory of modular relations.

The character analogue of the Lipschitz summation formula is known Chakraborty et al. (2009c, pp. 59–62), and so we may naturally treat a more general case.

The Lipschitz summation formula for quadratic fields is contained in Kanemitsu and Tsukada (2014), which is elucidation of Koshlyakov's results Koshlyakov (1954).

4.4 Plana's Summation Formula à La Koshlyakov

In this section we work in the situation of Example 4.1.1, i.e. $Z(s) = \zeta_\Omega(s) = \sum_{n=1}^{\infty} \frac{a_n}{n^s}$, $\sigma > 1$ and the equality (4.1.1). We prove the general Plana summation formula, Theorem 4.4.1 by the method of Koshlyakov et al.:

Theorem 4.4.1 (Plana's summation formula, (Koshlyakov 1954, p. 217))
Suppose that the Mellin transform $F(s)$ in (2.1.1) of the function $f(x)$ $(x > 0)$ satisfies the growth condition (Stirling formula)

$$F(s) = O\left(e^{-\frac{\pi}{2}|t|}\,|t|^{m\sigma-n}\right), \tag{4.4.1}$$

uniformly in $-d \le \sigma \le c$, where $-1 < -d < -\frac{1}{2}$, $1 < c$, and where $m > 0$, $n > 0$ are subject to the condition

$$\frac{m}{2} + n \ge 1,$$

and that $F(s)$ is regular in the strip $-d \le \sigma \le c$ except for a simple pole at $s = 0$ with residue $f(0)$. Then

$$Z_f(s) = \sum_{n=1}^{\infty} a_n f(n) = J(s) + \mathrm{P}(s), \tag{4.4.2}$$

where

$$\mathrm{P}(s) = \zeta_\Omega(0) f(0) + \rho \int_0^\infty f(x)\,\mathrm{d}x,$$

and where $I(x) = 2\sigma(x)$ ((4.1.4)) and $\sigma(x)$ is Koshlyakov's function (4.4.8) *and ρ is defined in (2.5.2). Equation* (4.4.2) *entails Plana's summation formula* (Srivastava and Choi 2001, p. 90, (9))

$$\sum_{n=1}^{\infty} f(n) = -\frac{1}{2} f(0) + \int_0^\infty f(x)\mathrm{d}x - 2\int_0^\infty \frac{f(ix) - f(-ix)}{2i} \frac{1}{e^{2\pi x} - 1}\mathrm{d}x, \tag{4.4.3}$$

under the conditions in Lemma 4.4.1 and the infinite series and the integral in (4.4.3) *are (absolutely) convergent.*

For quadratic fields $\Omega = \mathbb{Q}(\sqrt{|\Delta|})$, we have

$$\begin{aligned} J(s) &= \int_0^\infty \frac{f(ix) - f(-ix)}{2i} 2\sigma(x)\,\mathrm{d}x \qquad (4.4.4)\\ &= \begin{cases} \frac{2A}{\pi i} \sum_{n=1}^{\infty} a_n \left(K_0(2\bar{\varepsilon} A\sqrt{xn}) - K_0(2\varepsilon A\sqrt{xn})\right), & \Delta < 0 \\ \frac{4A}{\pi} \sum_{n=1}^{\infty} a_n \left(K_0(4\bar{\varepsilon} A\sqrt{xn}) + K_0(4\varepsilon A\sqrt{xn})\right), & \Delta > 0. \end{cases} \end{aligned}$$

Exercise 4.4.1 *Give details of the proof of the identity* (4.1.6) *(for $Z(s) = \zeta_\Omega(s)$).*

Solution. Application of HGT and the Cauchy residue theorem in the spirit of Example 2.1.1. □

Exercise 4.4.2 *In the situation of Theorem 4.4.1,* $\mathrm{P}(s) = R_0 + R_1$*, where* $R_j = \operatorname*{Res}_{s=j} F(s)\zeta_\Omega(s),\ j = 0,\ 1$*. Show that*

$$R_0 = \zeta_\Omega(0) f(0)$$

and

$$R_1 = \rho \lim_{s\to 1} F(s) = \rho \int_0^\infty f(x)\,\mathrm{d}x,$$

where ρ *is given by (2.5.3).*

Solution. These are found on Koshlyakov (1954, II, pp. 214–215) and can be readily verified. □

In order to transform the last integral in (4.1.6), we appeal to

Lemma 4.4.1 *We extend the domain* $\{\mathrm{Re}\, x > 0\}$ *of* $f(x)$ *to (2.1.80).*

Then $f(x)$ *is analytic in the domain (2.1.80) and we may speak of* $f(\pm ix)$*,* $x > 0$*. We have*

$$F(s) = \frac{1}{\sin\frac{\pi s}{2}} \int_0^\infty \frac{f(ix) - f(-ix)}{2i} x^{s-1}\mathrm{d}x. \tag{4.4.5}$$

The double integral in (4.4.6) *is absolutely convergent and the order of integrals may be changed.*

Proof The first assertion is proved on Koshlyakov (1954, II, pp. 215–216). By rotating the positive real axis by $\frac{\pi}{2}$, we obtain

$$F(s) = \int_0^\infty f(ix)(ix)^{s-1}\, i\mathrm{d}x = e^{\frac{\pi i}{2}s} \int_0^\infty f(ix)\, x^{s-1}\mathrm{d}x$$

and similarly

$$F(s) = e^{-\frac{\pi i}{2}s} \int_0^\infty f(-ix)\, x^{s-1}\mathrm{d}x.$$

Cf. Proposition 2.1.2. Hence (4.4.5) follows.

The second assertion is proved on Koshlyakov (1954, II, p. 216) based on the estimate $\sigma(x) = o\left(|z|^\delta\right)$ for every $\delta > 0$, where $\sigma(x)$ is defined by (4.4.8). In our case we estimate the integral in (4.4.5) by the left-hand side using the Stirling formula (4.4.1). Since the integral is absolutely convergent and is estimated as $O(|t|^{-\lambda})$, $\lambda > 1$, the outer integral $\int_{(-d)}$ is absolutely convergent. □

Proof of Theorem 4.4.1. Substituting (4.1.12) and (4.4.5) in $J(s)$ in (4.1.6), we deduce that

$$J(s) = \frac{1}{2\pi i} \int_{(-d)} A^{2s-1} G(z) \zeta_\Omega(1-z) \frac{1}{\sin \frac{\pi z}{2}} \left(\int_0^\infty \frac{f(ix) - f(-ix)}{2i} x^{z-1} \, \mathrm{d}x \right) \mathrm{d}z$$

$$= \int_0^\infty \frac{f(ix) - f(-ix)}{2i} I(x) \, \mathrm{d}x = \int_0^\infty I(x) \mathfrak{F}_F(x) \delta(x) \, \mathrm{d}x \tag{4.4.6}$$

on changing the order of integration. Here

$$\mathfrak{F}_F(x) = \frac{f(ix) - f(-ix)}{2i}, \quad \delta(x) = 1,$$

$$\begin{aligned} I(x) &= \frac{1}{2\pi i} \int_{(-d)} A^{2z-1} x^{z-1} G(z) \frac{1}{\sin \frac{\pi z}{2}} \zeta_\Omega(1-z) \, \mathrm{d}z \\ &= \frac{1}{2\pi i} \int_{(-d)} G(z) \frac{1}{\sin \frac{\pi z}{2}} \zeta_\Omega(1-z) \gamma(z) \, \mathrm{d}z \\ &= \sum_{n=1}^\infty a_n \frac{1}{2\pi i} \int_{(-d)} M^{(3)}(z) n^{z-1} \gamma(z) \, \mathrm{d}z = \sum_{n=1}^\infty a_n K(x, n), \end{aligned}$$

and

$$\begin{aligned} K(x, n) &= \frac{1}{2\pi i} \int_{(-d)} M^{(3)}(z) n^{z-1} \gamma(z) \, \mathrm{d}z = \frac{2A}{\pi} K_{r_1, r_2}(A^2 x n), \\ \gamma(z) &= A^{2z-1} x^{z-1}, \end{aligned} \tag{4.4.7}$$

where K_{r_1,r_2} is Koshlyakov's function defined by (4.4.9), and where

$$M^{(3)}(z) = G(z) \frac{1}{\sin \frac{\pi}{2} z} = \frac{2}{\pi} \Gamma(1-z) M^{(2)}(z).$$

Hence by (4.3.9),

$$\begin{aligned} M_{\mathbb{Q}}^{(3)}(z) &= \frac{2^z}{\sqrt{\pi}} \Gamma(1-z), \\ M_{\mathbb{Q}(\sqrt{|\Delta|})}^{(3)}(z) &= \frac{2}{\pi} \Gamma(1-z)^2 \cos \frac{\pi}{2} z, \quad \Delta < 0 \\ M_{\mathbb{Q}(\sqrt{\Delta})}^{(3)}(z) &= \frac{2^{2z}}{\pi} \Gamma(1-z)^2 \sin \frac{\pi}{2} z, \quad \Delta > 0. \end{aligned}$$

Hence the inner integral of (4.4.6) in the rational case amounts to

$$\frac{2A}{\sqrt{\pi}}\frac{1}{2\pi i}\int_{(-d)}\Gamma(1-z)(A^2xn)^{z-1}\,\mathrm{d}z = \frac{1}{2\pi i}\int_{(1+d)}\Gamma(z)(A^2xn)^{-z}\,\mathrm{d}z$$
$$= 2e^{-(A^2xn)} = 2e^{-2\pi xn}.$$

Hence, summing the geometric series, (4.4.6) leads to

$$J(s) = -2\int_0^\infty \frac{f(ix)-f(-ix)}{2i}\frac{1}{e^{2\pi x}-1}\,\mathrm{d}x,$$

which proves (4.4.3).

Equation (4.4.4) follows by substituting (4.4.10) in (4.4.7). □

Here $\sigma(x)$ is defined after (Koshlyakov 1954, I, (2.13)) by $(c > 1)$

$$\sigma(x) = \frac{1}{2\pi i}\int_{(c)}\frac{1}{2\cos\frac{\pi s}{2}}A^{1-2s}G(1-s)\zeta_\Omega(s)\,x^{-s}\,\mathrm{d}s = \frac{A}{\pi}\sum_{n=1}^\infty a_n K_{r_1,r_2}(A^2xn), \tag{4.4.8}$$

where (Koshlyakov 1954, (2.8))

$$K_{r_1,r_2}(x) = \frac{1}{2\pi i}\int_{(c)}\frac{\pi}{2\cos\frac{\pi s}{2}}G(1-s)x^{-s}\,\mathrm{d}s = \frac{1}{2\pi i}\int_{(c)}\frac{\pi}{2}M^{(3)}(1-z)x^{-z}\,\mathrm{d}z \tag{4.4.9}$$

is Koshlyakov's K-function. Indeed,

$$\sigma(x) = \frac{1}{2}I(x).$$

Exercise 4.4.3 *Check the following by (3.1.3).*

$$K_{1,0}(x) = \sqrt{\pi}e^{-2x}, \tag{4.4.10}$$
$$K_{0,1}(x) = \frac{1}{i}\left(K_0\left(2\bar{\varepsilon}\sqrt{x}\right) - K_0\left(2\varepsilon\sqrt{x}\right)\right),$$
$$K_{2,0}(x) = 2\left(K_0\left(4\bar{\varepsilon}\sqrt{x}\right) + K_0\left(4\varepsilon\sqrt{x}\right)\right).$$

This is stated in Koshlyakov (1954, p. 114, (2.5), (2.6)).

Corollary 4.4.1 *For*

$$f(x) = f(x; z, s, a) = z^x\,(x+a)^{-s}$$

Theorem 4.4.1 reduces to the generalized Hermite formula

$$\Phi_\Omega(z, s, w) = \frac{1}{2}w^{-s} + R_{r_1,r_2}\int_0^\infty \frac{z^x}{(x+w)^s}\mathrm{d}x \tag{4.4.11}$$

$$-2\int_0^\infty \left(x^2+w^2\right)^{-\frac{s}{2}} \sin\left(x\log z - s\arctan\frac{x}{w}\right)\sigma(t)\,\mathrm{d}t$$

valid for Re $w > 0$ *is a consequence of the functional Eq.* (4.1.9) *for the Dedekind zeta-function.*

Proof We note that

$$\frac{1}{2i}\left(f(ix)-f(-ix)\right) = \left(a^2+x^2\right)^{-\frac{s}{2}} \mathrm{I_m} e^{i\left(x\log z - s\arctan\frac{x}{a}\right)} \tag{4.4.12}$$
$$= \left(a^2+x^2\right)^{-\frac{s}{2}} \sin\left(x\log z - s\arctan\frac{x}{a}\right)$$

for $z > 0$ in the first instance and then for Re $z > 0$ by analytic continuation.
Substituting (4.4.12) in (4.4.3), we arrive at (4.4.11). □

We assemble corollaries to the above corollary which are proved in Li et al. (2012) directly.

Theorem 4.4.2 *For* Re $a > 0$ *we have*

$$\Phi(z,s,a) = \frac{1}{2}a^{-s} + \int_0^\infty \frac{z^x}{(x+a)^s}\mathrm{d}x$$
$$-2\int_0^\infty \left(x^2+a^2\right)^{-\frac{s}{2}} \sin\left(x\log z - s\arctan\frac{x}{a}\right)\frac{\mathrm{d}x}{e^{2\pi x}-1}.$$

Corollary 4.4.2 (Hermite's formula (Srivastava and Choi 2001, p. 91, (12))). *For* Re $a > 0$ *we have*

$$\zeta(s,a) = \frac{1}{2}a^{-s} + \frac{a^{1-s}}{s-1} + 2\int_0^\infty \left(a^2+x^2\right)^{-\frac{s}{2}} \sin\left(s\arctan\frac{x}{a}\right)\frac{\mathrm{d}x}{e^{2\pi x}-1}. \tag{4.4.13}$$

Corollary 4.4.3 (Binet's second expression (Srivastava and Choi 2001, p. 17, (30))). *For* Re $a > 0$ *we have*

$$\log\frac{\Gamma(a)}{\sqrt{2\pi}} = \left(a-\frac{1}{2}\right)\log a + 2\int_0^\infty \frac{\arctan\frac{x}{a}}{e^{2\pi x}-1}\mathrm{d}x \tag{4.4.14}$$

Proof of (2.1.15). In Corollary 4.4.2 we let $s \to 1$ after rewriting (4.4.13) as

$$\zeta(s,a) - \frac{1}{s-1} \tag{4.4.15}$$
$$= \frac{1}{2}a^{-s} + \frac{a^{1-s}-1}{s-1} + 2\int_0^\infty \left(a^2+x^2\right)^{-\frac{s}{2}} \sin\left(s\arctan\frac{x}{a}\right)\frac{\mathrm{d}x}{e^{2\pi x}-1}.$$

Then noting that $\sin\arctan\frac{x}{a} = \frac{x}{\sqrt{x^2+a^2}}$ and $\lim_{s\to 1}\frac{a^{1-s}-1}{s-1} = -\log a$, we deduce (2.1.15), completing the proof. □

4.5 Hamburger-Oberhettinger-Soni Modular Relations

Here we shall elucidate the genesis of some identities (esp. summation formulas) equivalent to the functional equations given by Hamburger (1922) (Corollary 4.5.1) and by Soni (1964), Soni (1966), Oberhettinger and Soni (1972) with emphasis on Hamburger's results, in the light of Corollary 4.1.1 and RHB correspondence.

Hamburger (1922) precedes Koshlyakov (1954) in the use of the partial fraction expansion (2.1.23) for the cotangent function. Although Chandrasekharan and Narasimhan (1961, p. 2) state that the equivalence "established constitutes an analogue, for Hecke's functional equation, of a result of Hamburger (1922) on Riemann's equation," analogous part is limited to the Bochner modular relation and Hardy's formula (2.12.46) as a counterpart of the partial fraction expansion (4.5.12), cf. Theorem 2.12.3 and Remark 4.5.1. The third equivalence is for the Riesz sum (2.12.44) but it does not reduce to (4.5.13), cf. Exercise 4.5.6. Berndt V (1971) mentions Hamburger's partial fraction expansion as a special case of Hardy's formula and Oberhettinger and Soni (1972) only quote this and other two results as (10): the Fourier expansion (4.5.13) for the 2nd periodic Bernoulli polynomial $\bar{B}_2(y)$ and (12): the Poisson summation formula (4.5.14). The latter is valid for a function of bounded variation *having an integral expression which is analytic in a vertical strip.* Oberhettinger and Soni (1972) is an extract from Soni (1964) and centers around the extension of Koshlyakov's results Koshlyakov (1934), Koshlyakov (1928/29a), and Koshlyakov (1928/29b), overlooking the more far-reaching Koshlyakov (1954). The main results of Soni (1966) and Oberhettinger and Soni (1972) are summation formulas of Sierpiński (a special case of Hecke type) and Voronoǐ in the case where the functions are a Mellin transform pair, which are essentially consequences of the theory of Koshlyakov's X-functions (4.5.5) and L-functions (4.5.6) (The X-function aspect has been explained in Kanemitsu and Tsukada (2014, pp. 97–100)). For the Voronoǐ summation formula, we refer, e.g. to Laurinčikas (1998). Here the *Principle* applies without processing: the simple change into the Mellin transform and interchange of integration, cf. Theorem 4.5.1. The novelty of Soni (1964)–Oberhettinger and Soni (1972) lies in the study of the functional equation of Hecke and Vononoǐ in analogy to Riemann's and establishing identities corresponding to the ones for the Riemann zeta-function, in the same sprit as Koshlyakov's study on Dedekind zeta-functions, where the line of reflection is $r = \frac{1}{2}$. The only difference occurs in the Vononoǐ case and the difference is in the residual function. Soni (1966) gives proof of two more identities on Soni (1964, p. 63) equivalent to the functional equation, which may be clarified through (4.5.9) (Koshlyakov 1954, I, p. 119, (4.7)). The Riesz sums as well as summation formulas of zeta-functions in Definition 4.1.1 are well-known and we shall be brief stating the results as in Exercise 4.5.6 and concentrate on the Poisson summation formula (Theorem 4.5.3).

Our main result on summation formula is the following (which is the culminating result stated in Liu et al. (2023)).

Theorem 4.5.1 *The Poisson summation formula* (4.5.14) *à la Hamburger is manifestation of the Principle with the Fourier transform pair. The Soni-Oberhettinger formula of the form*

$$\sum_{n=1}^{\infty} a_n f(\lambda_n) = \mathrm{P}(\cdot) + \sum_{n=1}^{\infty} b_n \hat{f}(\mu_n), \quad \hat{f}(y) = \int_0^{\infty} f(x) K(xy)\, \mathrm{d}x$$

is a consequence of Corollary 4.1.1 with $F(z)$ replaced by the Mellin transform (2.1.1).

Exercise 4.5.1 *Prove that the σ-function* (4.4.8) *admits the partial fraction expansion* (Koshlyakov 1954, I, (3.6))

$$2\pi i \sigma(iz) = -\pi i \rho - \frac{2\zeta_\Omega(0)}{z} + \sum_{n=1}^{\infty} a_n \left(\frac{1}{z-n} + \frac{1}{z+n} \right), \tag{4.5.1}$$

so that $\sigma(iz)$ is a meromorphic function with simple poles at every integer, whence except at poles,

$$\sigma(iz) = -\sigma(-iz) - \rho.$$

For Re $z \to \infty$,

$$\sigma(z) = o(|z|^{-\delta})$$

for any $\delta > 0$.

Solution. We apply Corollary 4.1.1 to the integral for $\sigma(x)$ in (4.4.8) by viewing $Z(s) = G(1-s)\zeta_\Omega(s)$ and applying the functional Eq. (4.1.12). Then

$$\sigma(x) = \frac{1}{2\pi i} \int_{(-d)} \frac{1}{2\cos\frac{\pi s}{2}} \zeta_\Omega(1-s) x^{-s}\, \mathrm{d}s + \mathrm{P}(x), \tag{4.5.2}$$

where $\mathrm{P}(x)$ is the sum of residues of the integrand. By the familiar procedure, the integral I in (4.5.2) becomes

$$I = \frac{1}{2\pi} \sum_{n=1}^{\infty} a_n \frac{1}{n} \frac{1}{2\pi i} \int_{(-d)} \Gamma\left(\frac{1-s}{2}\right) \Gamma\left(\frac{1+s}{2}\right) \left(\frac{x}{n}\right)^{-s} \mathrm{d}s,$$

which becomes on setting $z = \frac{s+1}{2}$,

$$I = \frac{1}{\pi} \sum_{n=1}^{\infty} a_n \frac{x}{\pi n^2} \frac{1}{2\pi i} \int_{\left(\frac{1-d}{2}\right)} \Gamma(1-z)\Gamma(z) \left(\frac{x^2}{n^2}\right)^{-z} \mathrm{d}z.$$

Noting that the resulting integral is the beta-function integral (2.5.18), we conclude that

$$I = \frac{x}{\pi}\sum_{n=1}^{\infty} a_n \frac{1}{x^2+n^2}. \tag{4.5.3}$$

On the other hand, noting that

$$\frac{1}{\cos\frac{\pi s}{2}} = \frac{-\frac{2}{\pi}}{s-1} + \cdots, \quad \zeta_\Omega(1-s) = \frac{-\rho}{s} + \cdots,$$

we see that

$$\mathrm{P}(x) = \mathrm{P}_K(x) = -\frac{1}{2}\rho - \frac{1}{\pi}\zeta_\Omega(0)x^{-1}. \tag{4.5.4}$$

Substituting (4.5.3) and (4.5.4) in (4.5.2), we conclude (4.5.1). Other assertions are left as exercise. Cf. Koshlyakov (1954, I, pp. 117–118). □

Following Koshlyakov, we define the X-function, $X(x) = X_{r_1,r_2}(x)$, to be $H_{0,\chi}^{\chi,0}\left(x \,\middle|\, \begin{matrix} - \\ \{(0,\frac{1}{2})\}_{j=1}^{r_1}, \{(0,1)\}_{j=r_1+1}^{r_2} \end{matrix}\right)$, i.e.

$$X_{r_1,r_2}(x) = \frac{1}{2\pi i}\int_{(c)} \Gamma^{r_1}\left(\frac{s}{2}\right)\Gamma^{r_2}(s)x^{-s}\,ds, \quad c > 0, \tag{4.5.5}$$

for $x > 0$ ($\operatorname{Re} x > 0$). It can be easily verified that

$$X_{1,0}(x) = 2e^{-x^2}, \quad X_{0,1}(x) = e^{-x}, \quad X_{2,0}(x) = 4K_0(2x).$$

In the case of an imaginary quadratic field, this is stated in Sect. 2.15. The Koshlyakov L-function is a relative of K-functions (4.4.9) and is related through

$$L_{r_1,r_2}(x) = \frac{1}{2}\Big(K_{r_1,r_2}(-ix) + K_{r_1,r_2}(ix)\Big) \quad \text{(Koshlyakov 1954), (7.11)).}$$

It is defined by (Koshlyakov 1954, (8.9))

$$\begin{aligned} L_{r_1,r_2}(x) &= \frac{1}{2\pi i}\int_{(c)} \frac{\pi}{2} G(1-s)\,x^{-s}ds \\ &= \frac{1}{2\pi i}\int_{(c)} \frac{\pi}{2}\frac{\Gamma^{r_1}\left(\frac{s}{2}\right)\Gamma^{r_2}(s)}{\Gamma^{r_1}\left(\frac{1-s}{2}\right)\Gamma^{r_2}(1-s)}x^{-s}ds, \quad c > 0,\ x > 0 \end{aligned}. \tag{4.5.6}$$

We have

$$L_{1,0}(x) = \pi x^{\frac{1}{2}} J_{\frac{1}{2}}(2x), \quad L_{0,1}(x) = \frac{\sin(x)}{x} = \mathrm{si}(x),$$

$$L_{2,0}(x) = \pi \left(\frac{2}{\pi} K_0(4\sqrt{x}) - Y_0(4\sqrt{x}) \right)$$

where $\mathrm{si}(x)$ the sinus cardinalis function. Proof is given in Kanemitsu and Tsukada (2014, pp. 105–106, Example 3.8).

We assume the setting in *Convention*. For each Koshlyakov's function $\mathfrak{K}(s)$ we let $\chi_{\mathfrak{K}}(s)x^{-s}$ denote its integrand:

$$\mathfrak{K}(s) = \frac{1}{2\pi i} \int_{(c)} \chi_{\mathfrak{K}}(s)x^{-s}\, \mathrm{d}s, \quad c > 1.$$

Hence

$$\chi_{\mathfrak{K}}(s)x^{-s} = \frac{\pi}{2\cos\frac{\pi s}{2}} G(1-s)x^{-s} \quad \text{for} \quad \mathfrak{K} = K_{r_1,r_2}$$

$$\chi_{\mathfrak{K}}(s)x^{-s} = \Gamma^{r_1}\left(\frac{s}{2}\right)\Gamma^{r_2}(s)x^{-s} \quad \text{for} \quad \mathfrak{K} = X_{r_1,r_2}$$

$$\chi_{\mathfrak{K}}(s)x^{-s} = \frac{\pi}{2} \frac{\Gamma^{r_1}\left(\frac{s}{2}\right)\Gamma^{r_2}(s)}{\Gamma^{r_1}\left(\frac{1-s}{2}\right)\Gamma^{r_2}(1-s)} x^{-s} \quad \text{for} \quad \mathfrak{K} = L_{r_1,r_2}.$$

We let $\mathrm{P}(x) = R_0 + R_1$ be the residual function in the strip $-d < \sigma < c, c > 1$, $0 < d < 1$ and

$$R_j = \mathrm{Res}\left(\chi_{\mathfrak{K}}(s)z^{-s}, s = j\right), \quad j = 0, 1.$$

Exercise 4.5.2 *Check the following.*

For $\mathfrak{K} = X_{r_1,r_2}$ *we have*

$$\mathrm{P}_X(x) = 2^{r_1}\zeta_\Omega^{(r)}(0) - 2^{r_1}\zeta_\Omega^{(r)}(0)x^{-1} \quad \textit{for} \quad \zeta_\Omega(s) \tag{4.5.7}$$

$$\mathrm{P}_X(x) = \frac{1}{4}\left(\gamma - \log\frac{4\pi}{x} - \frac{1}{x}(\gamma - \log 4\pi x)\right) \quad \textit{for} \quad \zeta(s)^2$$

while for $\mathfrak{K} = K_{r_1,r_2}$ *we have*

$$\mathrm{P}_K(x) = -\frac{1}{2}\rho - \frac{\zeta_\Omega(0)}{\pi}\frac{1}{x} \quad \textit{for} \quad \zeta_\Omega(s) \tag{4.5.8}$$

$$\mathrm{P}_K(x) = \frac{1}{2}\log x + \gamma + \frac{1}{4\pi x} \quad \textit{for} \quad \zeta(s)^2.$$

The first equality in (4.5.8) is verified in Exercise 4.5.1. Others can be similarly checked.

Theorem 4.5.2 (Koshlyakov) *The functional Eq.* (4.1.9) *(which we apply in the form* (4.1.12)*) is equivalent to each of the following.*

The Bochner modular relation

$$\sum_{n=1}^{\infty} a_n X_{r_1,r_2}(Axn) = \mathrm{P}_X(x) + x^{-1}\sum_{n=1}^{\infty} a_n X_{r_1,r_2}(Ax^{-1}n), \quad \mathrm{Re}\, x > 0. \tag{4.5.9}$$

The Fourier-Bessel expansion

$$\frac{x}{\pi}\sum_{n=1}^{\infty} a_n \frac{1}{x^2+n^2} + \mathrm{P}_K(x) = \sigma(x) = \frac{A}{\pi}\sum_{n=1}^{\infty} a_n K_{r_1,r_2}(A^2xn), \tag{4.5.10}$$

where the residual function is given by (4.5.7) *resp.* (4.5.8). *Koshlyakov's main theory is based on* (4.5.10).

Since Koshlyakov starts from the right-hand side of (4.5.10), this has been thought of as the partial fraction expansion, but the essence of his arguments depends on the (general) Fourier-Bessel expansion $G_{1,1}^{1,1} \leftrightarrow G_{0,2}^{2,0}$, cf. Kanemitsu and Tsukada (2014, Chap. 4). Deduction of (4.1.9) from (4.5.9) is made on Koshlykov (1954, I, p. 120, (3.4)). For (4.5.10), cf. Koshlykov (1954, I, p. 117, (3.4)) as modified in Kanemitsu and Tsukada (2014, p. 129, Theorem 4.7). It follows from (4.4.8) and (4.5.1). Deduction of (4.1.9) from (4.5.10) is done on Koshlykov (1954, II, pp. 225–226) based on the formula (Koshlyakov 1954, II, p. 225, (16.1)), which should read (supplementing the missing second integral on the right)

$$\zeta_\Omega(s) = \rho\frac{\alpha^{1-s}}{s-1} + \int_\alpha^{\alpha+i\infty} \sigma(-iz)z^{-s}\,\mathrm{d}z + \int_\alpha^{\alpha-i\infty} \sigma(iz)z^{-s}\,\mathrm{d}z, \quad 0<\alpha<1. \tag{4.5.11}$$

Formulas similar to this appears on Hamburger (1922, p. 137) resp. Soni (1966, p. 547) and Hamburger resp. Soni used it to deduce (4.5.14) from (4.5.12) resp. (4.5.10) from (4.5.9). We shall prove below that (4.5.11) is a special case of a version of the Plana summation formula, Lemma 4.5.1, and clarify the works of Hamburger and Soni.

Corollary 4.5.1 (Hamburger 1922) *The functional equation FER for the Riemann zeta-function is equivalent to the following equalities.*

- *The Bochner modular relation (the theta-transformation formula).*
- *The partial fraction expansion for the cotangent function* $i\cot\pi i z$.

$$1 + 2\sum_{n=1}^{\infty} e^{-2\pi n z} = \frac{1}{\pi z} + \frac{2z}{\pi}\sum_{n=1}^{\infty}\frac{1}{z^2+n^2}. \tag{4.5.12}$$

- *The Fourier expansion*

$$\sum_{n=1}^{[x]}(x-n) = \frac{x(x-1)}{2} - \frac{1}{2\pi^2}\sum_{n=1}^{\infty} n^{-2}(\cos 2\pi nx - 1). \tag{4.5.13}$$

- *The Poisson summation formula ($T > 0$)*

$$T\sum_{n=-\infty}^{\infty}\varphi(Tn) = \sum_{n=-\infty}^{\infty}\Phi\left(\frac{T}{2\pi}n\right) \tag{4.5.14}$$

valid for a function $\varphi(u)$ which is of bounded variation in any finite interval satisfying convergence conditions at infinity (p. 136) and such that

$$\Phi(z) := \int_{-\infty}^{\infty}\varphi(u)e^{zu}\,\mathrm{d}u \tag{4.5.15}$$

is analytic in a certain (vertical) strip.

The first two are contained in Theorem 4.5.2. Equation (4.5.14) is proved in Theorem 4.5.3 below. Proof of equivalence of (4.5.13) and (FER) is given in Liu et al. (2023). For modular relations for RZF viewed as an Epstein zeta-finction, cf. Exercise 2.12.7.

In its original form, the Fourier expansion (4.5.13) looks rather foreign compared with others. But it is $\frac{1}{2}\left(B_2(s) - \bar{B}_2(x)\right)$, a special case of (2.16.11) and amounts to a well-known result as shown in the next exercise.

Exercise 4.5.3 *Prove that* (4.5.13) *amounts to the familiar absolutely convergent Fourier series*

$$\bar{B}_2(x) = B_2(x-[x]) = \frac{1}{\pi^2}\sum_{n=1}^{\infty}\frac{\cos 2\pi nx}{n^2} \tag{4.5.16}$$

and that the left-hand side of (4.5.13) *is also the Riesz sum of order* 1

$$\frac{1}{2}\left(B_2(x) - \bar{B}_2(x)\right) = \sum_{n\le x}(x-n) = \frac{1}{2\pi^i}\int_{(c)}\frac{1}{s(s+1)}\zeta(s)x^{s+1}\,\mathrm{d}s \tag{4.5.17}$$

for $c > 1$. Cf. (4.5.22).

Solution. The left-hand side of (4.5.13) is equal to

$$\frac{1}{2}x(x-1) - \frac{1}{2}\bar{B}_2(x) + \frac{1}{12} = \frac{1}{2}\left(B_2(x) - \bar{B}_2(x)\right),$$

whence (4.5.16).

Equation (4.5.17) follows from the definition as remarked on Kanemitsu and Tsukada (2014, p. 170). Cf. Lemma 2.16.2. □

Around the same time, Walfisz (1923, 1925) remarks that the Fourier series for the 1st periodic Bernoulli polynomials is a consequence of the functional equation.

Exercise 4.5.4 *Write down the concrete form of* (4.5.9) *and* (4.5.10). *Prove* (4.5.13).

Solution. Equation (4.5.9) resp. (4.5.10) reads

$$\sum_{n=1}^{\infty} e^{-\pi n^2 x} = x^{-\frac{1}{2}} \sum_{n=1}^{\infty} e^{-\frac{\pi n^2}{x}} + \frac{1}{2}\left(x^{-\frac{1}{2}} - 1\right), \quad \operatorname{Re} x > 0,$$

which is (2.1.74), resp.

$$1 + 2\sum_{n=1}^{\infty} e^{-2\pi n z} = \frac{1}{\pi z} + \frac{2z}{\pi} \sum_{n=1}^{\infty} \frac{1}{z^2 + n^2}.$$

To prove (4.5.13) we integrate (2.1.20) from $\frac{1}{2}$ to x, $0 < x < 1$ to deduce that

$$\frac{1}{2\pi i}(\ell_1(x) + \log 2) = \frac{1}{2}\left(-\left(x - \frac{1}{2}\right) + \frac{i}{\pi}\log \sin \pi x\right).$$

Comparing the real and imaginary parts, we arrive at the Fourier series for the first Clausen function and the first periodic Bernoulli polynomial, respectively. Equation (4.5.13) follows on integrating them. Cf. Exercise 4.5.5 and Yamamoto (1977). □

The following exercise contains another application of Riesz sums in addition to those in Sect. 2.16.

Exercise 4.5.5 *Prove the equivalence of the Riemann functional equation and* (4.5.13).

Solution. First assume (4.5.13) and we recall the special case of Kanemitsu and Tsukada (2007, (3.50), p. 72)

$$\zeta(s) = \frac{1}{s-1} + \frac{1}{2} + \frac{1}{12}s - \frac{s(s+1)}{2}\int_1^{\infty} \bar{B}_2(t) t^{-s-2}\,\mathrm{d}t \tag{4.5.18}$$

valid for $\sigma > -1$. Noting that

$$-\frac{s(s+1)}{2}\int_0^{1} \bar{B}_2(t) t^{-s-2}\,\mathrm{d}t = \frac{1}{s-1} + \frac{1}{2} + \frac{1}{12}s, \tag{4.5.19}$$

we may *complete the integral* in (4.5.18) by (4.5.19) to obtain

$$\zeta(s) = -\frac{s(s+1)}{2}\int_0^\infty \bar{B}_2(t)t^{-s-2}\,\mathrm{d}t. \tag{4.5.20}$$

Substituting (4.5.16) in (4.5.20) and changing the order of integration and summation (by absolute convergence of both), we arrive at

$$\zeta(s) = -\frac{s(s+1)}{2\pi^2}\sum_{n=1}^{\infty}\frac{1}{n^2}(2\pi n)^{s+1}\int_0^\infty u^{-s-1}\cos u\,\frac{\mathrm{d}u}{u} \tag{4.5.21}$$

after transformation. We substitute the generalized Euler's formula valid for $|\alpha| \le \frac{\pi}{2}$, $\sigma > 0$

$$\Gamma(s)\cos\alpha s = \int_0^\infty t^s e^{-t\cos\alpha}\cos(t\sin\alpha)\,\frac{\mathrm{d}t}{t}$$

with $\alpha = \frac{\pi}{2}$ in (4.5.21) to deduce an asymmetric form of RFE.

Now we deduce (4.5.13) from Corollary 4.1.1 and FER. We generalize (4.5.17) as the Riesz sum of order 1 of a_n:

$$Z_f(x) := \sum_{n\le x}(x-n)a_n = \sum_{n=1}^{\infty} a_n f\left(\frac{n}{x}\right) = \frac{1}{2\pi i}\int_{(c)}\frac{1}{s(s+1)}\zeta_\Omega(s)x^{s+1}\,\mathrm{d}s, \tag{4.5.22}$$

where $f(y) = f_1(y)$ and

$$f_\varkappa(y) = \begin{cases} \frac{1}{\Gamma(\varkappa+1)}(1-y)^\varkappa, & (|y|<1) \\ \frac{1}{2}, & (y=1) \\ 0, & (|y|>1). \end{cases}$$

The second equality of (4.5.22) follow because the Mellin inversion of $f(y)$ is

$$f(y) = \frac{1}{2\pi i}\int_c \frac{\Gamma(s)}{\Gamma(s+2)}y^{-s}\,\mathrm{d}s = G_{1,1}^{1,0}\left(y\,\middle|\,\begin{matrix}2\\0\end{matrix}\right).$$

Now we apply Corollary 4.1.1 with

$$F(z) = F_f(z) = \mathfrak{G}_F(s,z) = \frac{\Gamma(s)}{\Gamma(s+2)} = \frac{1}{s(s+1)}.$$

Indeed, this procedure proves the general case (2.12.47). Then (4.1.21) reads

$$\begin{aligned} K(x,n) &= \frac{1}{2\pi i}\int_{(-d)} F(z)G(z)A^{2z-1}n^{z-1}x^{z+1}\,\mathrm{d}z \\ &= \frac{1}{2\pi i}\int_{(-d)}\frac{1}{z(z+1)}G(z)\gamma(z,x)\,\mathrm{d}z, \end{aligned}$$

where $\gamma(z,x) = \frac{1}{A(An)^2}(A^2xn)^{z+1}$.

For the rational case, $\mathrm{P}(x) = B_2(x)$, $A = \sqrt{\pi}$ and

$$\frac{1}{z(z+1)}G(z) = \frac{1}{\sqrt{\pi}}2^z \sin\frac{\pi}{2}i\Gamma(-z-1) = -\frac{1}{2i\sqrt{\pi}}2^z\left(e^{\frac{\pi}{2}iz} - e^{-\frac{\pi}{2}iz}\right)\Gamma(-z-1).$$

Hence

$$\begin{aligned}K(x,n) &= -\frac{1}{4i\pi^2n^2}\frac{1}{2\pi i}\int_{(-d)}\left(e^{\frac{\pi}{2}iz} - e^{-\frac{\pi}{2}iz}\right)\Gamma(-z-1)(2\pi xn)^{z+1}\,\mathrm{d}z\\ &= \frac{1}{4\pi^2n^2}\frac{1}{2\pi i}\int_{(1-d)}\left(e^{\frac{\pi}{2}iw} + e^{-\frac{\pi}{2}iw}\right)\Gamma(-w)(2\pi xn)^w\,\mathrm{d}w\end{aligned}$$

by the change of variable. Hence by Proposition 2.1.2,

$$J(x) = I(x) = \sum_{n=1}^{\infty} K(x,n) = \frac{1}{2\pi^2}\sum_{n=1}^{\infty}\frac{1}{n^2}\left(e^{2\pi ixn} + e^{-2\pi ixn}\right),$$

which leads to the Fourier series for $\bar{B}_2(x)$. □

Exercise 4.5.6 *In the quadratic field case, check that the expressions for the Riesz sum of order* 0 *are as follows. The corresponding results for* $Z_f(x)$ *in* (4.5.22) *are integrals thereof.*

In the case of an imaginary quadratic field,

$$K(x,n) = \sqrt{\frac{x}{n}}J_1(2A\sqrt{xn})$$

by (3.6.13), so that

$$Z_f(x) = \rho x + \zeta_\Omega(0) + \sqrt{x}\sum_{n=1}^{\infty}\frac{a_n}{\sqrt{n}}J_1\left(4\pi\sqrt{\frac{xn}{\sqrt{|\Delta|}}}\right),$$

(Walfisz 1961a, (19)), *which reduces to* Walfisz (1961a, (14)) *in the case* $\Delta = -4$.

In the case of a real quadratic field,

$$K(x,n) = -\sqrt{\frac{x}{n}}\left(Y_1(4A\sqrt{xn}) + \frac{2}{\pi}K_1(4A\sqrt{xn})\right)$$

by Walfisz (1925, (10.33)) *modified so as to* $J_{2,0}(w) = -Y_1(w) - \frac{2}{\pi}K_1(w)$ *in the notation of Watson. Hence*

$$Z_f(x) = \rho x + \zeta_\Omega(0) - \sqrt{x}\sum_{n=1}^{\infty}\frac{a_n}{\sqrt{n}}\left(Y_1\left(4\pi\sqrt{\frac{xn}{\sqrt{|\Delta|}}}\right) + \frac{2}{\pi}K_1\left(4\pi\sqrt{\frac{xn}{\sqrt{|\Delta|}}}\right)\right),$$

(Walfisz 1961a, (18)), *which reduces to the series in* Walfisz (1961a, (16)) *in the case* $\Delta = 1$. *The residual function is the well-known* $x\log x + (2\gamma - 1)x + \frac{1}{4}$ *coming from the double pole at* $s = 1$ *etc. and the coefficients are* $d(n)$.

Remark 4.5.1 (i) We remark that the partial fraction expansion for the cotangent function (4.5.10) is the special case of Watson's formula (2.12.50), i.e. by viewing FER as HTFE. On the other hand, as is elucidated on Kanemitsu and Tsukada (2014, pp. 241–243), (2.12.46) is a special case of (2.12.42) and follows from Lemma 2.12.4 with $A_1 = \frac{1}{2}$, i.e. by viewing HTFE as FER. Since the partial fraction expansion (4.5.10) is essential in Hamburger's determination of the Riemann zeta-function, we see a hidden relation to Hecke's method. Cf. the passages after Exercise 2.12.7. We remark ubiquity of the Fourier-Bessel expansion as the **partial fraction expansion** (4.5.10), Watson's formula and Hardy's formula (2.12.46).
(ii) The above deduction of (FER) from (4.5.13) is a simplified version of the first proof given in Titchmarsh (1951, pp. 13–17). The proof is given for $\bar{B}_1(x)$ and term-by-term integration is checked as on Titchmarsh (1951, p. 13) by integration by parts and it is seen that this corresponds to the integration of $\bar{B}_1(x)$ to arrive at $\bar{B}_2(x)$.

In Corollary 4.5.1, the Fourier expansion (4.5.13) may be replaced by

$$\sum_{n\le x}{}' 1 = B_1(s) + \frac{1}{\pi}\sum_{n=1}^{\infty}\frac{\sin 2\pi nx}{n}, \quad x \notin \mathbb{Z}.$$

The following lemma reveals the procedures adopted by Hamburger and Soni. Cf. Koshlyakov (1954, II, pp. 222–226).

Lemma 4.5.1 (A version of the Plana summation formula) *Let* $0 < \alpha < \min\{\lambda_1, 1\} < \beta$, *let* C *denote the rectangle with vertices at* $\alpha - iT$, $alpha + iT$, $\beta + iT$, $\beta - iT$, $T > 0$ *and let* $f(s)$ *be an arbitrary function analytic along and inside the contour* C *and satisfies the following limit conditions.*

$$\int_\alpha^\beta f(x+iT)\sigma(x-it)\,\mathrm{d}x \to 0, \quad \int_\alpha^\beta f(x-iT)\sigma(x+it)\,\mathrm{d}x \to 0 \quad as \quad T\to\infty \tag{4.5.23}$$

$$J(\beta,\infty) \to 0 \quad as \quad \beta\to\infty, \tag{4.5.24}$$

where

$$J(x,T) = \int_x^{x+iT} f(z)\sigma(-iz)\,\mathrm{d}z + \int_x^{x-iT} f(z)\sigma(iz)\,\mathrm{d}z.$$

Then

$$\Sigma_f(s) := \sum_{n=1}^{\infty} a_n f(n) = -\int_{(\alpha)} f(z)\sigma(iz)\,\mathrm{d}z \tag{4.5.25}$$

in the first instance. If (α) *is changed by the path* $(0+)$ *with indentation at the origin, then* (4.5.25) *becomes*

$$\Sigma_f(s) = -\frac{1}{2}a_0 - \int_{(0+)} f(z)\sigma_H(iz)\,\mathrm{d}z \tag{4.5.26}$$

where $a_0 = \mathrm{Res}_{z=0}(f(z)\sigma_H(iz)) = -2\zeta_\Omega(0)f(0)$.

$\Sigma_f(s)$ *has another expression* (Koshlyakov 1954, II, p. 223, IV)

$$\Sigma_f(s) = \rho\int_\alpha^\infty f(x)\,\mathrm{d}x + J(\alpha,\infty), \tag{4.5.27}$$

which amounts to a version of the Plana summation formula

$$\Sigma_f(s) = \sum_{n=1}^{\infty} a_n f(n) = \zeta_\Omega(0)f(0) + \rho\int_0^\infty f(x)\,\mathrm{d}x + J(0,\infty) \tag{4.5.28}$$

where

$$J(0,\infty) = \int_0^{i\infty} f(z)\sigma(-iz)\,\mathrm{d}z + \int_0^{-i\infty} f(z)\sigma(iz)\,\mathrm{d}z.$$

Proof By Exercise 4.5.1 and the Cauchy residue theorem, we have

$$\sum_{n=1}^{[\beta]} a_n f(n) = \int_C f(z)\sigma(is)\,\mathrm{d}z$$

which leads to (4.5.25) on letting $T \to \infty$, $\beta \to \infty$ in view of (4.5.23) and (4.5.24). Then (4.5.29) follows by Exercise 2.1.9, (i).

Using Exercise 4.5.1, (4.5.25) becomes

$$\sum_{n=1}^{[\beta]} a_n f(n) = \rho\int_\alpha^\infty f(x)\,\mathrm{d}x + J(\alpha,\infty) - J(\beta,\infty),$$

which leads to (4.5.27) by (4.5.24). By Exercise 4.5.1, the limiting case $\alpha \to 0$ of (4.5.27) needs the extra term $\zeta_\Omega(0)f(0)$ to arrive at (4.5.28). □

Theorem 4.5.3 *With the aid of a version of the Plana summation formula, Lemma 4.5.1, the Poisson summation formula* (4.5.14) *of Hamburger follows from the partial fraction expansion* (4.5.10), *which in turn follows from* (4.5.9) *and leads to the functional Eq.* (4.1.9).

Proof Equations (4.5.9)$\Longrightarrow$ (4.5.10).
It turns out that by coincidence, Soni's function $\sigma(z)$ is Koshlyakov's $\sigma(iz)$. If the upper part resp. lower part of the path (0+) in (4.5.29) is rotated by $-\frac{\pi}{2}$ resp. $\frac{\pi}{2}$, then it becomes the limit of the path of Soni. This leads to Soni (1966, (IV)), which leads to (4.5.10) in the case of the divisor problem. Other cases similarly follow.
Equations (4.5.10) $\Longrightarrow$ (4.5.14).
Corresponding to Hamburger we use (4.5.1) in the form

$$\sigma_H(iz) := \sigma(iz) + \frac{1}{2}\rho = -\frac{\zeta_\Omega(0)}{\pi}\frac{1}{iz} + \frac{1}{2\pi i}\sum_{n=1}^{\infty} a_n \left(\frac{1}{z-n} + \frac{1}{z+n}\right)$$

in place of $\sigma(iz)$, so that $\sigma_H(iz)$ is an odd function. Recall (4.5.26) as

$$\Sigma_f(s) = \sum_{n=1}^{\infty} a_n f(n) = \zeta_\Omega(0) f(0) - \int_{(0+)} f(z)\sigma(iz)\,\mathrm{d}z. \tag{4.5.29}$$

Similarly, for

$$\Sigma_{f^-}(s) := \sum_{n=1}^{\infty} a_{-n} f(-n) = \int_{(-\alpha)} f(z)\sigma_H(iz)\,\mathrm{d}z,$$

where $a_{-n} = a_n$, we have the limit case corresponding to (4.5.26):

$$\Sigma_{f^-}(s) = \sum_{n=1}^{\infty} a_n f(-n) = \zeta_\Omega(0) f(0) - \int_{(0+)} f(-z)\sigma(iz)\,\mathrm{d}z. \tag{4.5.30}$$

Hence adding (4.5.29) and (4.5.30), we have

$$\sum_{n=-\infty}^{\infty} a_n f(n) = -\int_{(0+)} \sigma_H(iz)(f(z) + f(-z))\,\mathrm{d}z,$$

where $a_0 = -2\zeta_\Omega(0) f(0)$. Hence we obtain

$$\sum_{n=-\infty}^{\infty} a_n f(n) = \int_C \sigma_H(iz)(f(z) + f(-z))\,\mathrm{d}z = \int_{-\infty}^{\infty} \sigma_H(ix)(f(x) + f(-x))\,\mathrm{d}x, \tag{4.5.31}$$

where C is the contour in Soni (1966, p. 547).
From (4.5.10), we have

$$\sigma_H(ix) = b_0 + \frac{A}{\pi}\sum_{n=1}^{\infty} a_n K_{r_1,r_2}(iA^2xn).$$

Substituting this in (4.5.31), we deduce that

$$\sum_{n=-\infty}^{\infty} a_n f(n) = \frac{A}{\pi} \sum_{n=-\infty}^{\infty} b_n \int_{-\infty}^{\infty} K_{r_1,r_2}(iA^2xn) f(x)\,\mathrm{d}x, \tag{4.5.32}$$

where $b_n = a_n$ for $n \neq 0$ and $b_0 = \frac{1}{2}\rho$. As in (4.5.15) we put .

$$\varphi(z) := \int_{-\infty}^{\infty} K_{r_1,r_2}(izx) f(x)\,\mathrm{d}x.$$

Substituting (4.4.10), Eq. (4.5.32) in the rational case, reads

$$\sum_{n=-\infty}^{\infty} f(n) = \frac{1}{\sqrt{\pi}} \sum_{n=-\infty}^{\infty} \int_{-\infty}^{\infty} e^{-2\pi xn} f(x)\,\mathrm{d}x. \tag{4.5.33}$$

With

$$\varphi(z) := \int_{-\infty}^{\infty} e^{izx} f(x)\,\mathrm{d}x,$$

Equation (4.5.33) translates into (4.5.14) with $T = 2\pi$.

In this case, it is simpler if we work with the rectangle C' with vertices at $\alpha - iT$, $\alpha + iT$, $-\alpha + iT$, $-\alpha - iT$, $T > 0$ and assuming convergence conditions, we count all the residues at in, $n \in \mathbb{Z}$.

Equations (4.5.14) $\Longrightarrow$ (4.1.9).

Equations (4.5.27) with $f(z) = z^{-s}$ leads to (4.5.11), in which we let $\alpha \to 0+$ for $\sigma < 1$ to arrive at a simplified form Koshlyakov (1954, II, (16.2)) of (4.5.28)

$$\zeta_\Omega(s) = J(0,\infty) = 2\sin\frac{\pi}{2}s \int_0^{\infty} x^{-s}\sigma(x)\,\mathrm{d}x \tag{4.5.34}$$

valid for $\sigma < r$. On the other hand, the Mellin inversion of (4.4.8), we have

$$\zeta_\Omega(s) = 2A^{1-2s}\cos\frac{\pi s}{2} G(s) \int_0^{\infty} \sigma(x) x^{-s}\,\mathrm{d}s, \quad \sigma > 1 - r. \tag{4.5.35}$$

Comparing (4.5.34) and (4.5.35) implies (4.1.9). □

References

Arai, T., Chakraborty, K., & Ma, J. (2015). Applications of the Hurwitz-Lerch zeta-function. *Pure and Applied Mathematics Journal, 4*(2–1), 30–35. https://doi.org/10.11648/j.pamj.s.2015040201.16.

Berndt, B. C. (1971). Identities involving the coefficients of a class of Dirichlet series V. *Transactions of the American Mathematical Society, 160*, 139–156.

Chakraborty, K., Kanemitsu, S., & Tsukada, H. (2009). *Vistas of special functions II*. Singapore etc: World Sci.

Erdélyi, A., Magnus, W., Oberhettinger, F., & Tricomi, F. G. (Eds.). (1953). *Higher transcendental functions* (Vol. 1–3). New York: McGraw-Hill.

Grosswald, E. (1984). *Topics from the theory of numbers*. Basel: Birkháuser.

Hamburger, H. (1922). Über einige Beziehungen, die mit der Funktionalgleichung der Riemannschen ζ-Funktion äquivalent sind. *Mathematische Annalen, 85*, 129–140.

Kanemitsu, S., & Tsukada, H. (2007). *Vistas of special functions*. Singapore etc: World Scientific.

Kanemitsu, S., Katsurada, M., & Yoshimoto, M. (2000). On the Hurwitz-Lerch zeta function. *Aequationes Mathematicae, 59*, 1–19.

Kanemitsu, S., & Tsukada, H. (2014). *Contributions to the theory of zeta-functions: the modular relation supremacy*. Singapore etc: World Sci.

Knopp, M., & Robbins, S. (2001). Easy proof of Riemann's functional equation for $\zeta(s)$ and of Lipschitz summation. *Proceedings of the American Mathematical Society, 129*, 1915–1922.

Koshlyakov, N. S. (1928/1929b). On Voronoĭ's sum formula. *Messenger of Mathematics, 58*, 30–32.

Koshlyakov, N. S. (1954). Investigation of some questions of analytic theory of the rational and quadratic fields, I-III (Russian). *Izvestiya Akademii Nauk SSSR, Seriya Biologicheskaya, 18*, 113–144, 213–260, 307–326; Errata: *ibid, 19* (1955), 271.

Koshlyakov, N. S. (1934). On a general summation formula and its applications. *Doklady Akademii Nauk SSSR, 4*, 187–190.

Laurinchikas, A., & Garunkstis, R. (2002). *The Lerch zeta-function*. Dordrecht-Boston-London: Kluwer Academic Publ.

Laurinčikas, A. (1998). On Voronoĭ summation formulae. *Voronoĭ's Impact on Modern Science*. Book I. Proc. Inst. Math. Natl. Acad. Sci. Ukr., Math. Appl, vol. 21 (1), P. Engel and H. Syta (Eds.), Inst, of Math., Kviv, pp. 117–136.

Li, H. -L., Kanemitsu, S., & Wang, X. -H. (2012). Plana's summation formula as a modular relation and applications. *Integral Transforms and Special Functions, 23*, 183–190.

Li, W. -B., Li, H. -Y., & Mehta, J. (2020). Around the Lipschitz summation formula. *Mathematical Problems in Engineering, 2020*, Article ID 5762823, 16. https://doi.org/10.1155/2020/5762823.

Li, F.-H., Wang, N.-L., & Kanemitsu, S. (2013). *Number theory and its applications*. London-Singapore-New Jersey: World Sci.

Lindelöf, E. (1905). *Calcul des résidus et ses applications a la théorie des fonctions*. Paris: Gauthier-Villars.

Liu, G. -D., Chakraborty, K., & Kanemitsu, S. (2023). A unifying principle in the theory of modular relations. *Mathematics,* 11030535. MDPI.

Morita, T. (1977). On the Hurwitz-Lerch L-functions. *Journal of the Faculty of Science, the University of Tokyo, Section IA, Mathematics, 24*, 29–43.

Oberhettinger, F., & Soni, K. L. (1972). On some relations which are equivalent to functional equations involving the Riemann zeta-function. *Mathematische Zeitschrift, 127*, 17–34.

Prudnikov, A. P., Bychkov, Y. A., & Marichev, D. I. (1986). *Integrals and series*. Nauka, Moscow: Supplementary chapters.

Rademacher, H. (1973). *Topics in analytic number theory*. Berlin-Heidelberg: Springer Verl.

Soni, K. (1964). *On some relations which are equivalent to the functional equations of type of Riemann's zeta function and tables of divisor transforms*. Ph.D. thesis, Oregon State University, Corvallis.

Soni, K. (1966). Some relations associated with an extension of Koshliakov's formula. *Proceedings of the American Mathematical Society, 17*, 543–551.

Srivastava, H. M., & Choi, J.-S. (2001). *Series associated with the zeta and related functions*. Dordrecht etc: Kluwer Acad. Publ.

Titchmarsh, E. C. (1951). *The Theory of the Riemann Zeta-Function* (Second edition revised by D. R. Heath-Brown). OUP 1986.

Walfisz, A. A. (1961). On the theory of a class of Dirichlet series. *ibid,27*, 9–16.

Walfisz, A. Z. (1923). *Über die summatorischen Funktionen einiger Dirichletscher Reihen*. Göttingnen: Inaugural Diss.

Walfisz, A. Z. (1925). Über das Piltzsche Teilerproblem in algebraishcen Zahlkörpern. *Mathematische Zeitschrift, 22*, 153–188.

Walfisz, A. A. (1961). On sums of coefficients of some Dirichlet series. *Soobšč. Akad. Nauk Grundz. SSR, 26*, 9–16.

Wolfram, S. (1998). *Mathematical functions site*. New York: Wolfram Research Inc.

Yamamoto, Y. (1977). Dirichlet series with periodic coefficients. In *Proceedings of the International Symposium on Algebraic Number Theory, Kyoto, 1976* (pp. 275–289). Tokyo: JSPS.

Chapter 5
Class Numbers of Abelian Fields and CPMD

This chapter essentially depends on the parity and is of algebraic nature. It is closely related to early sections of Chap. 2 and particularly, Sect. 2.6. These can be studied simultaneously. The main figures of this chapter are class numbers, Abelian fields, Dedekind determinant, characters, and their orthogonality.

5.1 Chowla's Problem and Maillet–Demyanenko Determinants: Odd Part

We treat both Chowla's problem and Maillet–Demyanenko determinants in a unified way in terms of the Dedekind determinant and $L(1, \chi) = \sum_{n=1}^{\infty} \frac{\chi(n)}{n} \neq 0$. The linkage can be perceived in a few papers. Milnor (1983) is another essential source on the linkage describing the structure of the Kubert space of functions and referring to Chowlas' problem. Funakura (1985) contains a prototype of Corollary 2.2.2 for the Laurent constant of a periodic Dirichlet series (which leads to the misleading title of Kronecker limit formula) but it has the other contribution to clarification of the Maillet determinants.

Definition 5.1.1 Chowla's problem seems to refer to an elementary proof of $L(1, \chi) \neq 0$ through an elementary proof of the linear independence of cotangent values or similar stuff.

But now the inverse problem of Chowla is referred to as Chowla's problem, i.e. to prove linear independence of circular functions at rational arguments from the non-vanishing of $L(1, \chi)$. We follow the tendency and refer to Chowla's problem (CP).

Milnor states Chowlas' problem to prove that every $\mathbb{Q}$-linear relation between $\zeta(2, x), 0 < x < 1, x \in \mathbb{Q}$ is a consequence of the Kubert identity.

K. Chakraborty et al., *Modular Relations and Parity in Number Theory*,
Infosys Science Foundation Series, https://doi.org/10.1007/978-981-96-6471-9_5

Livingston (1965) is one of the earliest references that refer to Chowla's problem (Chowla, 1952), but it is not mentioned in many of subsequent papers and mentioned only in Okada (1982), Adhikari et al. (2001), etc.

Baker et al. (1973) refer to and attack the 1964 form of Chowla's problem, cf., e.g. Chowla (1964): Whether there exists a rational-valued periodic function $f : \mathbb{Z} \to \mathbb{Q}$ with period p, p being a fixed prime such that

$$\sum_{n=1}^{\infty} \frac{f(n)}{n} = 0. \tag{5.1.1}$$

We will dwell on their results in Sect. 5.5.

Suppose $p \equiv 3 \bmod 4$ is a prime. Then Chowla (1970) proved that

$$\cot \frac{\pi a}{p}, \quad a = 1, \cdots, \frac{p-1}{2}$$

are linearly independent over $\mathbb{Q}$ is equivalent to

$$L(1, \chi) \neq 0 \tag{5.1.2}$$

for χ a non-principal Dirichlet character mod p, which implies the existence of infinitely many primes in an arithmetic progression. This holds for composite modulus, too.

Research on Maillet–Demyanenko determinants was started by Carlitz and Olson (1955) and reference to it in Ayoub (1975b) in which Ayoub also refers to circulants. Without noticing this, Endô (1990) used circulants (referring to the book Davis, 1979) to develop studies on similar determinants. Hence, although Ayoub (1975b) and Funakura (1990) (generalization of Fujisaki, 1981) are concerned with CP, they are the missing links that connect CP and MD. The reference on Iwasawa's paper (not published) in Ayoub (1975b) is misleading since he refers to his own paper (Ayoub, 1975a).

Iwasawa (1972, p. 14) states that there is no elementary proof known of (5.1.5) for an odd character χ to the modulus M. Cf. Sect. 2.16. Indeed, it seems that save for Baker et al. (1973), only Fujiwara (1978) ever tried to attack Chowla's original problem by the theory of n-gons on the grounds of (5.1.6).

In early studies on the Maillet determinants, the entries were expressed in terms of the least non-negative residue $R(a)$ of a modulo M, $a \equiv R(a) \mod M$, $0 \leq R(a) < M$. Then we may express it as

$$R(a) = M\left(\bar{B}_1\left(\frac{a}{M}\right) + \frac{1}{2}\right). \tag{5.1.3}$$

Only after this expression is introduced (Kanemitsu & Kuzumaki, 1998), the situation has become clearer and we are naturally motivated to study Maillet determinants with Clausen function entries.

Theorem 5.1.1 (Fujisaki, 1981) *Suppose M is an odd positive integer such that $(M, \varphi(M)) = 1$, where $\varphi(M)$ is Euler's function. Then the following are equivalent.*
(i)

$$\cot \frac{\pi a}{M}, \quad 1 \le a < \frac{M}{2}, \quad (a, M) = 1 \tag{5.1.4}$$

are linearly independent over $\mathbb{Q}$.
(i)' *Fix an integer c such that $(c, M) = 1$ (esp. $c = 2$). Then*

$$\cot \frac{\pi c a}{M}, \quad 1 \le a < \frac{M}{2}, \quad (a, M) = 1$$

are linearly independent over $\mathbb{Q}$.
(ii) (i), (i)' *hold with cotangent replaced by tangent.*
(iii) *For χ odd,*

$$\sum_{a=1}^{M-1} a\chi(a) \neq 0 \tag{5.1.5}$$

holds.
(iv) *For χ odd,*

$$\sum_{a=1}^{M-1} (-1)^a \chi(a) \neq 0 \tag{5.1.6}$$

holds.
(v) $\sin \frac{2\pi}{M}$ *is expressed as a linear combination of cotangents* (5.1.4) *with rational coefficients.*
(v') $\sin \frac{2\pi}{M}$ *is expressed as a linear combination of tangents* $\tan \frac{\pi c a}{M}$ *with rational coefficients.*
Suppose $M = p$ an odd prime. Then the above statements are equivalent to the following.
(vi) *The degree $\frac{p-1}{2}$ matrix*

$$\mathcal{M} = (2\overline{ab^{-1}} - p) = \left(2p\bar{B}_1\left(\frac{ab^{-1}}{p}\right)\right)$$

is a regular matrix, where a, b runs over $1, \cdots, \frac{p-1}{2}$ and $\overline{ab^{-1}} = c$ is equivalent to $a \equiv cb \bmod p$ and $1 \le c < p$.
(vii) *The degree $\frac{p-1}{2}$ matrix*

$$\mathcal{D} = ((-1)^{\overline{ab^{-1}}})$$

is a regular matrix, where the conditions are the same as in (vi).

Theorem 5.1.1 has been generalized as in the following two theorems by Funakura.

Theorem 5.1.2 (Funakura, 1985, Theorem 20-I) *For a positive integer M, the following are equivalent:*
(i)

$$\cot \frac{\pi a}{M}, \quad 1 \le a < \frac{M}{2}, \quad (a, M) = 1 \tag{5.1.7}$$

are linearly independent over $\mathbb{Q}$.
(ii) (i) *holds with cotangent replaced by tangent.*
(iii) *For every divisor d of M, $\sin \frac{2\pi}{d}$ is expressed as a linear combination of cotangents* (5.1.7) *with rational coefficients.*
(iv) (iii) *holds with cotangent replaced by tangent.*
(v)

$$L(1, \chi) \neq 0$$

for all odd characters mod M.

Theorem 5.1.3 (Funakura, 1985, Theorem 20-II) *Suppose one of the conditions in Theorem 5.1.2 holds. Then for an integer $M \ge 3$, the following are equivalent:*
(II-1) *For every prime divisor $p|M$ the exponent $M(p)$ is even, where the exponent is defined by*

$$M(p) = \begin{cases} 0 & M = p^a \\ \min\{m \in \mathbb{N} | p^m \equiv 1 \mod \frac{M}{p^a}\} & \textit{otherwise} \end{cases},$$

and where p^a is the highest power of p dividing M.
(II-2) *For χ odd,* (5.1.5) *holds.*
(II-3) *The degree $\frac{\varphi(M)}{2}$ matrix*

$$\mathcal{M} = (2\overline{ab^{-1}} - M) = \left(2M\bar{B}_1\left(\frac{ab^{-1}}{M}\right)\right) \tag{5.1.8}$$

is regular, where a, b run over $1 \le a, b < \frac{M}{2}$, $(a, M) = (b, M) = 1$.
Assume M is odd. Then the above conditions are equivalent to either of the following two conditions:
(II-4) *For χ odd,* (5.1.6) *holds.*
(II-5) *The degree $\frac{\varphi(M)}{2}$ matrix*

$$\mathcal{D} = ((-1)^{\overline{ab^{-1}}}) \tag{5.1.9}$$

is regular, where the conditions are the same as in (II-3).

In Funakura's definition, as with most of the papers on CPMD, the matrix $\mathcal{M}$ in (5.1.8) is given as $\mathcal{M} = (2c(a, b) - M)$, where for $1 \le a, b < \frac{M}{2}$, $(a, M) = (b, M) = 1$, the positive integer $c(a, b) < M$ is defined by

$$a \equiv c(a, b)b \mod M,$$

which is exactly the least positive residue mod M of ab^{-1}, so that the expression (5.1.8) is true (so does (5.1.9)).

Remark 5.1.1 (i) Theorems 5.1.2 and 5.1.3 are for the (inverse) Chowla problem. From the point of view of the original problem of Chowla in Definition 5.1.1, the statement of Theorem 5.1.3 is not satisfactory and Fujisaki's version is more relevant: Suppose $M > 1$ is odd and $(M, \varphi(M)) = 1$. Then statements (i)-(iv) of Theorem 5.1.2 are equivalent to statements (II-2) and (II-4) of Theorem 5.1.3. Maillet–Demyanenko determinant is restricted to the prime modulus, though.
(ii) It is to be noted that the determinant of the matrix in (II-3) resp. (II-5) is the Maillet resp. Demyanenko determinant. That these should be treated in a unified way is already stated in Fujisaki (1981). It was as late as 1998 that Hirabayashi (1998) unified them using the b-division vector method of Girstmair (1997) as mentioned in Kanemitsu and Kuzumaki (2007) which is one of the papers not listed in Yamamura (2009).

Girstmair (1986) is rather enlightening and clarifies the results of Okada (1981) and Wang (1982) on the rth derivative of cotangent functions are rather trivial save for $r = 0$ in which case (Baker et al., 1973) gives a more substantial non-vanishing result of $L(1, \chi)$. Bundschuh (1988) is equally informative including applications of Baker et al. (1973) and rather through references up to 1986, cf. also Sect. 5.4.

We turn to thc Maillet and Demyanenko determinants. People working on the subject are to be thankful to Professor Ken Yamamura for his tremendous and paintaking efforts toward the completion of the list of all related papers on the subject. He published a previous list up to 2008 in Yamamura (2009). Now he has published the updated list which may not be easily available and we add those which are published after 2009. They are Jakubec (2009), Yamamura (2009), Hirabayashi (2014), Hirabayashi 67 (2015), Jakubec (2017) in this order which are labeled as 64–68. In Yamamura (2009), the label 53 is repeated and so on p. 249, the labels are to be 54–63 and continue on to 64–68 in the updated list. Papers Ayoub (1975b), Shimada (1996), and Kanemitsu and Kuzumaki (2007) are to be added in the list. Although Ernvall (1990) is not directly connected to neither problem, it contains rich information of class numbers and generalized Bernoulli numbers (and their congruences) and should be included in the list. Wang (1982, 1984) seems to be the only researcher who worked in both Chowla's problem and Maillet determinant. On Wang (1984, pp. 306–307) he mentions the link between them but didn't go further. Wang (1982) should be included in the list, too.

Most of the papers rely on the Frobenius determinant relation which is to be called the Dedekind determinant relation, cf. Hawkins (1974), Lang (1978, pp. 89–91), so that we call it the Dedekind determinant.

Table 5.1 Merging of Chowla's problem and Maillet–Demyanenko determinant

Object	Chowla's problem	Maillet–Demyanenko
Merging	Ayoub (1975b)	Fujisaki (1981)
Ingredient	Dedekind det., Baker–Birch–Wirsing	Dedekind det.
Adv. meth.	Girstmair (1987)	Girstmair (1997)

Ishibashi (1995, 1998) made an effective use of Girtstmair's character coordinate method (Girstmair, 1987) to deduce linear independence results on the Estermann zeta-function, where character coordinates are due to Leopoldt (1959). We suspect that there might be a more unified method that connects both character coordinates (Girstmair, 1987) and b-division vectors (Girstmair, 1997) to give linear independence and determinantal expression (Table 5.1).

5.2 Merging of Chowla's Problem and Maillet–Demyanenko Determinant

Dirichlet's original motivation was that he could prove the non-vanishing of $L(1, \chi)$ by the Dirichlet class number formula (Theorem 2.6.1, (ii)) connecting the value $L(1, \chi) = \sum_{n=1}^{\infty} \frac{\chi(n)}{n}$ with the class number of the quadratic field which is a positive integer. As mentioned in the last section, this implies the infinitude of primes in an arithmetic progression (AP)—Dirichlet prime number theorem.

Chowla's (original) problem was to mean an elementary proof of non-vanishing of $L(1, \chi)$ and as preliminary attempts, he suggested linear independence of cotangent values. However, the problem has been understood to mean the reverse implication, i.e. proof of linear independence of cotangent values from $L(1, \chi) \neq 0$ and enormous amount of research has been made.

On the other hand, there is enormous amount of research on determinant expressions for the (relative) class number of (subfields of) a cyclotomic field, which is referred to as the Maillet–Demyanenko determinants.

These two streams of research have been conducted independently of each other and the connection went unnoticed. Our aim is to make these lines meet and streamline the heritage in proper perspective, *by unifying Chowla's problem and Maillet–Demyanenko determinants (CPMD) as different-looking expressions of the (relative) class number via the Dedekind determinant.* In the remaining of this section, we settle the odd part of CPMD, Theorem 5.2.2, following partially (Wang et al., 2023). We settle the even part case in Sect. 5.4. At the same time, we proceed further methodologically and show that the genesis of DFT as well as the Dedekind determinant is the characters of a finite Abelian group and its convolution map, thus revealing the ground of all these researches, i.e. CPMD boils down to analysis of the class number by group characters. In the light of class field theory, class numbers of Abelian extensions over the rationals have been settled, cf., e.g. Hasse (1952) and we add this related viewpoint.

One expression is in terms of the Hurwitz zeta-function which arises from the decomposition into residue classes mod M and the special case of $s = 1$ amounts to the Euler digamma function ψ in (2.1.16). For Chowla's problem, its odd part is relevant and the odd part of ψ is the cotangent function. This latter part follows from the functional equation or from the defining expression for ψ and the partial fraction expansion for the cotangent function. The other expression involves DFT of f and may be thought of as the base change formula or a consequence of the generalized inverse Eisenstein formula (which is a consequence of the odd part of the base change formula). For the Maillet–Demyanenko determinants, they are specified to Dirichlet L-functions, with imprimitive characters in general. They may be expressed in terms of the L-function with primitive character times the product arising from Eratosthenes sieve. $L(1, \chi)$ with primitive χ has the expression in terms of the periodic Bernoulli polynomial which leads to the one for the relative class number of a cyclotomic field $\mathbb{Q}(\zeta_M)$, with a primitive Mth root ζ_M of 1.

Although there is enormous amount of researches on these two subjects, most of them are exclusively devoted to only one of them, save for Wang (1982, 1984) who seems to be the only researcher that worked in both Chowla's problem and the Maillet determinant. On Wang (1984, pp. 306–307) he mentions the link between them but didn't go further. There is a good reason why no merging occurred. It is probably because most of researches on Maillet–Demyanenko determinants stick to the notation of the least positive residue modulo M, say, where M is the modulus or the conductor of a cyclotomic field. Cf. (5.1.3). And also because the use was not thorough enough of the discrete Fourier transform (DFT) of periodic arithmetic functions f which is a generalization of Dirichlet characters with orthogonality or vanishing $\hat{f}(M)$. It is the base change formula (2.2.18) with orthogonality that governs the whole situation. In this case, the Laurent constant is simply the value $L(1, f)$ and we obtain a universal expression. In view of this, it was possible to obtain equivalence of finite expressions for $L(1, \chi)$ for χ even or odd and Gauss's formula for ψ in Hashimoto et al. (2008).

It turns out that the zeta-symmetry travels incognito and DFT alone is not enough for merging of CPMD, cf. Remark 5.2.2. In addition to (2.2.21), we need separability of the general Gauss sum and the Eratosthese sieve together with the product formula for the relative class number (odd part) as the quotient of the class number h_M of the Mth cyclotomic field by the class number h_M^+ of its maximal real subfield (even part), i.e. multiplicative parity

$$h_M^- = \frac{h_M}{h_M^+}. \tag{5.2.1}$$

Cf. (5.3.2) for an Abelian field. As we see in Lemma 2.3.1, it is here that the zeta-symmetry takes place implicitly to assure primitivity, which implies separability, cf. the third equality in (5.2.17).

Most of the papers on CPMD rely on the Frobenius determinant (or the Dedekind–Frobenius determinant) relation which is to be called the Dedekind determinant relation, cf. Hawkins (1974), Lang (1978, pp. 89–91), Washington (1983), so that we call it the **Dedekind determinant** and state it as.

Lemma 5.2.1 (Dedekind determinant) *Let G be a finite Abelian group, let N be a subgroup of G, and let $T \subset G$ be a complete system of representatives of G/N. For a character λ of N, let Δ be the set of all characters of G whose restriction to N is λ. Then for any complex-valued function f on G with*

$$f(ah) = \lambda(h) f(a) \quad (a \in G, h \in N), \tag{5.2.2}$$

we have

$$\det(f(ab^{-1}))_{a,b\in T} = \prod_{\chi\in\Delta} \left(\sum_{a\in T} \bar{\chi}(a) f(a) \right), \tag{5.2.3}$$

and, if the principal character $\chi_0 \in \Delta$

$$\det(f(ab^{-1}) - f(a))_{\substack{a,b\in T \\ a,b\notin N}} = \prod_{\substack{\chi\in\Delta \\ \chi\neq\chi_0}} \left(\sum_{\substack{a\in T \\ a\notin N}} \bar{\chi}(a) f(a) \right). \tag{5.2.4}$$

For a more general account of group characters, cf. Hasse (1952, pp. 30–34). Lemma 5.2.1 is a special case of the following general theorem.

Theorem 5.2.1 *Let G be a finite Abelian group of order M written additively and $\hat{G}$ its character group. Consider a vector space of dimension M*

$$V = \{f : G \to \mathbb{C}\},$$

which is isomorphic to $\mathbb{C}[\hat{G}]$.

Let χ_a be the characteristic function

$$\chi_a(b) = \delta_{ab} = \begin{cases} 1 & a = b \\ 0 & a \neq b \end{cases}, \tag{5.2.5}$$

which means orthogonality

$$\chi_a(b) = \delta_{ab} = \frac{1}{M} \sum_{\varepsilon\in\hat{G}} \bar{\varepsilon}(a)\varepsilon(b). \tag{5.2.6}$$

Then for any function $f \in V$ we have the base change formula

$$\sum_{a\in G} f(a)\chi_a = f = \frac{1}{M}\sum_{\varepsilon\in\hat{G}}\sum_{a\in G} f(a)\varepsilon(-a)\varepsilon \tag{5.2.7}$$

and the relation between linear maps

$$\sum_{a\in G} f(a-b)\chi_a = T\chi_b, \quad T\varepsilon = \sum_{a\in G} f(a)\varepsilon(-a)\varepsilon. \tag{5.2.8}$$

Their matrix expressions entail

$$\det(f(a-b))_{a,b} = \prod_{\varepsilon\in\hat{G}}\sum_{a\in G}\varepsilon(a)f(-a). \tag{5.2.9}$$

Proof V has bases $\mathcal{E} = \{\varepsilon \in \hat{G}\}$ as well as $\mathcal{X} = \{\chi_a | a \in G\}$, where χ_a is defined by (5.2.5).

Suppose

$$f = \sum_{\varepsilon\in\hat{G}} c_\varepsilon(f)\varepsilon, \quad f(b) = \sum_{\varepsilon\in\hat{G}} c_\varepsilon(f)\varepsilon(b). \tag{5.2.10}$$

Multiplying by $\psi(b)$ and summing over $b \in G$, we deduce that

$$\sum_{b\in G}\psi(b)f(b) = \sum_{\varepsilon\in\hat{G}} c_\varepsilon(f)\sum_{b\in G}\psi\varepsilon(b).$$

The inner sum is $Mc_{\psi^{-1}}(f)$ and so

$$c_\varepsilon(f) = \frac{1}{M}\sum_{a\in G} f(a)\varepsilon^{-1}(a) = \frac{1}{M}\sum_{a\in G} f(a)\varepsilon(-a). \tag{5.2.11}$$

Hence (5.2.11) amounts to the definition of DFT (2.2.8) and (5.2.10) leads to the second expression in (5.2.7):

$$f = \frac{1}{M}\sum_{\varepsilon\in\hat{G}}\sum_{a\in G} f(a)\varepsilon(-a)\varepsilon. \tag{5.2.12}$$

Let T_a be the translation operator

$$T_a : V \to V; \quad T_a f(x) = f(x+a), \quad x, a \in G.$$

Then

$$T_a\varepsilon(x) = \varepsilon(x+a) = \varepsilon(a)\varepsilon(x),$$

so that

$$T_a\varepsilon = \varepsilon(a)\varepsilon. \tag{5.2.13}$$

For any fixed f, we define the linear operator $T = T(f)$

$$T = \sum_{a\in G} f(a)T_{-a}; \quad Tg(x) = \sum_{a\in G} f(a)g(x-a), \quad g \in V.$$

Hence

$$T\varepsilon = \sum_{a\in G} f(a)T_{-a}\varepsilon = \sum_{a\in G} f(a)\varepsilon(-a)\varepsilon,$$

i.e. (5.2.8), by (5.2.13). Let $\mathbb{T}$ be the matrix representation of T. Then $\mathbb{T}$ w.r.t. $\mathcal{E}$ is diagonal with entries $\sum_{a\in G} f(a)\varepsilon(-a)$, so that

$$\det \mathbb{T} = \prod_{\varepsilon\in\hat{G}} \left(\sum_{a\in G} f(a)\varepsilon(-a)\right).$$

Corresponding to (5.2.12), we have clearly

$$f = \sum_{a\in G} f(a)\chi_a.$$

This and (5.2.12) lead to (5.2.7).

We consider the representation of $\mathbb{T}$ w.r.t. $\mathcal{D}$. Similarly as above,

$$T_a\chi_b = \chi_{b-a}$$

and

$$T\chi_b = \sum_{a\in G} f(a)T_{-a}\chi_b = \sum_{a\in G} f(a)\chi_{a+b} = \sum_{a\in G} f(a-b)\chi_a,$$

i.e. (5.2.8). Hence

$$\mathbb{T} = (f(a-b))_{a,b}.$$

Comparing the value of $\det \mathbb{T}$, we conclude (5.2.9). □

Proof of Lemma 5.2.1. Fix $\psi \in \Delta$. Then for any $f : G \to \mathbb{C}$ satisfying (5.2.2), we have

$$f\psi^{-1} : G/N \to \mathbb{C}.$$

For any $a \in N$, we have $f\psi^{-1}(a) = f(1a)\lambda^{-1}(a) = 1$. Hence

$$\widehat{G/N} = \{\eta = \chi\psi^{-1} | \chi \in \Delta\}. \tag{5.2.14}$$

Writing $G/N = \{\bar{a} = a \mod N\}$, we apply Theorem 5.2.1 to obtain

$$\prod_{\eta\in\widehat{G/N}}\sum_{\bar{a}\in G/N}\eta(\bar{a})f\psi^{-1}(\bar{a}^{-1}) = \det(f\psi^{-1}(\bar{a}^{-1}\bar{b})_{a,b\in T}). \tag{5.2.15}$$

By (5.2.14), the LHS of (5.2.15) is

$$\prod_{\chi\in\Delta}\sum_{a\in T}\chi\psi^{-1}(a)f\psi^{-1}(a^{-1}) = \prod_{\chi\in\Delta}\sum_{a\in T}\chi(a)f(a^{-1}),$$

which is the LHS of (5.2.3). The RHS is

$$\det\left(Af(a^{-1}b)B\right),$$

where $A = \begin{pmatrix}\psi^{-1}(a_1^{-1}) & & O\\ & \ddots & \\ O & & \psi^{-1}(a_m^{-1})\end{pmatrix}$, say and B is similar with $\psi^{-1}(a^{-1})$ replaced by $\psi^{-1}(b)$. Hence

$$\det\left(Af(a^{-1}b)B\right) = \prod_{a\in T}\psi^{-1}(a^{-1})\det\left(f(a^{-1}b)\right)\prod_{b\in T}\psi^{-1}(b) = \det\left(f(a^{-1}b)\right), \tag{5.2.16}$$

which is the RHS of (5.2.3), thereby competing the proof. □

Remark 5.2.1 Theorem 5.2.1 may be applied to $G = \mathbb{Z}/M\mathbb{Z}$ and cover the theory of periodic functions of period M and (5.2.7) leads to the base change formula (2.2.17). According to the labeling (2.2.2), the sum over $j = 1, \cdots, M$ is the one over characters $\varepsilon \in \hat{G}$.

Lemma 5.2.1, which is a consequence of Theorem 5.2.1, is applied to $G = G_M := (\mathbb{Z}/M\mathbb{Z})^\times$ to handle the Maillet–Demyanenko determinants. In Kanemitsu and Kuzumaki (2001), a subgroup $-1 \notin H$ is considered (which is to be the congruence group by class field theory) and N is taken to be $H\{\pm 1\}$. In our application, we take $H = 1$ and $N = \{\pm 1\}$, in which case G/N is the Galois group of K/K^+.

Theorem 5.2.2 *Chowla's problem and Maillet–Demyanenko determinants merge on the grounds of the Dedekind determinant and the product of the special case of the odd part of the base change formula (2.2.21), which for the Dirichlet L-function with odd χ, is to be interpreted as*

$$\frac{\pi}{2M}\sum_{a=1}^{M}\chi(a)\cot\left(\pi\frac{a}{M}\right) = L(1,\chi) = P(1,\chi^*)L(1,\chi^*) \tag{5.2.17}$$

$$= i\frac{\pi\tau(\chi^*)}{f}P(1,\chi^*)\sum_{a=1}^{M-1}\chi^*(a)\bar{B}_1\left(\frac{a}{M}\right),$$

where χ^* is the primitive character with conductor f inducing χ and $P(s, \chi^*)$ is defined by (2.3.2).

The product over all odd primitive χ of (5.2.17) *leads to a constant multiple of the relative class number* (5.2.1).

Chowla's problem corresponds to Theorem 5.1.2 and Maillet–Demyanenko determinant to Theorem 5.1.3.

Proof In both cases, we apply Lemma 5.2.1 with $G = (\mathbb{Z}/M\mathbb{Z})^\times$ the group of reduced residue classes mod M, $N = \{\pm 1\}$, (5.2.18) as a complete system T of representatives of G/N and $\Delta = X^e$ resp. $\Delta = X^o$ the set of all even resp. odd Dirichlet characters.

To prove Chowla's problem part of Theorem 5.2.2, we modify the proof of Shirasaka (1989) a little and we may choose as T the complete set of representatives of $G/\{\pm 1\}$

$$T = \left\{a \middle| 1 \le a < \frac{M}{2},\ (a, M) = 1\right\}. \tag{5.2.18}$$

Suppose

$$0 = \sum_{a \in T} c_a \cot \frac{\pi a}{M} = i \sum_{a \in T} c_a \xi(a), \tag{5.2.19}$$

($c_a \in \mathbb{Q}$), say, where (cf. (5.5.3) for the even case)

$$\xi(a) := -i \cot \frac{\pi a}{M} = \frac{\zeta^{a/2} + \zeta^{-a/2}}{\zeta^{a/2} - \zeta^{-a/2}} \tag{5.2.20}$$

and where $\zeta = e^{\frac{2\pi i}{M}}$.

To accommodate this in the framework of Lemma 5.2.1, we form the system of linear equations from (5.2.19)

$$0 = \sum_{a,b \in T} c_a \xi(ab^{-1}), \tag{5.2.21}$$

by applying the automorphism σ_b^{-1} in (5.3.1) and consider the determinant $\det(\xi(ab^{-1}))_{a,b\in T}$. For an odd ψ we have

$$\det(\xi(ab^{-1}))_{a,b\in T} = \det(\psi(ab^{-1})\xi(ab^{-1}))_{a,b\in T}. \tag{5.2.22}$$

Indeed, the right-hand side is transformed into

$$\det(\psi(a)\bar{\psi}(b)\xi(ab^{-1}))_{a,b\in T} = \prod_{a\in T} \psi(a) \prod_{b \in T} \bar{\psi}(b) \det(\xi(ab^{-1}))_{a,b\in T}$$

and the coefficient $\prod_{a\in T} \psi(a) \prod_{b\in T} \bar{\psi}(b) = \psi(\prod_{a\in T} a)\bar{\psi}(\prod_{b\in T} b)$ may be expressed as $\psi(\prod_{a\in T} a)\bar{\psi}(\prod_{b\in -T} b) = 1$, whence (5.2.22) follows.

Now apply (5.2.3) with $f(a) = \psi(a)\xi(a)$ which is even, to deduce that

$$\det(\psi(ab^{-1})\xi(ab^{-1}))_{a,b\in T} = \prod_{\chi\in X^e}\left(\sum_{a\in T}\bar{\chi}(a)\psi(a)\xi(a)\right)$$
$$= \prod_{\chi\in X^e}\left(\sum_{a\in T}\bar{\chi}\psi(a)\xi(a)\right),$$

so that

$$\det(\xi(ab^{-1}))_{a,b\in T} = \prod_{\chi\in X^o}\left(\sum_{a\in T}\bar{\chi}(a)\xi(a)\right) \tag{5.2.23}$$

in view of (5.2.22).

We may express $\xi(a)$ as

$$\xi(a) = -\frac{4iM}{\pi\varphi(M)}\sum_{\psi\in X^o}\psi(a)L(1,\bar{\psi}) \tag{5.2.24}$$

from the first equality of (5.2.17) by orthogonality (2.3.1) (cf. Proposition 5.6.1). Substituting (5.2.24) in (5.2.23), we compute each factor

$$\sum_{a\in T}\bar{\chi}(a)\xi(a) = -\frac{4iM}{\pi\varphi(M)}\sum_{\psi\in X^o}L(1,\bar{\psi})\sum_{a\in T}\bar{\chi}(a)\psi(a).$$

Since the inner sum $\sum_{a\in T}\bar{\chi}(a)\psi(a)$ is

$$\frac{1}{2}\sum_{a\in T\cup(-T)}\bar{\chi}(a)\psi(a) = \frac{1}{2}\sum_{a=1}^{M}\bar{\chi}\psi(a),$$

which is $\frac{1}{2}\varphi(M)$ only when $\psi = \chi$ by orthogonality (2.3.1). Hence (5.2.23) amounts to

$$\det(\xi(ab^{-1}))_{a,b\in T} = \prod_{\chi\in X^o}-\frac{2iM}{\pi}L(1,\bar{\chi}). \tag{5.2.25}$$

Appealing to the non-vanishingness of $L(1,\chi)$, we conclude that the determinant of the system (5.2.21) of linear equations in unknowns c_a is non-zero. Hence the equation has only the trivial solution, i.e. cotangent values are linearly independent over $\mathbb{Q}$.

We turn to the proof of Maillet–Demyanenko determinant part of Theorem 5.2.2. The proof being almost verbatim to that of Kanemitsu and Kuzumaki (2001, Theorem 1), we reproduce the proof of just one case and suggest the steps for other cases. As is common in all the cases, we have by Lemma 5.2.1

$$\begin{aligned} D_k^0(0) := \det\left(\tilde{B}_k^0\left(\frac{\bar{a}\bar{b}^{-1}}{M}\right)\right)_{\bar{a},\bar{b}\in T} \\ = \prod_{\chi\in X^k}\sum_{\bar{a}\in T}\chi(\bar{a})\tilde{B}_k\left(\frac{\bar{a}}{M}\right) \\ = \prod_{\chi\in X^k}\sum_{\bar{a}\in T}\sum_{\alpha\in\bar{a}H}\chi(\alpha)\tilde{B}_k\left(\frac{R(\alpha)}{M}\right), \end{aligned} \tag{5.2.26}$$

where $R(\alpha)$ is the least non-negative residue mod M (5.1.3), $H = \{1\}$ and $X^k = X^e$ or $X^k = X^o$ according to k is even or odd. Hence

$$D_k^0(0) = \prod_{\chi\in X^k}\frac{1}{2}\sum_{a=1}^{M-1}\chi(a)B_k\left(\frac{a}{M}\right) = \prod_{\chi\in X^k}\frac{M^{1-k}}{2}B_{k,\chi^*}P(1-k,\chi^*), \tag{5.2.27}$$

$$P(1-k,\chi^*) = \prod_{\substack{p\mid M\\ p:prime}}\left(1-\chi^*(p)p^{k-1}\right)$$

by Lemma 2.3.2, where B_{k,χ^*} is the generalized kth Bernoulli number defined by (2.3.16). This corresponds to (5.3.3) and is a finite form of the relative class number.

Here we digress and go back to the product of $L(1,\chi)$ by appealing to the right-hand side of (5.2.17) (recall Lemma 2.3.2). For simplicity, we choose $k = 1$. Then $X^1 = X^o$ and we have

$$\begin{aligned} \frac{P(0,\bar{\chi}^*)}{P(1,\chi^*)}L(1,\chi) = P(0,\bar{\chi}^*)L(1,\chi^*) \\ = i\frac{\pi\tau(\chi^*)}{f}P(0,\bar{\chi}^*)\sum_{a=1}^{M-1}\chi^*(a)\bar{B}_1\left(\frac{a}{M}\right). \end{aligned}$$

Then (5.2.27) becomes

$$\begin{aligned} D_1^0(0) = \prod_{\chi\in X^o}\frac{f}{2\pi i\tau(\chi^*)}\frac{P(0,\bar{\chi}^*)}{P(1,\chi^*)}L(1,\chi) \\ = \left(-\frac{1}{2\pi i}\right)^{\frac{\varphi(M)}{2}}\delta_1\prod_{\chi\in X^o}\tau(\bar{\chi}^*)L(1,\chi), \end{aligned}$$

where slightly correcting (Kanemitsu & Kuzumaki, 2001, Theorem 3), we understand δ_k to mean

$$\delta_k = \prod_{\chi \in X^o} \frac{P(1-k, \bar{\chi}^*)}{P(k, \chi^*)}.$$

Therefore the essential ingredient is again the product $\prod_{\chi \in X^o} L(1, \chi)$. If all $\chi \in X^o$ are primitive, the situation will be much simpler.

Remark 5.2.2 (i) As we already remarked, substituting (2.2.8) and using the (inverse) Eisenstein formula (2.2.28), the right-hand side of (2.2.21) just returns to the left-hand side.

It follows that the DFT formula alone does not lead to merging.

(ii) If we apply the functional equation (twice, cf. below)

$$\begin{aligned} D(1, f^o) &= i\frac{\pi}{\sqrt{M}} D(0, \hat{f}^o) \qquad (5.2.28) \\ &= i\frac{\pi}{\sqrt{M}} \left(\sum_{k=1}^{M} \hat{f}(k)\zeta\left(0, \frac{k}{M}\right)\right) = -i\frac{\pi}{\sqrt{M}} \sum_{k=1}^{M-1} \hat{f}(k)\bar{B}_1\left(\frac{k}{M}\right) \end{aligned}$$

and we come back to the same expression as is expected.

In Remark 5.2.2, (i) we applied the (inverse) Eisenstein formula (2.2.28), which is the special case of (2.2.18) with $f(a) = \chi_b(a)$, which in turn is contained in (2.2.19), i.e. the base change formula.

In Remark 5.2.2, (ii) we used the formulas

$$D(1, f^o) = i\frac{\pi}{\sqrt{M}} D(0, \hat{f}^o), \qquad (5.2.29)$$

and (2.1.48).

Both follow from the functional equation in the generalized Euler identity form, a trivial variant of (2.2.38).

5.3 Cyclotomic Fields and Class Numbers

We quote from Takagi (1971, pp. 171–172).

Prime Ideal Theorem in an Arithmetic Progression (PNTAP). It is a consequence of class field theory that given a congruence group H in an algebraic number field k, there exists an Abelian extension K/k which is a class field over H. However, as is stated in the proof of Theorem 12.4 in Sect. 12.4, if there exists a Galois extension K/k corresponding to H, K not necessarily being a class field over H, then the following theorem holds.

Theorem 5.3.1 (Theorem 1) *For every congruence classification in an algebraic number field k, each congruence class contains infinitely many (absolute) prime ideals of degree* 1, *so that for the (Hecke) L-function corresponding to the classification, we have* $L(1, \chi_i) \neq 0$, $\chi_i \neq \chi_1$.

Thus $L(1, \chi_i) \neq 0$ is assured rather excessively by class field theory.

[Historical remarks]. It was Hecke* who first proved $L(1, \chi_i) \neq 0$ by analytic methods. On the other hand, it had been long known in the tradition of Dirichlet that if there exists (an Abelian extension) K/k, then the Dirichlet PNTAP for H holds true, whence $L(1, \chi_i) \neq 0$. It may be Weber who first claimed it explicitly in literature. The existence of K, however, being established by class field theory in 1915**, as its byproduct, both PNTAP and $L(1, \chi_i) \neq 0$ were established at the same time. Since around that period, communication was shut down because of WWI, it was not the case that Hecke gave a new proof of the already established theorem ***. Moreover, even if PNTAP is established by class field theory, a direct proof of $L(1, \chi_i) \neq 0$ is desirable. What is interesting is that although the original Dirichlet PNTAP was established in 1837 by the theory of quadratic forms and its analytic proof was given by Mertens in 1897, both were established simultaneously in the general case.

Footnote

$*$ =Hecke (1917a).

$**$ =Takagi (1915)

$***$ =Landau (1918). Landau, pp. 52–53, says something about minor slips in the aforementioned proof of Hecke.

Fundamental references on class numbers of an Abelian field are Hasse (1952) and Meyer (1957). For the relative class number (also referred to as the first factor), cf. e.g. Ernvall (1990). For cyclotomic fields, we refer to Lang (1978) and Washington (1983) (Table 5.2).

The Mth cyclotomic polynomial $\Phi_M(X) \in \mathbb{Q}[X]$ is the degree $\varphi(M)$ monic polynomial whose roots are primitive Mth roots ζ_M, of unity. The field $\mathbb{Q}_M = \mathbb{Q}(\zeta_M)$ is the splitting field of $\Phi_M(X)$, called the **Mth cyclotomic field**. We may assume that M is not of the form $2a$ with a odd. $\mathbb{Q}_M$ is an Abelian extension of $\mathbb{Q}$ with its Galois group $G_M = \mathrm{Gal}(\mathbb{Q}_M/\mathbb{Q}) \cong (\mathbb{Z}/M\mathbb{Z})^\times$. Here the right-hand side is the multiplicative group of reduced residue classes mod M and the isomorphism φ is given by

Table 5.2 Expressions for $L(1, \chi)$; universal and deformed

Theory	Class field theory	Cyclotomic fields
Ingredient	Congruence classification	Adjoint of prim. root of 1
Special function	Euler digamma	polylog.
Expression	Universal	DFT

$\varphi(\sigma_s) = s \mod M$, each $\sigma \in \mathrm{Gal}(K/\mathbb{Q})$ being a raising-to-the-power $s \mod M$ map σ_s, $(s, M) = 1$:

$$\zeta^\sigma = \zeta^s. \tag{5.3.1}$$

By the Kronecker–Weber theorem, every Abelian extension K of $\mathbb{Q}$ is a subfield of $\mathbb{Q}_M$ and conversely.

Let K be an Abelian extension of $\mathbb{Q}$ of degree n. For the class number $h = h_K$ of K resp. h_K^+ of the maximal real subfield K^+ of K, Hasse (1952, p. 11, (3)) reads

$$h = h_K^+ h_K^-, \tag{5.3.2}$$

where

$$h_K^+ = \frac{1}{R^+} \prod_{\chi \neq \chi_0 \in X^e} \sum_{a \in T} \chi(a) A_1\left(\frac{a}{f_\chi}\right),$$

where R^+ is the regulator of K^+ resp.

$$h_K^- = Qw \prod_{\chi \in X^o} \left(-\frac{1}{2f_\chi}\right) \sum_{a=1}^{f_\chi - 1} a\chi(a) = Qw \prod_{\chi \in X^o} \left(-\frac{1}{2} B_{1,\chi}\right), \tag{5.3.3}$$

and where the product is over all even resp. odd primitive characters χ with conductor f_χ. The data are $w = |W|$ and $Q = [E : WE^+]$, and where W is the group of roots of unity of K, $E = \mathfrak{O}_K^\times$ resp. $E^+ = \mathfrak{O}_{K^+}^\times$ is the unit group of K resp. K^+. As in Hasse (1952), if K is real, then $h = h_K^+$, $h_K^- = 1$. (5.2.1) is on Hasse (1952, p. 11, (3b)) and reads as in (5.3.3).

This follows from analytic class number formulas for K and its maximal real subfield K^+, which follows by comparison of the residue in two ways as with the Dirichlet class number formula, Theorem 2.6.1. One side follows from the decomposition of the Dedekind zeta-function (which in turn is a consequence of the decomposition theorem in class field theory), cf. Hasse (1952, pp. 4–11), Lang (1978, pp. 75–76), Gut (1929, p. 186, (345)):

$$\zeta_K(s) = \prod_{\chi} L(s, \chi) = \zeta(s) \prod_{\chi \neq \chi_0} L(s, \chi), \tag{5.3.4}$$

where χ runs through all primitive characters induced by those of the Galois group $\mathrm{Gal}(K/\mathbb{Q})$, i.e. primitive to the modulus f_χ. Cf. Lemma 2.6.1 for a quadratic field. This implies (cf. Theorem 2.6.1)

$$L(1, \chi) = \begin{cases} \frac{2^n}{\sqrt{d}} hR, & K \text{ real}, \\ \frac{(2\pi)^{\frac{n}{2}}}{w\sqrt{|d|}} h, & K \text{ imaginary}. \end{cases} \tag{5.3.5}$$

Combining (5.3.2), (5.3.5) and Washington (1983, pp. 41–42), we have the relative class number formula

$$\prod_{\chi\in X^o} L(1,\chi)=\frac{(2\pi)^{\frac{n}{2}}h_K^-}{Qw\sqrt{\left|\frac{d_K}{d_{K^+}}\right|}}.$$

Cf. (2.6.2) for an imaginary quadratic field.

5.4 Even Part of (CPMD)

For Maillet determinants with Clausen functions, Bundschuh is the first to introduce them. Bundschuh (1988, Lemma 5) and Funakura (1985, Formula (9)) give

$$\det\left(\frac{k!}{(-2\pi i)^k}A_{1-k}\left(\frac{ab^{-1}}{M}\right)\right)=\prod_{\substack{\chi\in(\mathbb{Z}/\hat{M}\mathbb{Z})^\times\\ \chi(-1)=(-1)^k}}\left(-\frac{k!}{(2\pi i)^k}M^kL(k,\bar\chi)\right), \tag{5.4.1}$$

where $A_k(x)$ is the Clausen function of order k defined by (2.1.35). The negative order Clausen functions are indeed the derivatives of the Lerch zeta-function, which in turn are derivatives of the cotangent function and (5.4.1) amounts to the results of Okada and Wang. Bundschuh proceeds further to give quantitative version of these types of theorems.

Then the extensive study of Kanemitsu and Kuzumaki (2007) appeared.

In addition to Kanemitsu et al. (2007), Liu et al. (2010), Maier (1936), Kanemitsu and Kuzumaki (2001), Chakraborty et al. (2018), we refer to Deninger (1984) which gives determinant expression for h_M^+ as well as a generalization of Gut's results (Gut 1929).

Along with (generalized) Bernoulli numbers (2.3.16) we define the value of $A_k(0)$ (the kth **Clausen number**, so to say) as in Kanemitsu and Kuzumaki (2001) by

$$A_k(0)=\begin{cases}\infty, & \text{for } k=1\,,\\ 0, & \text{for } k \text{ even},\\ \frac{k!}{(2\pi i)^{k-1}}\zeta(k), & \text{for } k \text{ odd}\,,\neq 1\,,\end{cases} \tag{5.4.2}$$

and define the kth generalized Clausen number $A_{k,\chi}$ for primitive even χ with conductor f by

$$A_{k,\chi}=f^{k-1}\sum_{a=0}^{f-1}\chi(a)A_k\left(\frac{a}{f}\right)\qquad\text{for}\qquad \chi\neq\chi_0$$

and

$$A_{k,\chi_0} = A_k(0).$$

For a positive odd integers k

$$L(k,\chi) = \left(\frac{2\pi i}{f}\right)^{k-1} \frac{1}{k!\tau(\bar{\chi})} A_{k,\bar{\chi}}, \tag{5.4.3}$$

which entails

$$L(1,\chi) = 2\frac{\tau(\chi)}{f} L'(0,\bar{\chi}) = \frac{\tau(\chi)}{f} A_{1,\bar{\chi}} = \frac{1}{\tau(\bar{\chi})} A_{1,\bar{\chi}}.$$

Note that the first equality is the limiting case of the functional equation (2.3.11) and the third by (2.3.8).

Corresponding to Lemma 2.3.2, we have

Lemma 5.4.1 *Let $\chi \neq \chi_0$ be a Dirichlet character* mod M *with conductor $f = f_\chi$ and let $P(s,\chi^*)$ be the product in (2.3.2). Then for odd $k \in \mathbb{N}$,*

$$\begin{aligned} M^{k-1} \sum_{a=0}^{M-1} \chi(a) A_k\left(\frac{a}{M}\right) &= A_{k,\chi^*} P(1-k,\chi^*) \\ &= \left(\frac{f}{2\pi i}\right)^{k-1} k!\tau(\chi^*) P(1-k,\chi^*) L(k,\bar{\chi}^*). \\ &= \left(\frac{f}{2\pi i}\right)^{k-1} k!\tau(\chi^*) \frac{P(1-k,\chi^*)}{P(k,\bar{\chi}^*)} L(k,\bar{\chi}). \end{aligned} \tag{5.4.4}$$

Proof Proof is similar to that of Lemma 2.3.2. The first equality follows from Eratosthenes sieve, the second from (5.4.3), and the third from (2.3.2).

We define a typical even type Maillet determinant with Clausen function entries $D_k^1(0)$ à la Kanemitsu and Kuzumaki (2001, Definition 2) : For an odd positive integer k let

$$\tilde{B}_k^1\left(\frac{\bar{a}}{m}\right) = \sum_{\alpha \in aH} B_k^1\left(\frac{R(\alpha)}{m}\right) = \sum_{\alpha \in aH} A_k\left(\frac{R(\alpha)}{m}\right). \tag{5.4.5}$$

There are similar types of Maillet and Demyanenko determinants for which we have similar results. Cf. Kanemitsu and Kuzumaki (2001, Theorem 3).

Theorem 5.4.1 *With* (5.4.5) *and notation as in the proof of Theorem 5.2.2, we have*

$$\begin{aligned} D_k^1(0) :&= \det\left(\tilde{B}_k^1\left(\frac{\bar{a}\bar{b}^{-1}}{M}\right)\right)_{\bar{a},\bar{b}\in T} \\ &= \left(\frac{k!}{2(2\pi i)^{k-1}}\right)^g \delta_k' \prod_{\chi\in X^e} \left(\frac{f_\chi}{M}\right)^{k-1} \tau(\bar{\chi}^*) L(k,\chi) \end{aligned} \tag{5.4.6}$$

for k odd, where $g = \sharp T = \frac{\varphi(M)}{2}$ *and*

$$\delta'_k = \prod_{\chi \in X^e} \frac{P(1-k, \bar{\chi}^*)}{P(k, \chi^*)}. \tag{5.4.7}$$

Proof By Lemma 5.2.1

$$\begin{aligned} D_k^1(0) &= \prod_{\chi \in X^{k+1}} \sum_{\bar{a} \in T} \chi(\bar{a}) \tilde{B}_k^1 \left(\frac{\bar{a}}{M}\right) \\ &= \prod_{\chi \in X^{k+1}} \sum_{\bar{a} \in T} \sum_{\alpha \in \bar{a} H} \chi(\alpha) A_k \left(\frac{R(\alpha)}{M}\right) \\ &= \prod_{\chi \in X^{k+1}} \frac{1}{2} \sum_{a=1}^{M-1} \chi(a) A_k \left(\frac{a}{M}\right). \end{aligned} \tag{5.4.8}$$

Substituting from (5.4.4), we conclude that

$$D_k^1(0) = \prod_{\chi \in X^{k+1}} \left(\frac{k!}{2(2\pi i)^{k-1}}\right) \frac{P(1-k, \chi^*)}{P(k, \bar{\chi}^*)} \left(\frac{f_\chi}{M}\right)^{k-1} \tau(\chi^*) L(k, \bar{\chi}). \tag{5.4.9}$$

Factoring $\frac{k!}{2(2\pi i)^{k-1}}$ out as the gth power and rewriting χ for $\bar{\chi}$, we prove (5.4.6).

Therefore the essential ingredient is again the product $\prod_{\chi \in X^e} L(1, \chi)$.

This theorem corresponds to Theorem 5.2.2 for the odd part. Since the even part of $\psi\left(\frac{a}{M}\right)$ (and $\ell_1\left(\frac{a}{M}\right)$) leads to the same expression $L(1, \chi) = \frac{1}{\sqrt{M}} \sum_{a=1}^{M-1} \hat{\chi}(a) A_1\left(\frac{a}{M}\right)$, merging of CPMD takes slightly different form. That is, the determinant expression in Theorem 5.4.1 and (multiplicative) linear independence of values of $A_1\left(\frac{a}{M}\right)$ known as Ramachandra units, which we describe in the next section.

5.5 BBW Theorem; Ramachandra Units

We clarify the results of Baker–Birch–Wirsing (BBW) and of Ramachandra about linear independence of the Clausen function values at rational arguments.

Baker et al. (1973) proved the following. Let $\bar{\mathbb{Q}}$ denote the field of all algebraic numbers.

Theorem 5.5.1 (Baker–Birch–Wirsing) *If* $f : \mathbb{Z} \to \bar{\mathbb{Q}}$ *is a non-vanishing periodic function of period* M *such that* (i) $f(r) = 0$ *if* $1 < (r, M) \le M$, (ii) *The* M*th cyclotomic polynomial* Φ_M *is irreducible over* $K := \mathbb{Q}(f(1), \cdots, f(M))$, *then*

$$D(1, f) = \sum_{n=1}^{\infty} \frac{f(n)}{n} \neq 0. \tag{5.5.1}$$

Although Theorem 5.5.1 does not cover Dirichlet's result (5.1.2) fully and indeed the proof uses it, it does cover certain cases (p. 226).

As a subspace of $C(M)$ in (2.2.1), let

$$C^a(M) = \{f : \mathbb{Z} \to \bar{\mathbb{Q}} | f(n+M) = f(n), D(1, f) = 0, \text{(i)}\}. \tag{5.5.2}$$

be the vector space of periodic functions $f : \mathbb{Z} \to \bar{\mathbb{Q}}$ of period M satisfying (i) $f(r) = 0$ if $(r, M) > 1$ and the negation of (5.5.1). Baker et al. (1973) also proved that it consists of odd functions, whence Theorem 5.5.1 is valid for even functions. Although even functions are to be considered, (5.6) is not mentioned.

It is instructive to look at some of their results in the light of DFT, esp. Theorem 2.2.1. They consider all Fourier transforms (2.2.8) of $f \in C^a(M)$ denoted $\hat{C}^a(M)$ but do not employ the associated Dirichlet series (2.2.4) fully. Under DFT, the vanishing condition $\hat{f}(M) = 0$ is kept, implying convergence of $\sum_{n=1}^{\infty} \frac{f(n)}{n}$ while $D(1, f) = 0$ implies $\hat{f}(M) = 0$.

The second equality of (2.2.20) explains Lemma 1 (which means Baker et al., 1973, Lemma 1 and the same for other citations to the end of this section) giving an equivalent condition for g to belong to $C^a(M)$. The odd part of the first equality is (2.3.20) which explains the condition for an odd function on p. 228. In case f is rational valued, this amounts to the inverse problem of Chowla.

As is stated on p. 231, Condition (ii) means that the composite field $K(\zeta_M)$ is an Abelian extension of K and the Galois group $\mathrm{Gal}(K(\zeta)/K)$ is $\widehat{(\mathbb{Z}/M\mathbb{Z})^\times}$ and its elements σ_a are raising-to-the-power-a maps (5.3.1).

They can be extended to a K-automorphism of $\mathbb{Q}$. If f is rational valued, then $K = \mathbb{Q}$ and a runs through all reduced residue classes and leads to the inverse problem of Chowla.

The key result is Lemma 2 which shows linear dependence of g-values $g \in \hat{C}^a(M)$ by choosing the maximal set of linearly independent members over $\mathbb{Q}$ of $\ell_1\left(\frac{a}{M}\right)$, $a = 1, \cdots, M-1$, i.e. of $A_1\left(\frac{a}{M}\right)$. Thus this is closely related to a maximal set of multiplicatively independent units, cf. Sect. 5.4. If the maximal number coincides with the order of $\mathrm{Gal}(K(\zeta)/K)$, then the Maillet determinant may be used.

In the extensive list of Yamamura (2009) on Maillet determinants, Kanemitsu and Kuzumaki (2001) is referred to as one centering around the even part. It turns out that Siegel (1964, pp. 183–207) considered the class numbers of cyclotomic fields $\mathbb{Q}_p$ with p an odd prime and deduced determinant expressions including Maillet determinants with Bernoulli polynomial entries, developing the work of Kummer.

Later Ramachandra (1964, 1966, 1969, 1993) considered Maillet-like determinants with entries related to the Clausen function in his study on multiplicative independence of units in cyclotomic fields with general modulus, generalizing Siegel (1964, Satz 19.6).

Ramachandra (1966) states a motivation, cf. (5.2.20). For the primitive Mth root $\zeta = e^{2\pi i \frac{1}{M}}$ of unity, $\zeta^a = e^{2\pi i \frac{a}{M}}$, $(a, M) = 1$ are roots of unity and

$$u_b = \frac{1-\zeta^{ab}}{1-\zeta^a}, \quad 1 < b < \frac{M}{2}, \quad (b, M) = 1 \tag{5.5.3}$$

are units in $\mathbb{Q}_M$. Milnor (previously Kummer) asked if u_b and $\pm\zeta$ form a basis of the group of units in $\mathbb{Q}_M$ and for this, multiplicative independence of u_b is necessary or linear independence of their logarithms

$$\log u_b = \log(1-\zeta^{ab}) - \log(1-\zeta^a) = -\left(\ell_1\left(\frac{ab}{M}\right) - \ell_1\left(\frac{a}{M}\right)\right).$$

In view of (2.1.24), this amounts to linear independence of $A_1\left(\frac{ab}{M}\right) - A_1\left(\frac{a}{M}\right)$. This can be accommodated in (5.2.4) by writing b^{-1} for b, i.e. one considers (5.2.4) with $N = \{\pm 1\}$, $\Delta = X^o$ (all even characters mod M) and T the half-interval (5.2.18) and $f(a) = A_1\left(\frac{a}{M}\right)$

$$\det\left(A_1\left(\frac{ab^{-1}}{M}\right) - A_1\left(\frac{a}{M}\right)\right)_{a,b\in T} = \prod_{\substack{\chi\in\Delta \\ \chi\neq\chi_0}} \left(\sum_{a\in T} \bar{\chi}(a) A_1\left(\frac{a}{M}\right)\right). \tag{5.5.4}$$

Then the argument is the same as in Kanemitsu and Kuzumaki (2001).

In the formulation of Washington (1983, p. 147), we have

Theorem 5.5.2 (Ramachandra, 1966) *Let $\zeta = \zeta_M$ be a primitive Mth root of unity and let $M = \prod_{i=1}^k p_i^{a_i}$ be the prime decomposition. Let I run through all subsets of $\{1, \cdots, k\}$ except for itself and let $n_I = \prod_{\{i\in I\}} p_i^{a_i}$. For $b = 1, \cdots, \left[\frac{M}{2}\right]$, $(b, M) = 1$ let*

$$\xi_b = \zeta^{d_b} \prod_{n_I} \frac{1-\zeta^{bn_I}}{1-\zeta^{n_I}}, \quad d_b = \frac{1}{2}(1-b) \sum_{I\subset\{1,\cdots,k\}} n_I. \tag{5.5.5}$$

Then $\{\xi_b\}$ forms a multiplicatively independent units in $\mathbb{Q}(\zeta_M)^+$.

Cf. Washington (1983, pp. 147–152), where the units in Theorem 5.5.2 are called **Ramachandra units**. Thus, Ramachandra's work may be described as the even part of Chowla's problem with Clausen function argument. In some cases, the definition of the regulator may be thought of as the Maillet determinant with Clausen function argument.

5.6 Dirichlet Series with Periodic Coefficients Revisited

We recall Sect. 2.2, esp. Theorem 2.2.1 for periodic Dirichlet series.

Exercise 5.6.1 Prove that Theorem 2.2.1 is a consequence of Theorem 5.2.1 viewed as $\mathbb{C}[\hat{G}]$.

Let G be a finite Abelian group. Character theory of G has been developed in Gal (1961) and Li et al. (2013, pp. 109–111). As with the inner product in $C(M)$, we recall the inner product (ξ, χ) which also works as the Fourier transform $\hat{\xi}$ of $\xi \in L^2(G)$ defined by

$$(\xi, \chi) = \hat{\xi}(\chi) = \int_G \xi\bar{\chi} = \frac{1}{\sqrt{g}} \sum_{x \in G} \bar{\chi}(x)\xi(x).$$

Theorem 5.6.1 *The Fourier inversion formula holds.*

$$\xi(x) = \hat{\hat{\xi}}(x^{-1}). \tag{5.6.1}$$

Corollary 5.6.1 (Gal, 1961, p. 150, Lemma 1) *Let* $\chi \in \hat{G}$, $\eta_\chi \in \hat{\hat{G}}$. *Then the system of linear equations*

$$\sum_{x \in G} \chi(x)\xi(x) = \eta(\bar{\chi}) =: \eta_\chi, \quad \chi \in \hat{G}, \tag{5.6.2}$$

say, is uniquely solved as

$$\xi(x) = \frac{1}{g} \sum_{\chi \in \hat{G}} \bar{\chi}(x)\eta_\chi \quad x \in G. \tag{5.6.3}$$

Exercise 5.6.2 Prove Theorem 5.6.1 and Corollary 5.6.1.

Solution. For (5.6.1) we have

$$\begin{aligned}\hat{\hat{\xi}}(x^{-1}) &= \hat{\hat{\xi}}(\bar{x}) = (\hat{\xi}, \bar{x}) = \frac{1}{\sqrt{g}} \sum_{\chi \in \hat{G}} \hat{\xi}(\chi)x(\chi) = \frac{1}{\sqrt{g}} \sum_{\chi \in \hat{G}} \hat{\xi}(\chi)\chi(x) \\ &= \frac{1}{g} \sum_{y \in G} \xi(y) \sum_{\chi \in \hat{G}} \bar{\chi}(y)\chi(x) = \xi(y)\delta_{yx} = \xi(x),\end{aligned}$$

where $x(\chi) = \chi(x)$ is duality. Compare this proof with that of (2.2.9).

Given $\eta \in \hat{G}$, we find a ξ such that (5.6.2) holds:

$$\sum_{x \in G} \chi(x)\xi(x) = \eta(\bar{\chi}) = \hat{\hat{\eta}}(\bar{\chi}) = a\hat{\xi}(\bar{\chi}) = a\frac{1}{\sqrt{g}} \sum_{x \in G} \chi(x)\xi(x). \tag{5.6.4}$$

Hence choosing $a = \sqrt{g}$, we have

$$\hat{\xi} = \frac{1}{\sqrt{g}}\eta,$$

which implies by (5.6.1)

$$\xi(x) = \frac{1}{\sqrt{g}}\hat{\eta}(\bar{x}). \tag{5.6.5}$$

Equation (5.6.4) leads to (5.6.2) while (5.6.5) to

$$\xi(x) = \frac{1}{\sqrt{g}}\hat{\eta}(\bar{x}) = \frac{1}{g}\sum_{\chi\in\hat{G}} x(\chi)\eta(\chi) = \frac{1}{g}\sum_{\bar{\chi}\in\hat{G}} x(\bar{\chi})\eta(\bar{\chi}) = \frac{1}{g}\sum_{\chi\in\hat{G}} \bar{\chi}(x)\eta_\chi,$$

i.e. (5.6.3). The proof also implies that the map $L^1(G) \to L^\infty(\hat{G})$; $\xi \to \hat{\xi}$ is an isomorphism where in the former the product is defined by convolution and in the latter by pointwise multiplication. □

Let $M \geq 3$ be an integer and we consider the setting modulo M. For $f = \chi$, a non-principal Dirichlet character mod M, we recall (2.2.35) as

$$L(1,\chi) = -\frac{1}{M}\sum_{\substack{a=1\\(a,M)=1}}^{M-1} \chi(a)\psi\left(\frac{a}{M}\right). \tag{5.6.6}$$

By Theorem 2.3.3, (5.6.6) amounts to (2.3.20), (2.3.21), respectively.

Proposition 5.6.1 *The system of linear equations* (5.6.6)

$$\eta_\chi := -ML(1,\chi) = \sum_{a\in G_M} \chi(a)\tilde{\psi}\left(\frac{a}{M}\right), \quad \chi \in \widehat{G_M}, \tag{5.6.7}$$

where $\tilde{\psi}$ is the periodic extension of period M and η_{χ_0} is defined by (5.6.12) *is solved for $1 \leq b \leq M-1$ as*

$$\begin{aligned}\psi\left(\frac{b}{M}\right) &= \frac{1}{\varphi(M)}\sum_{\chi\in\hat{G}_M} \bar{\chi}(b)\eta_\chi \qquad (5.6.8)\\ &= -\frac{M}{\varphi(M)}\sum_{\chi\neq\chi_0} \bar{\chi}(b)L(1,\chi) - \chi_0(b)(\log M + \gamma + \frac{1}{\varphi(M)}\log N_M).\end{aligned}$$

And for $\chi \neq \chi_0$ even, the system (5.6.6)

$$\eta^e_\chi := -ML(1,\chi) = -\sqrt{M}\sum_{a=1}^{M-1}\hat{\chi}(a)A_1\left(\frac{a}{M}\right) \tag{5.6.9}$$

$$= -\sum_{b\in G_M}\chi(b)\sum_{a=1}^{M-1}\varepsilon_b(-a)A_1\left(\frac{a}{M}\right)$$

is solved as

$$\left(\sqrt{M}\hat{A}_1\left(\frac{b}{M}\right)=\right)\sum_{a=1}^{M-1}\varepsilon_b(-a)A_1\left(\frac{a}{M}\right) = \frac{1}{\varphi(M)}\sum_{\substack{\chi\in\hat{G}_M\\ \chi\ even}}\bar{\chi}(b)\eta^e_\chi, \tag{5.6.10}$$

$$= -\frac{M}{\varphi(M)}\sum_{\substack{\chi\neq\chi_0\\ \chi\ even}}\bar{\chi}(b)L(1,\chi)+\chi_0(b)\log N_M.$$

The odd part

$$\cot\frac{b}{M}\pi = \frac{2M}{\pi\varphi(M)}\sum_{\chi odd}\bar{\chi}(b)L(1,\chi) \tag{5.6.11}$$

also follows. Cf. (5.2.24).

Proof η_{χ_0} is defined by the second equality of (5.6.7), i.e.:

$$\eta_{\chi_0} = \sum_{\substack{a=1\\ (a,M)=1}}^{M-1}\psi\left(\frac{a}{M}\right) = -\varphi(M)(\log M+\gamma)-\log N_M, \tag{5.6.12}$$

where

$$\log N_M = \sum_{d|M}\varphi(d)\Lambda\left(\frac{M}{d}\right) = \varphi(M)\sum_{p|M}\frac{\log p}{p-1}$$

by Lehmer (1975, p. 136) and Kanemitsu and Tsukada (2007, p. 176, Theorem 8.3). And $\eta^e_{\chi_0}$ is found in Kanemitsu and Tsukada (2007, p. 182, (8.52))

$$\eta^e_{\chi_0} = \sum_{b=1}^{M-1}\chi_0(b)\sum_{a=1}^{M-1}\varepsilon_b(-a)\log\left(2\sin\frac{a}{M}\pi\right) = -\log N_M.$$

Hence

$$\eta_{\chi_0} = \eta_{\chi_0^e} - \varphi(M)(\log M+\gamma). \tag{5.6.13}$$

We apply Corollary 5.6.1. In solving (5.6.9), we note that the fourth term is twice the quantity over the half-interval of length $\frac{\varphi(M)}{2}$ which corresponds to the group of all even characters whose order is $\frac{\varphi(M)}{2}$. Hence (5.6.3) is to be applied with $\varphi(M)$.

Adding (5.6.8) and (5.6.10), we deduce for $b \not\equiv 0 \bmod M$ that

$$\psi\left(\frac{b}{M}\right) - \sqrt{M}\hat{A}_1\left(\frac{b}{M}\right) = -\log M - \gamma - \frac{M}{\varphi(M)} \sum_{\chi \text{ odd}} \bar{\chi}(b) L(1, \chi). \tag{5.6.14}$$

Recalling Gauss' first formula (2.2.24), we deduce (5.6.11).

Unlike (5.6.8) or (5.6.10), (5.6.11) is not a direct consequence of (2.3.20) and (2.3.21) (which is stated as Zhang, 1996, Lemma 1) by the Fourier inversion, since odd characters do not form a group. It is a manifestation of the parity principle to the effect that "odd=whole−even." (5.6.11) is Zhang (1996, (7)), which is proved in a rather computational way.

Partial proof of Theorem 2.3.3. We may deduce Gauss' first formula (2.2.24) from (2.3.20) and (2.3.21). In view of Proposition 5.6.1, it suffices to prove (5.6.11) through (5.6.14). Substituting (2.3.20) in the right-hand side of (5.6.14), we prove it by orthogonality of odd characters which is a consequence of the parity principle. □

References

Adhikari, S. D., Saradha, N., Shorey, T. N., & Tijdeman, R. (2001). Transcendental infinite series. *Indagationes Mathematicae, N.S., 12*, 1–14.

Ayoub, R. (1975b). On a theorem of Iwasawa. *Journal of Number Theory, 7*, 108–120.

Ayoub, R. (1975a). On a theorem of Chowla. *Journal of Number Theory, 7*, 105–107.

Baker, A., Brich, B. J., & Wirsing, E. A. (1973). On a problem of Chowla. *Journal of Number Theory, 5*, 224–236.

Bundschuh, P. (1988). On the linear independence of certain algebraic numbers. *Keio Seminar on Mathematical Science No, 12*, 11–30.

Carlitz, L., & Olson, F. R. (1955). Maillet's determinant. *Proceedings of the American Mathematical Society, 6*, 265–269.

Chakraborty, K., Kanemitsu, S., & Kuzumaki, T. (2018). Seeing the invisible: around generalized Kubert functions. *Annales Universitatis Scientiarum Budapestinensis Sectio Computatorica, 47*, 185–195.

Chowla, S. (1952). The Riemann zeta and allied functions. *Bulletin of the American Mathematical Society, 58*, 287–305.

Chowla, S. (1964). On a special infinite series. *Norke Vid. Selsk. Forhand., 17*, 85–87.

Chowla, S. (1970). The nonexistence of nontrivial relations between the roots of a certain irreducible equation. *Journal of Number Theory, 2*, 120–123.

Davis, Ph. J. (1979). *Circulant matrices*. New York etc: Wiley.

Deninger, C. (1984). On the analogue of Chowla-Selberg formula for real quadratic fields. *Journal für die Reine und Angewandte Mathematik, 351*, 171–191.

Endô, A. (1990). The relative class numbers of certain imaginary abelian number fields and determinants. *Journal of Number Theory, 34*, 13–20.

Ernvall, R. (1979). Generalized Bernoulli numbers, generalized irregular primes, and class number. *Annales Universitatis Turkuensis, A Series mathematics, 178*, 72.

Fujisaki, G. (1981). On Chowla's theorem. *RIMS Kokyuroku, 411*, 199–202.

Fujiwara, M. (1978). On linear relations between roots of unity. *RIMS Kokyuroku, 334*, 71–73.

Funakura, T. (1985). On characterization of Dirichlet L-functions. *Acta Arithmetica, 126*, 305–315.
Funakura, T. (1990). On Kronecker's limit formula for Dirichlet series with periodic coefficients. *Acta Arithmetica, 55*, 59–73.
Gal, I. (1961). *Lectures on number theory*. Minneapolis: Jones Letter Service.
Girstmair, K. (1986). Letter to the editor. *Journal of Number Theory, 23*, 405.
Girstmair, K. (1987). Character coordinates and annihilators of cyclotomic numbers. *Manuscripta Mathematica, 59*, 375–389.
Girstmair, K. (1997). Class number factors and distribution of residues. *Abhandlungen aus dem Mathematischen Seminar der Universität Hamburg, 67*, 65–104.
Gut, M. (1929). Die Zetafunktion, die Klassenzahl und die Kronecker'shce Grenzformel eines beliebiges Kreiskörpers. *Commentarii Mathematici Helvetici, 1*, 160–226.
Hashimoto, M., Kanemitsu, S., & Toda, M. (2008). On Gauss' formula for ψ and finite expressions for the L-series at 1. *Journal of the Mathematical Society of Japan, 60*, 219–236.
Hasse, H. (1952). *Über die Klassenzahl abelscher Zahlkörper* (2nd ed). Akademie. (with Appendix by J. Martinet).
Hawkins, T. (1974). New light on Frobenius' creation of the theory of group characters. *Archive for History of Exact Sciences, 12*, 217–243.
Hecke, E. (1917). Über die L-Funktionen und den Dirchletschen Primzahlsatz für beliebigen Zahlkörper, Nachr. König. Ges. Wiss. Góttingen, Math.-phys. Kl. 1917, 299–323=Math. Werke, Vandenhoeck u. *Ruprecht, Göttingen,1970*, 178–197.
Hirabayashi, M. (2015). A generalization of Jakubec's formula. *Mathematica Slovaca65*, 215-227.
Hirabayashi, M. (1998). A generalization of Maillet and Demjanenko determinants. *Acta Arithmetica, 83*, 391–397.
Hirabayashi, M. (2014). A determinant formula for the relative class number formula for an imaginary abelian number field. *Communications in Mathematics, 22*, 133–140.
Ishibashi, M. (1995). The value of the Estermann zeta functions at $s = 0$. *Acta Arithmetica, 73*, 357–361.
Ishibashi, M. (1998). $\mathbb{Q}$-linear relations of special values of the Estermann zeta function. *Acta Arithmetica, 86*, 239–244.
Iwasawa, K. (1972). *Lectures on p-adic L-functions*. Princeton UP, 13–20
Jakubec, S. (2009). On some new estimates for $h^-(\mathbb{Q}(\zeta_p))$. *Acta Arithmetica, 137*, 43–50.
Jakubec, S. (2017). Connection between multiplication theorem for Bernoulli polynomials and first factor h_p^-. *Mathematica Slovaca, 67*, 345–348.
Kanemitsu, S. and Kuzumaki, T. (1998). On a generalization of the Maillet determinant. In K. Györy et al. (Eds.),*Number Theory, Proceedings of a Eger Conference, 1997*, 271–287. de Gruyter Berlin.
Kanemitsu, S., & Kuzumaki, T. (2001). On a generalization of the Maillet determinant II. *Acta Arithmetica, 99*, 343–361.
Kanemitsu, S., & Kuzumaki, T. (2007). On a generalization of the Demyanenko determinant. *Abhandlungen aus dem Mathematischen Seminar der Universität Hamburg, 77*, 25–38.
Kanemitsu, S., & Tsukada, H. (2007). *Vistas of special functions*. Singapore etc: World Scientific.
Landau, E. (1918). Über Ideale und Primideale in Idealklassen. *Mathematische Zeitschrift,2*, 52–154=Collected Works. *Thales Verl. Essen,7*, 11–113 (1985).
Lang, S. (1978). *Cyclotomic fields*. Berlin etc: Springer.
Lehmer, D. H. (1975). Euler constants for arithmetic progressions. *Acta Arithmetica,27*, 125–142; Selected Papers of D. H. Lehmer, Vol. II, 591-6oe, Charles Babbage Res. Center, Manitoba 1981.
Leopoldt, H.-W. (1959). Über die Hauptordnung der ganzen Elements eines abelschen Zahlkörpers. *Journal fur die Reine und Angewandte Mathematik, 201*, 119–149.
Liu, G.-D., Wang, N.-L., & Wang, X.-H. (2010). The discrete mean square of Dirichlet L-function at integral arguments. *Proceedings of the Japan Academy, Series A Mathematical Sciences, 86*, 149–153.
Livingston, A. E. (1965). The series $\sum_1^\infty f(n)/n$ for periodic f. *Canadian Mathematical Bulletin, 8*, 413–432.

Li, F.-H., Wang, N.-L., & Kanemitsu, S. (2013). *Number theory and its applications*. London-Singapore-New Jersey: World Sci.

Meyer, C. (1957). *Die Berechnung der Klassenzahl Abelscher Körper über quadratischen Zahlkörpern*. Berlin: Akademie.

Milnor, J. (1983). On polylogarithms, Hurwitz zeta functions, and the Kubert identities. *Enseignement des Mathématiques, 2*(29), 281–322.

Okada, T. (1981). On an extension of a theorem of S. *Chowla. Acta Arithmetica, 38*, 341–345.

Okada, T. (1982). On a certain infinite series for a periodic arithmetical function. *Acta Arithmetica, 40*, 1430153.

Ramachandra, K. (1964). Some applications of Kronecker's limit formulas. *Annals of Mathematics, 2*(80), 104–148.

Ramachandra, K. (1969). On the class number of relative abelian fields. *The Journal für die Reine und Angewandte Mathematik, 236*, 1–10.

Ramachandra, K. (1966). On the units of cyclotomic fields. *Acta Arithmetica, 12*, 165–173.

Ramachandra, K. (1993). On a theorem of E. E. Kummer. *RIMS Kokyuroku, 837*, 48–56.

Shimada, T. (1996). Fermat quotient of cyclotomic units. *Acta Arithmetica, 76*, 335–358.

Shirasaka, S. (1989). On a theorem of S. Chowla. *Memoirs of the Faculty of Science, Kyushu University, Series A: Mathematics, 43*, 21–23.

Siegel, C. L. (1964). *Analytische Zahlentheorie II*. Göttingen: University of Göttingen.

Takagi, T. (1915). Zur Theorie der relativ-Abel'schen Zahlkörper, II. Proceedings of the Physico-Mathematical Society of Japan *8*, 243–254=*The Collected Papers of Teiji Takagi* (1973), Iwanami Shoten, Tokyo, 51–67.

Takagi, T. (1971). *Algebraic number theory—number theory and class field theory*, 2nd edn. Tokyo: Iwanami-shoten; 2nd ed. 1947, ditto.

Wang, N. -L., Chakraborty, K. and Kanemitsu, S. (2023). Unification of Chowla's problem and Maillet-Demyanenko determinants, *Mathematics*, 2146241 MDPI.

Wang, K. (1982). On a theorem of S. Chowla. *Journal of Number Theory, 15*, 1–4.

Wang, K. (1984). On Maillet's determinant. *Journal of Number Theory, 18*, 305–312.

Washington, L. (1983). *Introduction to cyclotomic fields*. New York etc: Spring.

Yamamura, K. (2009). Bibliography on determinantal expressions of relative class numbers of imaginary abelian number fields. In T. Aoki, S. Kanemitsu, J. -Y. Liu (Eds.), *Dreaming in Dreams-Proceedings of the 5th China-Japan Seminar on Number Theory Held at Kinki University, Higashi-osaka, Japan*. World Sci., 244–250. Chinese translation, Harbin Technology University Press 2017.

Zhang, W.-P. (1996). On the mean values of Dedekind sums. *Journal de Théorie Nombres Bordeaux, 8*, 429–442.

Chapter 6
Historical Remarks

We assemble some interesting historical remarks and elucidation of heritage (every 50 years).

- Kubert identity or distribution property. This has been touched on only fragmentarily in the book, cf. Table 2.2. Milnor (1983) is the best reference on the subject and Chakraborty et al. (2018) gives generalizations. Arai et al. (2015), Lang (1978, pp. 65–68), Romanoff (1951), etc. contain related information.
- The Lerch–Chowla–Selberg formula (for an imaginary quadratic field).

  We add two more references Lepka (1995) and Yamamura (2010–18). The former tries to restore Lerch's work on number theory but remains the first step and needs the sequel. For example, Review [23] on pp. 19–20 lacks mentioning the Lerch–Chowla–Selberg formula. It does succeed in restoring (Lerch, 1905, 1906) which were developed by Y. Yamamoto and then by Wang et al. (2020). It lacks the papers [4], [6], [7], [8] of Lerch on Weil (1975, p. 68) about Lerch's formula (2.1.46) and the Hurwitz–Lerch zeta-function. The latter contains a complete list of literature on Kronecker's limit formula, which is a laborious work and is very useful if it is available on Internet.

  As Theorem 2.5.1 asserts, the Kronecker limit formula is a consequence of the CSIF, which in turn is equivalent to the HFE and the other ingredient $L'(1, \chi)$ is expressed in finite form by the Kummer's Fourier series, which is a consequence of the HFE, Corollary 2.1.1. We contend that LCSF is a consequence of functional equations up to the decomposition theorem as stated in Corollary 1.2.1.

  The Kronecker limit formula plays an important role and in the imaginary quadratic case it has been studied extensively since Kronecker and there is enormous amount of literature. Siegel (1961) is one of the most useful references on the Kronecker limit formula, etc. (where the discriminant is assumed to be 1) and Kumagai (2002) is supplementing the second limit formula. Equally useful and detailed accounts are Chowla and Selberg (1949), Lang (1973), Selberg and Chowla (1967), Zhang

K. Chakraborty et al., *Modular Relations and Parity in Number Theory*,
Infosys Science Foundation Series, https://doi.org/10.1007/978-981-96-6471-9_6

and Williams (1995), etc. and references therein. For other fields, the Kronecker limit formula is studied, cf., e.g. Bump and Goldfeld (1987), Novikov (1980).

- For further developments of the theory of modular relations, one can see Dixit and his school. Cf., e.g. Dixit and Gupta (2021), Dixit and Kumar (2021), and related papers.

- The Riemann–Siegel formula (RSF) and the Ewald expansion. Both of these are formula which hold in the critical strip $0 < \sigma < 1$. For the original RSF, Edwards (1974) is a good reference. The RSF given in Sects. 2.14 and 2.15 is for zeta-functions associated with modular forms and is more or less in the spirit of Guthman (1996, 1997a, b, 1999). For the RSF for $\zeta^2(s)$, Motohashi (1987) is a good reference. RSF should be studied further since there are applications to other fields including quantum information, cf., e.g. Feiler and Schleich (2013).

  For the Ewald expansion, we refer to Lavrik (1968, 1990), Terras (1985) and Kanemitsu and Tsukada (2014, Chap. 5) for well-known results and Bringmann et al. (2017), Bruinier and Funke (2004) for harmonic Maass forms.

  For the incomplete gamma functions, Erdélyi et al. (1953, I, p. 266, (26), (27)) and Erdélyi et al. (1953, II, p. 133, (1)–(4)) give

$$\Gamma(a,x)=\int_x^\infty t^{a-1}e^{-t}\,dt=e^{-a}\Psi(1-a,1-a;x)=e^{-a}x^a\Psi(1,1+a;x) \tag{6.0.1}$$

  and

$$\begin{aligned}\gamma(a,x)&=\int_0^z t^{a-1}e^{-t}\,dt=\Gamma(a)-\Gamma(a,x)\\&=a^{-1}x^a\Phi(a,1+a;x)=a^{-1}x^ae^{-x}\Phi(1,1+a;x),\quad \operatorname{Re}a>0.\end{aligned}$$

  (6.0.1) is also expressed as

$$\Gamma(a,x)=e^{-\frac{x}{2}}x^{\frac{1-a}{2}}W_{\varkappa,\mu}(x),\quad \varkappa=\frac{a-1}{2},\quad \mu=-\frac{a}{2}.$$

  Hence we are motivated to unify the theory of Ewald expansion and the Fourier–Whittaker expansion.

- The spectral theory of automorphic forms may be said to begin with the discovery of Maass (wave) forms (Maass, 1949). In pursuit of extending Hecke's result on the L-functions associated to *holomorphic* automorphic forms, Maass went backward from the Dirichlet series with Grössen characters attached to real quadratic fields and introduced *non-holomorphic* automorphic forms, eigen-functions of the hyperbolic Laplacian Δ, and the Maass (wave) forms. One of the main objectives of this book is to give genesis of Maass theory as that of the theory of zeta-functions satisfying the ramified function equation which are in correspondence with those which have Fourier–Whittaker expansions.

Selberg (1956) and Roelcke (1956) independently developed the L^2-theory of automorphic forms and established spectral expansions. Combined with the development of representation theory notably by Gelfand et al. (1966), the theory has been vastly advanced.

On Baker (2003, p. 2), R. Baker mentions "The idea is to make a difficult subject a bit easier for beginners." There are books devoted to the subject: Bump (1984), Iwaniec (1995–2002), Motohashi (1997), Venkov (1982), Venkov (1990), but none of them are easy to read. In Agarwal et al. (2023), we concentrate on the Kuznetsov sum formula as opposed to the Kuznetsov trace formula Kuznetsov (1981) and pave a light-hearted promenade to understanding of the subject to such an extent that one can go on reading more advanced material and possibly come to the front to make research. Selberg introduced Selberg (1956) a very important class of series, known as Poincaré series which led him to introduce a new class of zeta-functions, the Kloosterman(–Selberg) zeta-functions. Kuznetsov established the trace formula by equating two expressions for the inner product of two Poincaré series, one by the unfolding method and the other by the Parseval formula. The unfolding method is ubiquitous starting probably from Petersson's use Petersson (1932) in computation of the Fourier coefficients of a modular form of weight k.

References

Agarwal, P., Kanemitsu, S., & Kuzumaki, T. (2023). Automorphic L-functions by way of the Kuznetsov sum formula. *The Advances in Computational Mathematics and Modelling, 1*, 47–72.

Arai, T., Chakraborty, K. and Ma, J. (2015). Applications of the Hurwitz-Lerch zeta-function. *Pure and Applied Mathematics Journal,4*(2-1), 30–35. https://doi.org/10.11648/j.pamj.s.2015040201.16

Baker, R.C. (2003). *Kloosterman sums and Maass forms*, Vol. I, Kendrick Press.

Bringmann, A., Folsom, A., Ono, K., & Rolen, L. (2017). *Harmonic Maass forms and mock modular forms: Theory and applications*. Providence: AMS.

Bump, D. (1984). *Automorphic forms of* GL(3, $\mathbb{R}$). Lecture Notes in Mathematics 1083. Berlin etc: Springer.

Bump, D., & Goldfeld, D. (1987). A Kronecker limit formula for cubic fields. In Modular Forms (Ed.), Rankin, R. (pp. 43–49). Ellis Harwood.

Chakraborty, K., Kanemitsu, S., & Kuzumaki, T. (2018). Seeing the invisible: Around generalized Kubert functions, Annales Univ. Sci. Budapest. *Sect. Comp., 47*, 185–195.

Chowla, S., & Selberg, A. (1949). On Epstein's zeta-function (I). *Proceedings of the National Academy of Sciences of the United States of America,35*, 371–374; Collected Papers of Atle Selberg I, Springer, 1989, 367–370. *The Collected Papers of Sarvadaman Chowla II, CRM,1999*, 719–722.

Dixit, A., & Gupta, R. (2021). Koshliakov zeta functions I. Modular relations. *Advances in Mathematics,393*, Paper No. 108093.

Dixit, A., & Kumar, R. (2021). Superimposing theta structure on a generalized modular relation. *Research in the Mathematical Sciences,8*(3), Paper No. 41, 83.

Edwards, H. M. (1974). *Riemann's Zeta-function*. New York-London: Academic.

Erdélyi, A., Magnus, W., Oberhettinger, F., & Tricomi, F. G. (ed.), (1953). *Higher transcendental functions*, Vol. I–III. New York: McGraw-Hill.

Feiler, C., & Schleich, W. P. (2013). Entanglement and analytical continuation: An intimate relation told by the Riemann zeta-function, *New Journal of Physics,15*, 063009 (31pp).

Gelfand, I. M., Graev, M. I. and Piatetski-Shapiro, I. (1966). *Representation theory and automorphic functions* (English transl. Saunders, 1969).

Guthman, A. (1996). The Riemann-Siegel integral formula for Dirichlet series associated to cusp forms. In *Analytic and Elementary Number Theory*, Vienna, 53–69.

Guthman, A. (1999). Asymptotic expansions for Dirichlet series associated to cusp forms. *Publications de l'Institut Mathématique (Beograd)* (N.S.) *65* (79), 69–96.

Guthman, A. (1997a). Die Riemann-Siegel-Integralformel für die Mellintransformation von Spitzenformen. (German). *Archiv der Mathematik (Basel), 69*, 391–402.

Guthman, A. (1997b). New integral representations for the square of the Riemann zeta-function. *Acta Arithmetica, 82*, 309–330.

Iwaniec, H. (1995). *Introduction to the spectral theory of automorphic forms*, Biblioteca de la Revista Matemática Iberoamericana, Madrid. Second edition: *Spectral methods of automorphic forms*. Grad. Stud. Math. Vol. 53. American Mathematical Society and Revista Matemática Iberoamericana, Providence, 2002.

Kanemitsu, S., & Tsukada, H. (2014). *Contributions to the theory of zeta-functions: The modular relation supremacy*. Singapore etc: World Scientific Publishing.

Kumagai, H. (2002). On unified Kronecker limit formulae, Kyushu. *Journal of Mathematics, 56*, 41–51.

Kuznetsov, N. V. (1980). Petersson's conjecture for cusp forms of weight zero and Linnik's conjecture. Sums of Kloosterman sums. *Mathematics of the USSR-Sbornik* (N.S.), **111 (153)**, 334–383, 479 (in Russian). English translation: Math. USSR Sbornik, **39** (1981), 299–342.

Lang, S. (1973). *Elliptic functions*. New York, etc.: Addison-Wesley.

Lang, S. (1978). *Cyclotomic fields*. Berlin etc: Springer.

Lavrik, A. F. (1968). An approximate functional equation for the Dirichlet L-function. Trudy Moskov. Mat. Obšč. 18, 91–104 =Trans. *Moscow Mathematical Society,18*, 101–115.

Lavrik, A. F. (1990). Functional equations with a parameter for zeta-functions (Russian). *Izvestiya Akademii Nauk SSSR. Seriya Matematicheskay,***54** , 501–521; English translation in *Mathematics of the USSR-Izvestiya,36*, 519–540 (1991).

Lepka, K. (1995). *Matyžs Lerch's work on number theory*. Brno: Faculty of Science Masaryk University.

Lerch, M. (1905). Essais sur le calcul de nombre des classes de formes quadratiques binaires aux coefficients entiers. *Acta Mathematica, 29*, 333–424.

Lerch, M. (1906). Essais sur le calcul de nombre des classes de formes quadratiques binaires aux coefficients entiers. *Acta Mathematica, 30*, 203–293.

Maass, H. (1949). Über eine neue Art von nichtanalytischen automorphen Funktionen und die Bestimmung Dirichletscher Reihen durch Funktionalgelichungen. *Mathematische Annalen, 121*, 141–183.

Milnor, J. (1983). On polylogarithms, Hurwitz zeta functions, and the Kubert identities. *Enseignement des Mathématiques, 2*(29), 281–322.

Motohashi, Y. (1987). *Lectures on the Riemann-Siegel formula*. Boulder: University Press of Colorado.

Motohashi, Y. (1997). *Spectral theory of the Riemann zeta-function*. Cambridge etc: Cambridge UP.

Novikov, A. P. (1980). Kronecker's limit formula in a real quadratic field. *Izvestija Akademii Nauk SSSR Serija Matematiceskaja., 44*, 147–176.

Petersson, H. (1932). Über die Entwicklungskoeffizienten der automorphen Formen. *Acta Mathematica, 58*, 160–215.

Roelcke, W. (1956). Über die Wellengleichung bei Grenzkreisgruppen erster Art, *Sitz.-Ber. Heidelberg Akad. Wiss. Math-Naturwiss. Kl.***1953/1955**, 159–267.

Romanoff, N. P. (1951). Hilbert spaces and the theory of numbers. II. *Izvestiya Akademii Nauk SSSR. Seriya Matematicheskaya, 15*, 131–152.

Selberg, A., & Chowla, S. (1967). On Epstein's zeta-function. *Journal für die reine und angewandte Mathematik,227*, 86–110; (1989) *Collected Papers of Atle Selberg I*, 521–545. Springer; (1999) *The Collected Papers of Sarvadaman Chowla II*, 1101–1125. CRM.

Selberg, A. (1956). Harmonic analysis and discontinuous groups in weakly symmetric Riemannian spaces with applications to Dirichlet series. *The Journal of the Indian Mathematical Society, 20*, 47–87.

Siegel, C. L. (1961). *Lectures on advanced analytic number theory*. Bombay: Tata Institute.

Terras, A. (1985). *Harmonic analysis on symmetric spaces and applications I*. Berlin etc: Springer.

Venkov, A. B. (1982). Spectral Theory of automorphic functions. *Trudy Matematicheskogo Instituta imeni V.A. Steklova,153*(1981), 3–171; English transl. in *Proceedings of the Steklov Institute of Mathematics* (4), 1–163 (1982).

Venkov, A. B. (1990). *Spectral Theory of automorphic function and its applications*. Dordrecht-Boston-London: Kluwer Academic Publishers.

Wang, X.-H., Mehta, J., & Kanemitsu, S. (2020). The boundary Lerch zeta-function and short character sums à la Y. Yamamoto. *Kyushu Journal of Mathematics, 74*, 313–335. https://doi.org/10.2206/kyushujm.74.313

Weil, A. (1975). *Elliptic functions according to Eisenstein and Kronecker*. Berlin-Heidelberg-New York: Springer.

Zhang, N. -X., & Williams, K. (1995). On the Epstein zeta function. *Tamkang Journal of Mathematics,26*, 165–176.

Index

K. Chakraborty et al., *Modular Relations and Parity in Number Theory*,
Infosys Science Foundation Series, https://doi.org/10.1007/978-981-96-6471-9

GPSR Compliance

The European Union's (EU) General Product Safety Regulation (GPSR) is a set of rules that requires consumer products to be safe and our obligations to ensure this.

If you have any concerns about our products, you can contact us on ProductSafety@springernature.com

In case Publisher is established outside the EU, the EU authorized representative is:

Springer Nature Customer Service Center GmbH
Europaplatz 3
69115 Heidelberg, Germany

Batch number: 09288666

Printed by Printforce, the Netherlands